Steel Structures

Practical Design Studies

Fourth Edition

Steel Structures

Practical Design Studies

Fourth Edition

Hassan Al Nageim

CRC Press is an imprint of the
Taylor & Francis Group, an **informa** business
A SPON PRESS BOOK

CRC Press
Taylor & Francis Group
6000 Broken Sound Parkway NW, Suite 300
Boca Raton, FL 33487-2742

Printed at CPI on sustainably sourced paper
Version Date: 20160707

International Standard Book Number-13: 978-1-4822-6355-8 (Paperback)

Library of Congress Cataloging-in-Publication Data

Names: Al Nageim, H. (Hassan), author.
Title: Steel structures : practical design studies / Hassan Al Nageim.
Description: Fourth edition. | Boca Raton : Taylor & Francis, CRC Press, 2017. | Includes bibliographical references and index.
Identifiers: LCCN 2016028953 | ISBN 9781482263558 (pbk. : alk. paper)
Subjects: LCSH: Structural analysis (Engineering) | Building, Iron and steel.
Classification: LCC TA645 .A43 2017 | DDC 693/.71--dc23
LC record available at https://lccn.loc.gov/2016028953

Visit the Taylor & Francis Web site at
http://www.taylorandfrancis.com

and the CRC Press Web site at
http://www.crcpress.com

Printed and bound by CPI Group (UK) Ltd, Croydon, CR0 4YY

Contents

Preface to the fourth edition

The structural engineer works as a member of a team, and to operate successfully requires flair, sound knowledge and judgement, experience and the ability to exercise great care. His or her role may be summarized as planning, design, supervision of the preparation of drawings and tender documents and supervision of construction. He or she makes decisions about materials, structural form and design methods to be used. He or she recommends acceptance of tenders, inspects, supervises and approves fabrication and construction. He or she has an overall responsibility for safety and must ensure that the consequences of failure due to accidental causes are limited in extent.

The designer's work, which is covered partially in this book, is one part of the structural engineer's work.

The main purpose of the fourth edition remains, again, to present:

- Up-to-date principles and relevant consideration in the field of structural design
- New and revised practical design examples for some of the major types of steel-framed buildings
- New chapters on the structural design of major types of structural elements to Eurocodes of practice such as
    - Plate girder
    - Cased steel columns
    - Composite floors and composite steel beams

All buildings can be framed in different ways with different types of joints and analysed using different methods. Member design for ultimate conditions is specified. Projects are selected to show alternative designs for the same structure.

Designs are now to confirm to limit state theory according to the British steel codes of practice and the new Eurocodes of practice. Design principles are set out briefly and designs made to the British codes of practice and to the Eurocodes of practice in most cases for comparison. Many more design calculations and checks are required for the limit state codes of practice than for the previous elastic codes of practice and thus not all load cases or details checks can be carried out for every design project. However, further necessary design work is indicated in these cases.

Though computer methods, mainly for analysis, but also increasingly used for the member and connection design are now the design office procedural norm, approximate, manual methods are still of great importance. These are required mainly to obtain sections for computer analysis and to check final design.

The book, as in the case of the third edition, is aimed at final year students, candidates in master's degree courses in structural engineering and young engineers in industry. Fundamental knowledge of the methods of structural analysis and design from a basic design course is assumed.

The preparation of this edition of this book has not been confined to a single person, as the title page may suggest. In this work, I have been helped by many, both directly and indirectly, and my thanks are due and gratefully given to them all and in particular to John Ellis, Academic Programme Manager, Liverpool John Moores University and David Tinker, Liverpool John Moores University for providing the figures and drawings and Professor Aldo Cauvin Giuseppe Stagnitto, Department of Structural Mechanics, University of Pavia, Italy for providing comprehensive material used as the main bases for Chapter 1. None of this, however, would have been possible without the generous cooperation of my wife Shadha and sons Dr. Haydar and Dr. Yassier. I am, therefore, extremely grateful for their continuous support and help.

I hope that many students and engineers will find this current edition as helpful in their studies as the first, second and third editions were to me and my students. Then all the effort put into this work will have been well worthwhile.

Professor Hassan Al Nageim
Professor of Structural Engineering
Liverpool John Moores University
Liverpool, United Kingdom

Preface to the third edition

The main purpose of the third edition remains again to present principles, relevant considerations and sample designs for some of the major types of steel-framed buildings. All buildings can be framed in different ways with different types of joints and analysed using different methods. Member design for ultimate conditions is specified. Projects are selected to show alternative designs for the same structure.

Designs are now to confirm to limit state theory according to the British steel codes of practice and the new Eurocodes of practice. Design principles are set out briefly and designs made to the British codes of practice and to the Eurocodes of practice in most cases for comparison. Many more design calculations and checks are required for the limit state codes of practice than for the previous elastic codes of practice, and thus not all load cases or details checks can be carried out for every design project. However, further necessary design work is indicated in these cases.

Though computer methods, mainly for analysis, but also increasingly used for the member and connection design, are now the design office procedural norm, approximate, manual methods are still of great importance. These are required mainly to obtain sections for computer analysis and to check final design.

The book, as in the case of the second edition, is aimed at final year students, candidates of master's degree courses in structural engineering and young engineers in industry. Fundamental knowledge of the methods of structural analysis and design from a basic design course is assumed.

The preparation of this edition of this book has not been confined to a single person, as the title page may suggest. In this work, I have been helped by many, both directly and indirectly, and my thanks are due and gratefully given to them all and in particular to John Ellis, Academic Programme Manager, Liverpool John Moores University; Paul Hodgkinson, Liverpool John Moores University for providing the figures and drawings; and Professor Aldo Cauvin Giuseppe Stagnitto, Department of Structural Mechanics, University of Pavia, Italy for providing comprehensive material used as the main bases for Chapter 1.

None of this, however, would have been possible without the generous cooperation of my wife Shadha and sons Haydar and Yassier. I am, therefore, extremely grateful for their continuous support and help. I hope that many students and engineers will find this current edition as helpful in their studies as the first and second editions were to my students and myself. Then all the efforts put into this work would have been well worthwhile.

Professor Hassan Al Nageim
Professor of Structural Engineering
Liverpool John Moores University
Liverpool, United Kingdom

Preface to the second edition

The main purpose of the second edition is again to present principles, relevant considerations and sample designs for some of the major types of steel-framed buildings. All buildings can be framed in different ways with different types of joints and analysed using different methods. Member design for ultimate conditions is specified. Projects are selected to show alternative designs for the same structure.

Designs are now to conform to limit state theory – the British steel codes of practice and the new Eurocodes of practice. Design principles are set out briefly and designs made to the British codes of practice only. Reference is made to the Eurocodes of practice in one special case. Many more design calculations and checks are required for the limit state codes of practice than for the previous elastic codes of practice, and thus not all load cases or detailed checks can be carried out for every design project. However, further necessary design work is indicated in these cases.

Though computer methods, mainly for analysis, but also increasingly used for member and connection design, are now the design office procedural norm, approximate, manual methods are still of great importance. These are required mainly to obtain sections for computer analysis and to check final designs.

The book, as in the case of the first edition, is aimed at final year students, candidates on master's degree courses in structural engineering and young engineers in industry. Fundamental knowledge of the methods of structural analysis and design from a basic design course is assumed.

Professor Hassan Al Nageim
Professor of Structural Engineering
Liverpool John Moores University
Liverpool, United Kingdom

Preface to the first edition

The purpose of the book is to present the principles and practice of design for some of the main modern structures. It is intended for final year degree students to show the application of structural engineering theory and so assist them to gain an appreciation of the problems involved in the design process in the limited time available in college. In such a presentation, many topics cannot be covered in any detail.

Design is a decision-making process where engineering judgement based on experience, theoretical knowledge, comparative design studies, etc., is used to arrive at the best solution for a given situation. The material in the book covers the following:

(a) Discussion of conceptual design and planning.
(b) Presentation of the principles and procedures for the various methods of analysis and design.
(c) Detailed analysis and design for selected structures. Preliminary design studies are made in other cases where the full treatment of the problem is beyond the scope of this book.

In detailed design, the results are presented in the form of sketches showing framing plans, member sizes and constructional details.

Although the book is primarily concerned with the design of steel structures, important factors affecting both the overall design and detail required are discussed briefly. These include the choice of materials, type of foundations used, methods of jointing and the fabrication process and erection methods. Other design considerations such as fatigue, brittle fracture, fire resistance and corrosion protection are also noted.

The use of computers in design is now of increasing importance. Where required, computer programs are used in the book for analysis. While examples of computer-aided design have not been included, a project on this topic is listed at the end of the book. It is felt that the student must thoroughly understand design principles before using design programs.

In college, student are instructed through formal lectures backed by reading from textbooks and journals and by consultation with staff and fellow students. The acquisition of knowledge and the exchange of ideas help them to develop their expertise and judgement and to make sound decisions. However, the most important part of the learning process is the carrying out of practical design work where the students are given selected coursework exercises, which cover the stages in the design process. Such exercises have been included at the end of most chapters. These, generally, consist of making designs for given structures including framing plans, computer analysis, design and detailed drawings.

In many first degree courses, student are also required to undertake a project for which they may choose a topic from the structural engineering field. This gives them the opportunity to make a study in a particular area of interest in greater depth than would be possible through the normal lectures. Some suggestions for projects are given at the end of the book. These may be classified as follows:

(a) Comparative design studies
(b) Computer-aided design projects
(c) Construction and testing of structural models and presentation of results in report form

The intention of the book is to help equip the young engineers for their role in structural engineering in industry. It is important to foster interest in structural engineering in industry. It is important to foster interest in structural design where this is shown by a student. It is hoped that this book will go some way towards this goal.

Professor Hassan Al Nageim
Professor of Structural Engineering
Liverpool John Moores University
Liverpool, United Kingdom

Chapter 1

Introduction to structural design

The meaning, the purpose and the limits of structural design – general

1.1 INTRODUCTION

To clarify the meaning of what we call structural design, it is important to avoid any confusion, as it often unfortunately happens, between architectural structural design, structural elements design and analysis.

Analysis is a verification process, using knowledge of applied mechanics and technical tools, of the dimensions of a project, which in a more or less definitive way has already been defined by a design process.

It is clear that structural analysis procedures can be taught in schools as every well-defined and formalized subject can. In fact, structural analysis is a branch of applied physics.

However, it is less clear to understand to which extent a correct design procedure can be acquired by more or less conventional teaching procedures because design implies an ill-definable quality called creativity.

In this section, the basic steps of structural design and analysis are explained first, together with the tools that are needed to execute them, including the use of databases and expert systems.

1.2 PHASES OF STRUCTURAL DESIGN

In structural design we can distinguish three basic phases:

1. Conceptual design, creative and dimensioning phase, in which the structural form is created. The structural type, materials and the basic dimensions of the members are chosen.

   This phase is immediately followed by a preliminary verification of the choices, using approximate analysis and cost estimate procedures.
2. Planning or development phase of the preliminary design, in which details are defined and the final modelling and analysis of the structure are made.
3. Documentation phase, in which the final drawing, cost estimates and contracts are prepared.

In the flow chart of Figure 1.1, the various phases of structural design and the related tools are described in some detail.

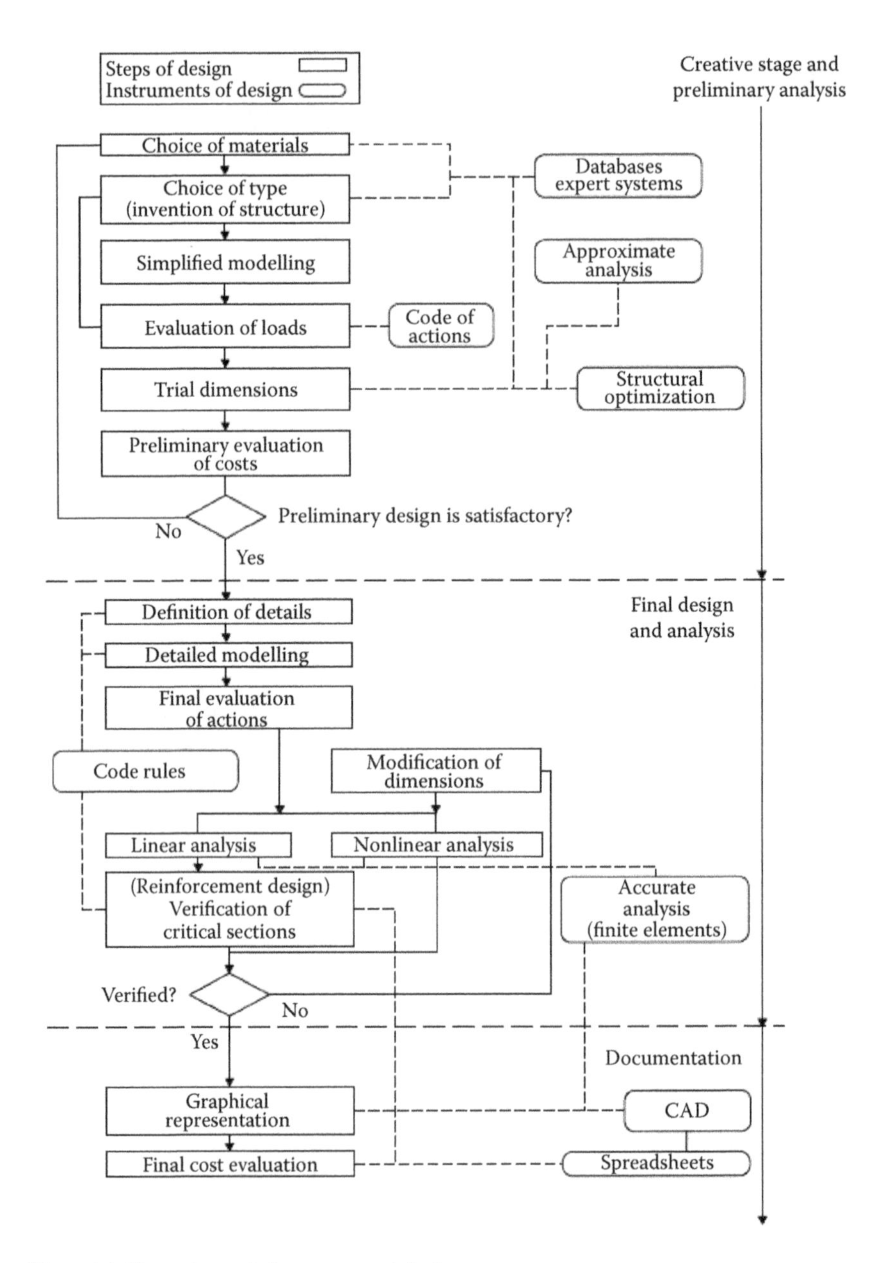

Figure 1.1 Flow-chart of the structural design process.

1.2.1 Basic considerations concerning the structural design process

Documentation phase is a logical and essentially mechanical consequence of the first two phases.

Very often the 'creative' phase is left completely to the architect (with little involvement of the structural engineer), who, in most cases, lacks the necessary sensitivity and knowledge to conceive a statically sound structure. In addition, the architect is often unable to perform the preliminary verifications necessary to validate his choices of structural type, material and dimensions.

The structural engineer often performs his work from phase two onwards by verifying the dimensions of the structure and making sure that the structure complies with stability, safety and sustainability. This situation does not produce relevant 'inconveniencies' whenever the structure plays a secondary role in architecture and the architectural and structural designer can proceed independently from one another. This is the case of low-rise small span buildings where the structure is determined by the exterior form of the building, which can be designed without much concern about statics. In this case, there is often little to invent from the structural point of view and the role of the structural engineer is not fundamental for architectural expression.

However, in the case of long span structures (bridges, buildings with long span roofs and floors) and tall buildings, structure cannot be concealed but on the contrary it becomes the main architectural means of expression and a relevant part of the budget too. This is also true in the extreme case of bridges and tower structures, where forms and structure are coincident. In this case, we have to deal with structural art.

It is interesting to note that an ill-conceived structure is unsatisfactory not only for technical and economical reasons but often also from an aesthetics point of view because it tends to convey, even to the inexperienced observer, an unpleasant sense of instability.

In the author's opinion, it is important that the structural engineer be involved in the project from the very beginning, that is, from what we have called the creative phase, provided that he is prepared to understand the importance of the boundary conditions that are determined by aesthetics and function and not only by statics.

1.3 THE MEANINGS OF STRUCTURAL DESIGN

Structural design means artistic invention and dimensioning. Invention is the creation of a structural form, dimensioning is to assign to every structural member adequate dimensions for stability, serviceability, suitability and sustainability.

Dimensioning is usually obtained through a 'trial and error' procedure, involving repeated analysis of the structure. Only in some particular cases can this process be rationalized using an optimization algorithm.

The starting point in this process is usually the adoption of dimensions derived from similar projects.

Therefore, structural design intended as dimensioning is strictly related to analysis and prediction of the sizes for the structural elements: it can be reduced to repeated analysis and, therefore, to an essentially mechanical process.

The confusion between design and analysis stems from this limited view of the design process in which analysis plays such an essential role.

1.4 CAN STRUCTURAL DESIGN BE TAUGHT?

If we give to structural design the first meaning, that is, artistic invention of new structural forms, the answer is basically no. We can, however, as we will see later, help inventive people to better express their qualities.

In fact a subject can only be taught what can be formalized in some way and can be expressed by rules and objective principles.

We could say that all kinds of teaching can be reduced to three basic categories:

1. Scientific teaching where the laws and principles that govern nature are taught.
2. Technical teaching where the techniques and tools (often but not necessarily derived from scientific principles) that interact with nature are taught and described.
3. Historical teaching (in the broader sense) where what humans have done in the past (also the very near past) is described and commented on.

The third category can be particularly useful in a purely creative job.

In fact, nothing is created from nothing: creative people use examples they like and appreciate to inspire them in the invention of new forms. In other words, a new creation is the result of the elaboration, modification and adaptation of old forms in such a complete way that it can be considered something entirely new and not an imitation of the old.

The purpose of a school in the field of art (and structural design is art) is to propose and comment on examples and not to teach how the invention must take place.

In fact, there have been many attempts to formalize and therefore to teach what is beautiful in art and what is not.

However, in most cases, these attempts to express the 'rules' of the work of art resulted in hampering creativity rather than enhancing it.

As a consequence the purpose of a school, as far as creative structural design is concerned, is to propose examples of what has been done in the field with particular reference (but not only) to contemporary examples.

In addition to this, as structural design is not only art but also (and not only) science and technique, a school must give, as is obviously possible and necessary, a sound scientific preparation.

We could say, with reference to the situation in most countries, where architects and engineers seem to live in different worlds, although working on the same projects, that architects should be more 'engineers' and engineers should be more 'architects', so that a common ground for agreement could be more easily found. The gap between the two cultures should be filled as much as possible, so that structural design will be seen as good art and good technique at the same time.

It is, however, true that in engineering schools, where lecturers are more frequently researchers in the field of structural mechanics rather than architectural designers, they do not help very much in transmitting to the students a correct philosophy of structural design from architectural points of view.

In the best situation, the lecturer of structural design should have both the sensibility of an artist and the skill of a good structural engineer.

1.5 DATABASES AND EXPERT SYSTEMS IN STRUCTURAL DESIGN

Suitable examples are important in the generation of structural creativity. They should be organized in databases where both the relevant graphic documentation and numerical data and indexes can be found easily.

From these databases, knowledge base systems, such as neural network systems, can be derived, organized in rules and introduced in an expert system, which, according to a suitable inference mechanism, can help in choosing the most suitable structural type that satisfies given boundary, loading and environmental conditions.

Figure 1.2 shows a very simplified example of a neural network used in the choice of suitable structural schemes for bridges.

Instruments, such as neural network systems and databases, are tools that help in selecting a reasonable structural type to suit given external objective boundary conditions.

1.6 THE IMPORTANCE OF THE COMPUTER MODELLING PROCESS

Digital computers make both simple and complex structural matrix analysis possible. The structure can be analysed as a single entity, possibly including

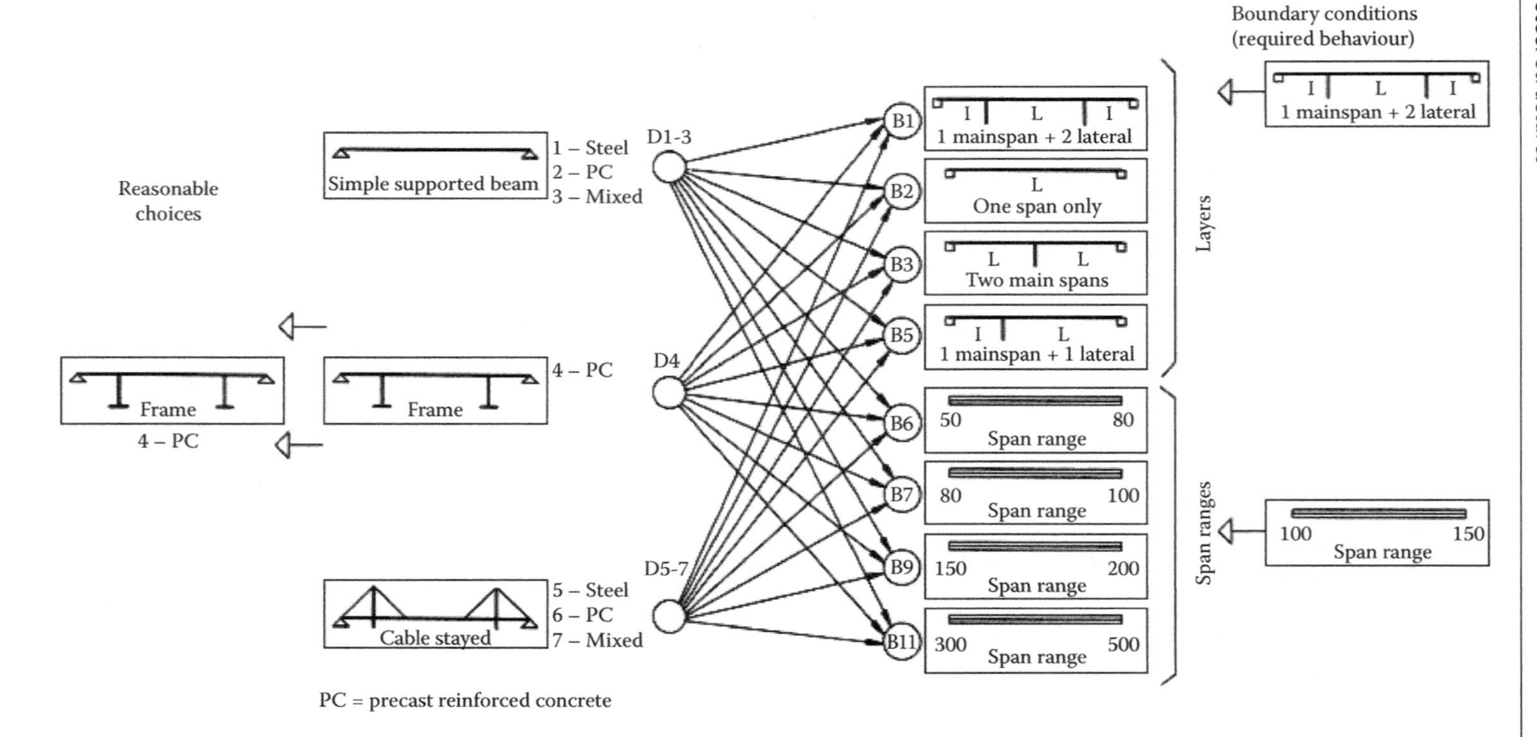

Figure 1.2 Simplified example of application of neural networks to structural design.

the subsoil, modelled as an assemblage of finite elements. To get realistic results, therefore, correct modelling for these structures becomes imperative.

Engineers are always aiming for simple structural forms, which gives a qualitative understanding of the structural behaviour (which is sometimes lacking in more sophisticated holistic approaches), and results that are on the safe side. The designer is looking for a sense of control on the design process, which may not become available when sophisticated computer programs are used. A careful examination of the results of the structural analysis must be performed especially with reference to displacements.

Computer methods of analysis and, in particular, the finite element method have permitted the designer to analyse very complex three-dimensional structural schemes, taking into account, sometimes, the soil structure interaction.

There are, however, limits to this possibility. These limits are determined by

- The intrinsic complexity of the structural geometry;
- The degree of detail required in the results;
- The required degree of accuracy;
- The intrinsic complexity of the constitutive laws of the materials in nonlinear analysis;
- The number or type of loading conditions.

However, the following are important in modelling structures for analysis:

- The choice of plane or space scheme;
- The modelling of the influence of joint dimensions in frames;
- The realistic modelling of supports, for example the introduction of soil structure interaction by modelling the soil as an elastic medium, or, more accurately, as an assemblage of nonlinear finite elements having the stress deformation characteristics of the layers involved;
- The taking into account of nonintentional imperfections and existing fractures in the structural material;
- The taking into account of construction phases.

In the case of two-dimensional, three-dimensional and nonlinear analysis, the following are important too:

- The choice of the types of finite elements;
- The fineness of the mesh to be used in the various zones of the structure according to the stress gradients;
- The choice of the type of analysis;
- The choice of the constitutive laws of the materials.

Chapter 2

Steel structures – structural engineering

2.1 NEED FOR AND USE OF STRUCTURES

Structures are one of humankind's basic needs next to food and clothing, and are a hallmark of civilization. Humans' structural endeavours to protect themselves from the elements and from their own kind, to bridge streams, to enhance a ruling class and for religious purposes go back to the dawn of humanity. Fundamentally, structures are needed for the following purposes:

- To enclose space for environmental control;
- To support people, equipment, materials, etc., at required locations in space;
- To contain and retain materials;
- To span land gaps for transport of people, equipment, etc.

The prime purpose of structures is to carry loads and transfer them to the ground.

Structures may be classified according to use and need. A general classification is

- Residential – houses, apartments, hotels;
- Commercial – offices, banks, department stores, shopping centres;
- Institutional – schools, universities, hospitals;
- Exhibition – churches, theatres, museums, art galleries, leisure centres, sports stadia, etc.;
- Industrial – factories, warehouses, power stations, steelworks, aircraft hangers, etc.

Other important engineering structures are

- Bridges – truss, girder, arch, cable suspended, suspension;
- Towers – water towers, pylons, lighting towers, etc.;
- Special structures – offshore structures, carparks, radio telescopes, mine headframes, etc.

Each of the structures listed above can be constructed using a variety of materials, structural forms or systems. Materials are discussed first and then a general classification of structures is set out, followed by one of steel structures. Though the subject is steel structures, steel is not used in isolation from other materials. All steel structures must rest on concrete foundations and concrete shear walls are commonly used to stabilize multistorey buildings.

2.2 STRUCTURAL MATERIALS – TYPES AND USES

From earliest times, naturally occurring materials such as timber, stone and fibres were used structurally. Then followed brickmaking, rope-making, glass and metalwork. From these early beginnings the modern materials manufacturing industries developed.

The principal modern building materials are masonry, concrete (mass, reinforced and prestressed), structural steel in rolled and fabricated sections and timber. All materials listed have particular advantages in given situations, and construction of a particular building type can be in various materials, for example, a multistorey building can be loadbearing masonry, concrete shear wall or frame or steel frame. One duty of the designer is to find the best solution that takes account of all requirements – economic, aesthetic and utilitarian.

The principal uses, types of construction and advantages of the main structural materials are as follows:

- Masonry – Loadbearing walls or columns in compression and walls taking in-plane or transverse loads. Construction is very durable, fire resistant and aesthetically pleasing. Building height is moderate, say to 20 storeys.
- Concrete – Framed or shear wall construction in reinforced concrete is very durable and fire resistant and is used for the tallest buildings. Concrete, reinforced or prestressed, is used for floor construction in all buildings, and concrete foundations are required for all buildings.
- Structural steel – Loadbearing frames in buildings, where the main advantages are strength and speed of erection. Steel requires protection from corrosion and fire. Claddings and division walls of other materials and concrete foundations are required. Steel is used in conjunction with concrete in composite and combined frame and shear wall construction.

Structural steels are alloys of iron, with carefully controlled amounts of carbon and various other metals such as manganese, chromium, aluminium, vanadium, molybdenum, neobium and copper. The carbon content is less than 0.25%, manganese less than 1.5% and the other elements are in trace amounts. The alloying elements control grain size and hence steel properties,

Table 2.1 Strengths of steels used in structures

Steel type and use	*Yield stress (N/mm²)*
Grade S275 – structural shapes	275
Grade S355 – structural shapes	355
Quenched and self-tempering	500
Quenched tempered–plates	690
Alloy bars – tension members	1030
High carbon hard-drawn wire for cables	1700

giving high strengths, increased ductility and fracture toughness. The inclusion of copper gives the corrosion resistant steel Cor-ten. High-carbon steel is used to manufacture hard drawn wires for cables and tendons.

The production processes such as cooling rates, quenching and tempering, rolling and forming also have an important effect on the microstructure, giving small grain size, which improves steel properties. The modern steels have much improved weldability. Sound full-strength welds free from defects in the thickest sections can be guaranteed.

A comparison of the steels used in various forms in structures is given in Table 2.1. The properties of hot-rolled structural steels are given in Chapter 3 (Table 3.3).

Structural steels are hot-rolled into shapes such as universal beams and columns. The maximum size of a universal column in the United Kingdom is 356 × 406 UC, 634 kg/m, with 77 mm-thick flanges. Trade ARBED in Luxembourg roll a section 360 × 401 WTM, 1299 kg/m, with 140 mm-thick flanges. The heavy rolled columns are useful in high-rise buildings where large loads must be carried. Heavy built-up H, I and box sections made from plates and lattice members are needed for columns, transfer girders, crane and bridge girders, etc. At the other end of the scale, lightweight cold-rolled purdins are used for roofing industrial buildings. Finally, wire, rope and high-strength alloy steel bars are required for cable-suspended and cable-girder roofs and suspended floors in multistorey buildings.

2.3 TYPES OF STRUCTURES

2.3.1 General types of structures

The structural engineer adopts a classification for structures based on the way the structure resists loads, as follows:

1. Gravity masonry structures – Loadbearing walls resist loads transmitted to them by floor slabs. Stability depends on gravity loads.
2. Framed structures – A steel or concrete skeleton collects loads from plate elements and delivers them to the foundations.

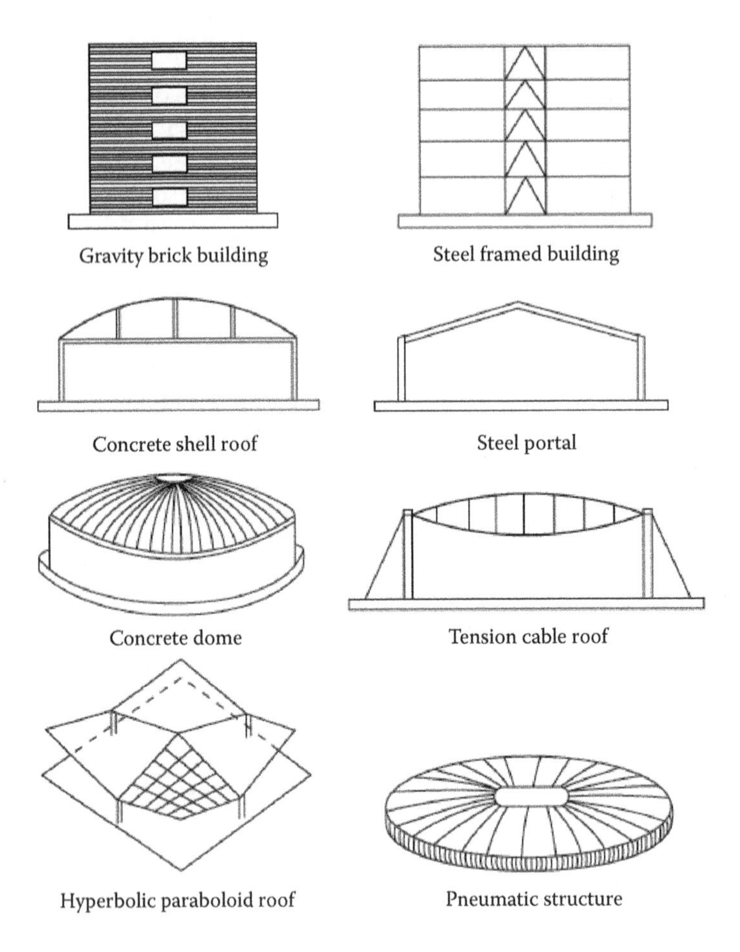

Figure 2.1 General types of structures.

3. Shell structures – A curved surface covers space and carries loads.
4. Tension structures – Cables span between anchor structures carrying membranes.
5. Pneumatic structures – A membrane sealed to the ground is supported by internal air pressure.

Examples of the above structures are shown in Figure 2.1.

2.3.2 Steel structures

Steel-framed structures may be further classified into the following types:

1. Single-storey, single- or multibay structures which may be of truss or stanchion frames or rigid frame of solid or lattice members;

2. Multistorey, single- or multibay structures of braced or rigid frame construction – many spectacular systems have been developed;
3. Space structures (space decks, domes, towers, etc.) – space decks and domes (except the Schwedler dome) are redundant structures, while towers may be statically determinate space structures;
4. Tension structures and cable-supported roof structures;
5. Stressed skin structures, where the cladding stabilizes the structure.

As noted above, combinations with concrete are structurally important in many buildings. Illustrations of some of the types of framed steel structures are shown in Figure 2.2. Braced and rigid frame and truss roof and space deck construction are shown in the figure for comparison. Only framed

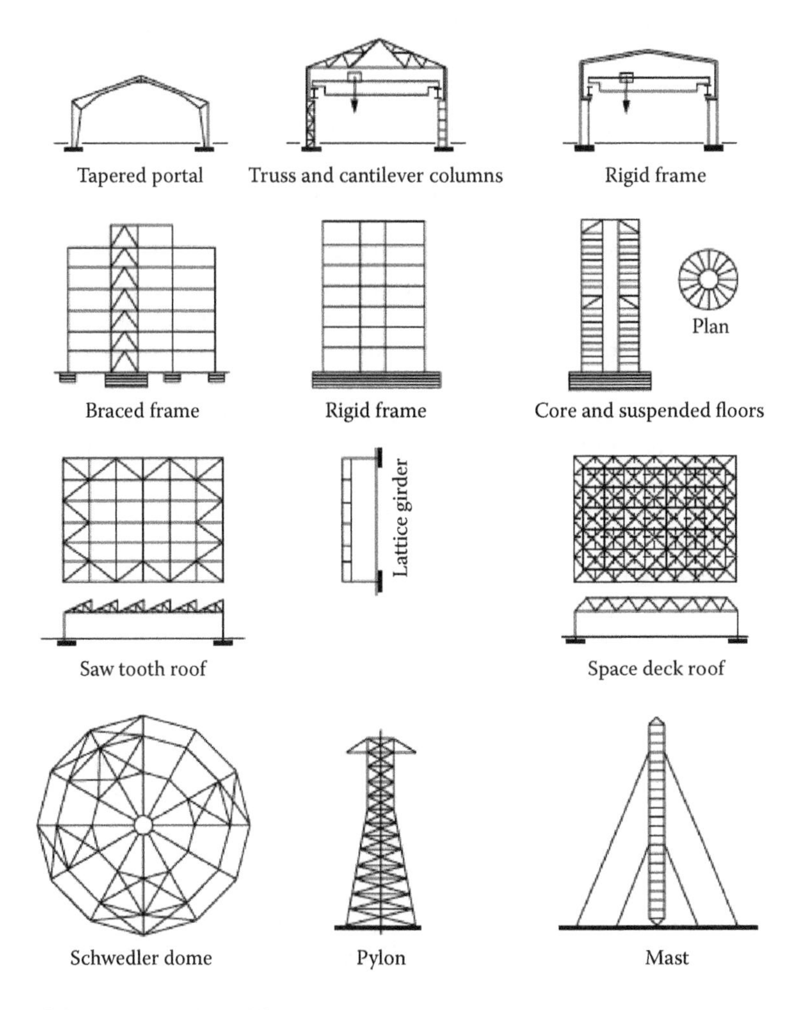

Figure 2.2 Examples of steel-framed structures.

structures are dealt with in the book. Shell types, for example, tanks, tension structures and stressed skin structures, are not considered.

For the framed structures the main elements are the beam, column, tie and lattice member. Beams and columns can be rolled or built-up I, H or box. Detailed designs including idealization, load estimation, analysis and section design are given for selected structures.

2.4 FOUNDATIONS

Foundations transfer the loads from the building structure to the ground. Building loads can be vertical or horizontal and cause overturning and the foundation must resist bearing and uplift loads. The correct choice and design of foundations is essential in steel design to ensure that assumptions made for frame design are achieved in practice. If movement of a foundation should occur and has not been allowed for in design, it can lead to structural failure and damage to finishes in a building. The type of foundation to be used depends on the ground conditions and the type of structure adopted.

The main types of foundations are set out and discussed briefly, as follows:

1. Direct bearing on rock or soil. The size must be sufficient to ensure that the safe bearing pressure is not exceeded. The amount of overall settlement may need to be limited in some cases, and for separate bases differential settlement can be important. A classification is as follows:

    - Pad or spread footing used under individual columns;
    - Special footings such as combined, balanced or tied bases and special shaped bases;
    - Strip footings used under walls or a row of columns;
    - Raft or mat foundations where a large slab in flat or rubbed construction supports the complete building;
    - Basement or cellular raft foundations; this type may be in one or more storeys and form an underground extension to the building that often serves as a carpark.

2. Piled foundations, where piles either carry loads through soft soil to bear on rock below or by friction between piles and earth. Types of piles used vary from precast driven piles and cast-in-place piles to large deep cylinder piles. All of the above types of foundations can be supported on piles where the foundation forms the pile cap.

Foundations are invariably constructed in concrete. Design is covered in specialist books. Some types of foundations for steel-framed buildings are shown in Figure 2.3. Where appropriate, comments on foundation design are given in worked examples.

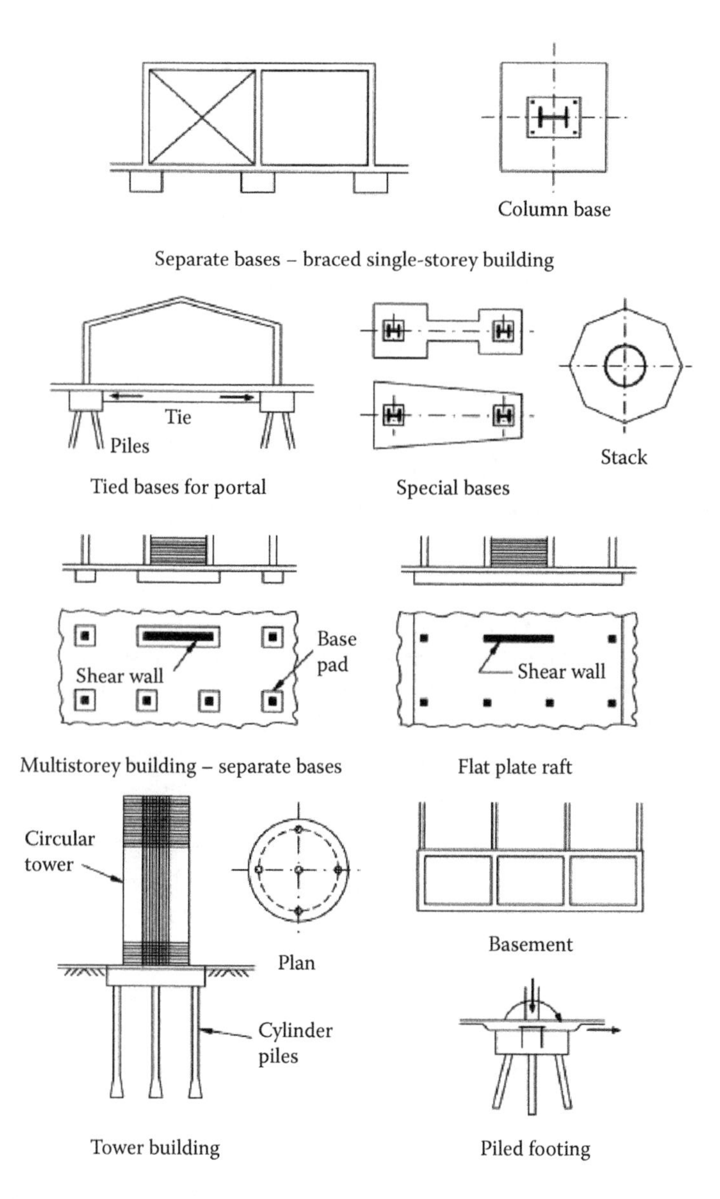

Figure 2.3 Types of foundations.

2.5 STRUCTURAL ENGINEERING

2.5.1 Scope of structural engineering

Structural engineering covers the conception, planning, design, drawings and construction for all structures. Professional engineers from a number of disciplines are involved and work as a team on any given project under the overall control of the architect for a building structure. On engineering structures such as bridges or powerstations, an engineer is in charge.

Lest it is thought that the structural engineer's work is mechanical or routine in nature, it is useful to consider his or her position in building construction where the parties involved are

- The client (or owning organization), who has a need for a given building and will finance the project;
- The architect, who produces proposals in the form of building plans and models (or a computer simulation) to meet the client's requirements, who controls the project and who engages consultants to bring the proposals into being;
- Consultants (structural, mechanical, electrical, heating and ventilating, etc.), who carry out the detail design, prepare working drawings and tender documents and supervise construction;
- Contractors, who carry out fabrication and erection of the structural framework, floors, walls, finishes and installation of equipment and services.

The structural engineer works as a member of a team and to operate successfully requires flair, sound knowledge and judgement, experience and the ability to exercise great care. His or her role may be summarized as planning, design preparation of drawings and tender documents and supervision of construction. He or she makes decisions about materials, structural form and design methods to be used. He or she recommends acceptance of tenders, inspects, supervises and approves fabrication and construction. He or she has an overall responsibility for safety and must ensure that the consequences of failure due to accidental causes are limited in extent.

The designer's work, which is covered partially in this book, is one part of the structural engineer's work.

2.5.2 Structural designer's work

The aim of the structural designer is to produce the design and drawings for a safe and economical structure that fulfils its intended purpose. The steps in the design process are as follows:

1. Conceptual design and planning. This involves selecting the most economical structural form and materials to be used. Preliminary designs are often necessary to enable comparisons to be made. Preliminary design methods are discussed in Chapter 4.
2. Detailed design for a given type and arrangement of structure, which includes:

    - Idealization of the structure for analysis and design;
    - Estimation of loading;

- Analysis for the various load cases and combinations of loads and identification of the most severe design actions;
- Design of the foundations, structural frames, elements and connections;
- Preparation of the final arrangement and detail drawings.

The materials list, bill of quantities and specification covering welding, fabrication, erection, corrosion protection and fire protection may then be prepared. Finally, the estimates and tender documents can be finalized for submission to contractors.

The structural designer uses his or her knowledge of structural mechanics and design, materials, geotechnics and codes of practice and combines this with his or her practical experience to produce a satisfactory design. He or she takes advice from specialists, makes use of codes, design aids, handbooks and computer software to help him or her in making decisions and to carry out complex analysis and design calculations.

2.6 CONCEPTUAL DESIGN, INNOVATION AND PLANNING

Conceptual design in the structural engineering sense is the function of choosing a suitable form or system or framing arrangement to bring the architectural solution into being. The building layout, limits and parameters have often been determined solely by the architect. In such cases, the structural engineer may not be able to select the optimum structural solution. Ideally, conceptual design should result from a team effort, where architect, structural engineer and service engineers contribute to the final solution. Modern architectural practices take this multidisciplinary approach.

The architectural decisions are based on functional, aesthetic, environmental and economic considerations. Any of these factors may control in a given case. For example, for an industrial plant it is the functional requirement, whereas for an exhibition building it is the aesthetic aspect. Financial control is always of paramount importance and cost over-runs lead to many legal and other problems.

Novelty and innovation are always desirable and we seem to strive after these goals. Architects, engineers and builders always push existing forms of construction to the limits possible with materials available and within the state of knowledge at the time. Structural failures determine when limits are reached and so modifications are made and new ideas developed. Often it is not a new solution that is required, but the correct choice and use of a well-proven existing structural system that gives the best answer. The engineer continually seeks new and improved methods of analysis, design and construction, and the materials scientist continually seeks to improve material properties and protection systems through research and development. These advances lead to safer and more economical structures. Much of recent structural research has centred on the use of computers in all aspects of the work from architectural and structural modelling and design for construction and building finishing control.

The following are instances of recent structural engineering innovation:

1. Analysis – Elastic matrix and finite element analysis, second order analysis, cable net analysis, plastic analysis;
2. Design – Plastic design, limit state design, computer-aided design, structural optimization and neural network systems;
3. Construction – Space decks, geodesic domes, tension structures, box girder bridges, high-rise tube buildings, etc.

Planning may be described as the practical expression of conceptual design. The various proposals must be translated from ideas and sketches into drawings consisting of plans and elevations to show the layout and functions and perspective views to give a realistic impression of the finished concept. Computer drafting software is now available to make this work much quicker than the older manual methods. Three-dimensional computer simulation with views possible from all directions gives great assistance in the decision-making process. A scale model of the complete project is often made to show clearly the finished form. The preparation and presentation of planning proposals are very important because the final approval for a scheme often rests with nontechnical people such as city councillors or financiers.

The engineer must also consider construction in any of the major materials – masonry, concrete, steel or timber, or again some combination of these materials for his or her structures and then make the appropriate selection. A list of factors that need to be considered at the conceptual and planning stage would include:

1. Location of the structure and environmental conditions;
2. Site and foundation conditions including problems with contaminated sites, toxicity pollutants, chemical attack, legislation for remediation, effects of changes in the ground and environmental conditions either naturally or as a result of the works or on adjacent works;
3. Weather conditions likely during construction;
4. Availability of materials;
5. Location and reputation of fabrication industry;
6. Transport of materials and fabricated elements to site;
7. Availability and quality of labour for construction;
8. Degree of supervision needed for construction;
9. Measures needed to give protection against corrosion and fire;

10. Likelihood of damage or failure due to fatigue or brittle fracture;
11. Possibility of accidental damage;
12. Maintenance required after completion;
13. Possibility of demolition in the future.

The final decision on the form and type of structure and construction method depends on many factors and will often be taken on grounds other than cost, though cost often remains the most important.

2.7 COMPARATIVE DESIGN AND OPTIMIZATION

2.7.1 General considerations

Preliminary designs to enable comparisons and appraisals to be made will often be necessary during the planning stage in order to establish which of the possible structural solutions is the most economical. Information from the site survey is essential because foundation design will affect the type of superstructure selected as well as the overall cost.

Arrangement drawings showing the overall structural system are made for the various proposals. Then preliminary analyses and designs are carried out to establish foundation sizes, member sizes and weights so that costs of materials, fabrication, construction and finishes can be estimated. Fire and corrosion protection and maintenance costs must also be considered. However, it is often difficult to get true comparative costs and contractors are reluctant to give costs at the planning stage. Preliminary design methods are given in the book and worked examples have been selected to show design comparisons using different structural systems or design methods.

By optimization is meant the use of mathematical techniques to obtain the most economical design for a given structure. The aim is usually to determine the topology of the structure, arrangement of floors, spacing of columns or frames or member sizes to give the minimum weight of steel or minimum cost. Though much research has been carried out and sophisticated software written for specific cases, the technique is not of general practical use at present. Many important factors cannot be satisfactorily taken into account.

The design of individual elements may be optimized, for example, plate girders or trusses. However, with optimum designs the depths are often some 50% greater than those normally adopted and the effect of this on the total building cost should be considered. Again, in optimizing member costs it is essential to rationalize sizes, even if this may lead to some oversized items. Floor layouts and column spacings should be regular and, as a consequence, fabrication and erection will be simplified and cost reduced.

2.7.2 Aims and factors considered in design comparison

The aim of the design comparison is to enable the designer to ascertain the most economical solution that meets the requirements for the given structure. All factors must be taken into consideration. A misleading result can arise if the comparison is made on a restricted basis. Factors to be taken into account include:

1. Materials to be used;
2. Arrangement and structural system and flooring system to be adopted;
3. Fabrication and type of jointing;
4. Method of erection of the framework to be used;
5. Type of construction for floor, walls, cladding and finishes;
6. Installation of ventilating/heating plant, lifts, water supply, power, etc.;
7. Corrosion protection required;
8. Fire protection required;
9. Operating and maintenance costs.

Aesthetic considerations are important in many cases and the choice of design may not always be based on cost alone.

Most structures can be designed in a variety of ways. The possible alternatives that may be used include:

1. The different methods of framing that will achieve the same structural solution;
2. Selection of spacing for frames and columns;
3. Flooring system to be used, for example, *in situ* concrete, precast concrete or profile steel sheeting;
4. The various methods that may be used to stabilize the building and provide resistance to horizontal loading;
5. The different design methods that may be applied to the same structural form, for example, simple design or semirigid design or rigid design using either elastic or plastic theory;
6. Design in different materials, for example, mild steel or high strength steels. The weight saving may be offset by the higher cost of the stronger material.

It should be noted that often no one solution for a given structure ever appears to dominate to the exclusion of all other alternatives. Though the rigid pinned base portal has almost entirely replaced the truss and stanchion frame for single-bay buildings, lattice girder roofs are used in many single-storey multibay buildings.

2.7.3 Specific basis of comparisons for common structures

In the following sections a classification is given on which design comparisons for some general purpose structures may be made. More detailed design comparisons are given in later chapters.

(a) *Single-storey, single-bay buildings*

For a given plan size, the designer can make the following choices.

(i) TYPE OF BUILDING AND DESIGN METHOD (FIGURE 2.4a)

The alternatives are:

1. Truss and stanchion frame with cantilever columns on knee-braces with pinned or fixed bases using simple design;

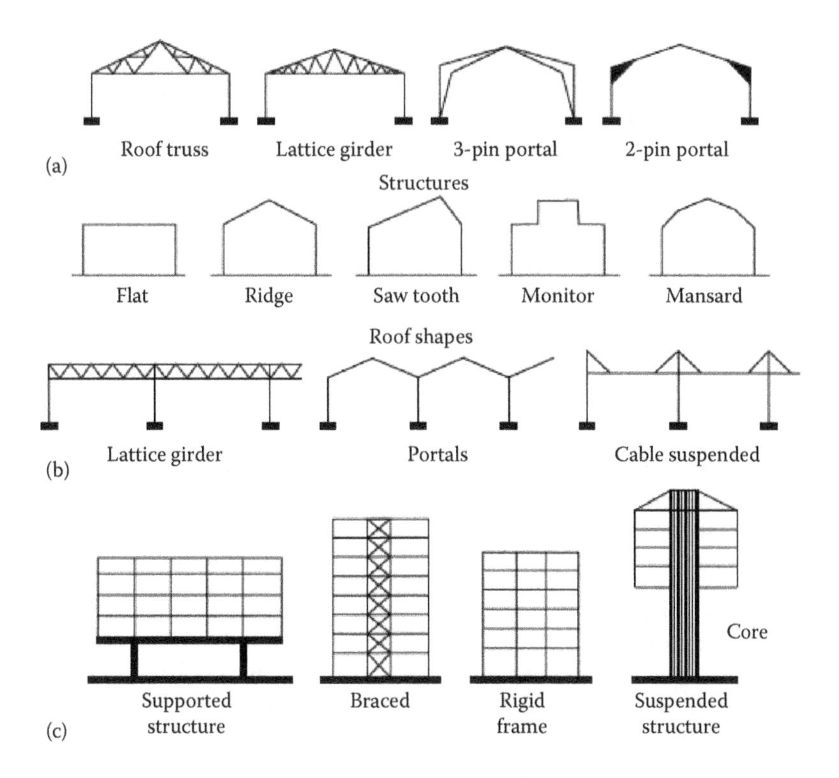

Figure 2.4 Comparison of designs for various structures: (a) single-storey, single-bay structures; (b) single-storey, multibay structures; (c) multistorey structures.

2. Three-pinned portal of I-section or lattice construction using simple design;
3. Rigid portal with pinned or fixed base constructed in

   - Rolled I – or hollow section designed using elastic or plastic theory;
   - Built-up I – tapered or haunched sections designed using elastic theory;
   - Lattice construction designed using elastic theory.

The design may be fully welded or with rigid joints mode using high-strength bolts.

(ii) DESIGN VARIABLES

The basic variable is column spacing which governs the size of purlins, sheeting rails, main frame members and foundations. Designs may be made with various column spacings to determine which gives the most economical results. Various roof shapes are possible such as flat, ridge, sawtooth, monitor or mansard (Figure 2.4a). The roof slope is a further variable; the present practice is to use flatter slopes. In the longitudinal direction, these buildings are in braced simple design. The gable ends are normally simple design.

(b) *Single-storey, multibay buildings*

Three common types of single-storey, multibay buildings are the lattice girder roof, multibay portal and cable suspended roof (Figure 2.4b). The comments from (a) above apply. For wide-span buildings, the sawtooth or space deck roof shown in Figure 2.2 is used.

(c) *Multistorey buildings*

Many different systems are used and many parameters can be varied in design. Some important aspects of the problem are as follows.

(i) OVERALL FRAMING

The column spacing can be varied in both directions. The locations of the liftshaft/staircase can be varied. Not all columns may be continuous throughout the building height. Plate girders can be used to carry upper columns over clear areas. Economy can be achieved if the bottom storey columns are set in, allowing girders to cantilever out. All columns can carry load, or the outer ends of floor beams can be suspended from an umbrella girder supported by the core (Figure 2.2).

(ii) FLOORING

The type of flooring and arrangement of floor framing affect the overall design. The main types of flooring used are cast-*in situ* concrete in one- or two-way spanning slabs or precast one-way floor slabs. The cast-*in situ* slabs can be constructed to act compositively with the steel floor beams. Flat slab construction has also been used with steel columns where a special steel shear head has been designed.

(iii) STABILITY

Various systems or framing arrangements can be used to stabilize multi-storey buildings and resist horizontal loads. The building may be braced in both directions, rigid one way and braced the other or rigid in both directions. Alternatively, concrete shear walls or liftshafts can be used to provide stability. Tube construction is used for very tall buildings (Chapter 11).

(iv) DESIGN METHOD

For a given framing system, various design methods can be used. The methods given in BS5950 and EC3 are simple, semirigid or rigid design. Rigid design can be carried out using either elastic or plastic methods. More accurate methods taking secondary effects into account are possible with elastic analysis. Analysis and design methods are discussed more fully in the next chapter.

(v) FIRE PROTECTION

This is necessary for all steel-framed buildings, and solid casing of beams and columns may be taken into account in design. However, light-weight hollow or sprayed-on casing is generally used in modern practice. Methods have been developed for assessment of fire resistance for steel members.

(vi) FOUNDATIONS

Types of foundations used for steel-framed buildings were set out above. The type selected for prevailing soil conditions can affect the choice of superstructure. One common case is to use pinned bases in poor soil conditions because fixity would be expensive to achieve. Again, where provision must be made for differential settlement, buildings of simple design perform better than those of rigid design. If a monolithic raft or basement foundation is provided, the super structure can be designed independently of the foundation.

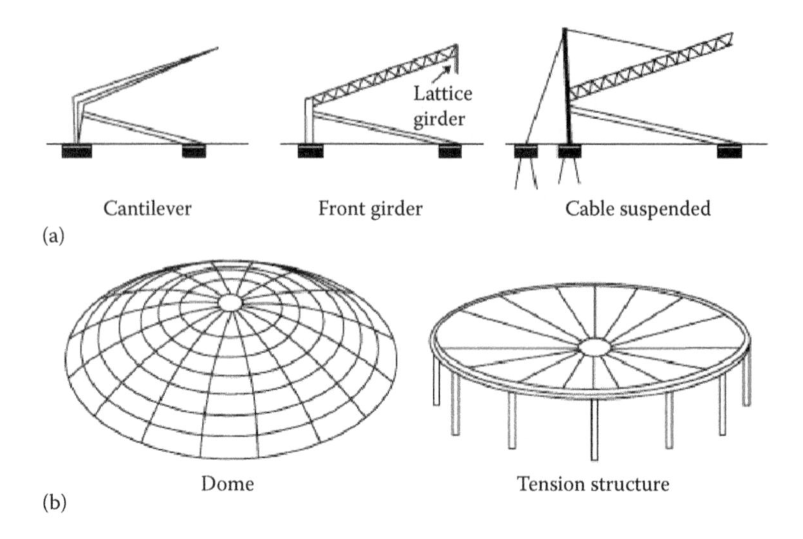

Figure 2.5 Special purpose structures: (a) grandstand structures; (b) covered areas.

(d) *Special purpose structures*

In some structures there may be two or more entirely different ways of framing or forming the structure while complying with all requirements for the finished project. One such structure is the sports grandstand. Three solutions for the structure are as follows:

1. Cantilever construction is used throughout.
2. End columns support a lattice girder which carries the front of the roof.
3. The roof is suspended on cables.

Another example is the dome and circular suspended roof structure (Figure 2.5).

2.8 LOAD PATHS, STRUCTURAL IDEALIZATION AND MODELLING

2.8.1 Load paths

Loads are applied to surfaces, along members and to points along members. Surfaces, that is, roof and floor slabs and walls, transfer loads to members in skeletal structures. Transfer of load from surface to member and member to member, such as slab to beam, wall to column, beam to beam and beam to column to foundation, are the load paths through the structure. Each individual slab, member or frame must be strong enough to carry its loads.

Members and frames must be spaced and arranged to carry its loads (loads carried by that individual element) in the most efficient manner. Standard framing arrangements have been developed for all types of buildings.

2.8.2 Structural idealization

Structures support loads and enclose space and are of three-dimensional construction. Idealization means breaking the complete structure down into single elements (beams, columns, trusses, braced or rigid frames) for which loads carried are estimated and analysis and design made. It is rarely possible to consider the three-dimensional structure in its entirety. Examples of idealization are as follows:

1. Three-dimensional structures are treated as a series of plane frames in each direction. The division is made by vertical planes. For example, a multistorey building (Figure 2.6a) readily divides into transverse rigid frames and longitudinal braced frames. The tower structure in Figure 2.6b could be analysed, using software, as a space frame. Alternatively, it is more commonly treated as a series of plane frames to resist horizontal loads and torsion due to wind.
2. The structure is divided into vertically separated parts by horizontal planes into say, roof, walls and foundations. These parts are designed separately. The reactions from one part are applied as loads to the next part. The horizontal division of a truss and stanchion frame is shown in Figure 2.6c, while a portal frame is treated as a complete unit. A domed-roof stadium, similarly divided, is shown in Figure 2.6d. The domed roof may be designed as a three-dimensional unit or, in the case of specially framed domes, as a series of arched ribs.

2.8.3 Modelling

Another idealization method is modelling for analysis. In this case, the structure is changed so that the analysis of a different form of structure that is more convenient to carry out can be made. This often means modelling the structure for analysis using a plane frame program. Two examples shown in Figure 2.7a and b are

- A ribbed deck idealized as a grid;
- Connected concrete shear walls modelled as a plane frame.

These examples can also be modelled for more accurate finite element analysis.

The modelling problem with steel structures also often includes composite action of steel and concrete elements. Examples are as follows:

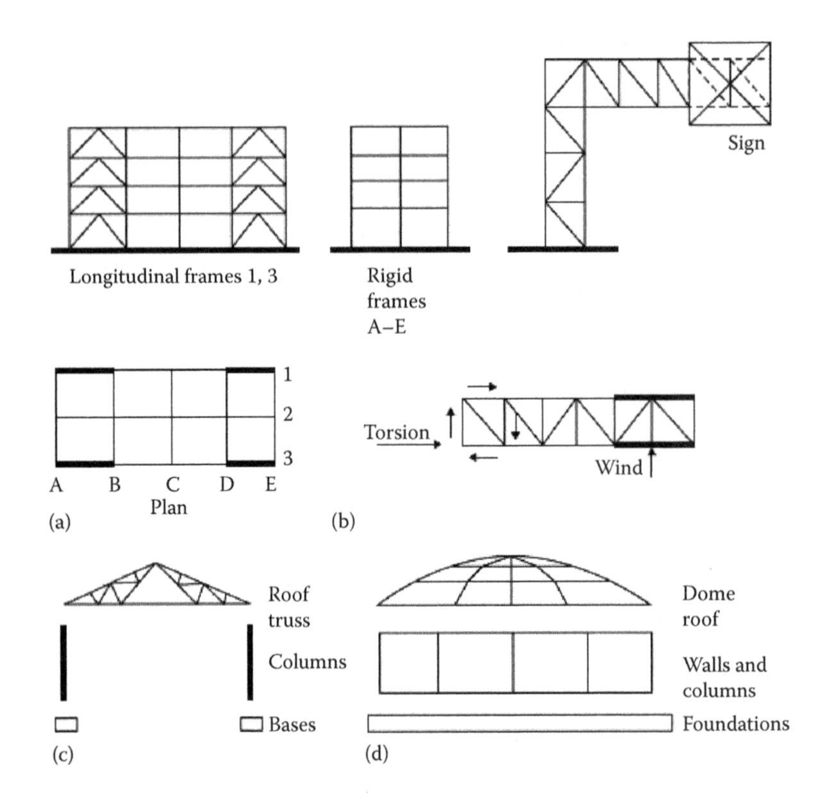

Figure 2.6 Idealization of structures: (a) multistorey building; (b) tower structure; (c) truss and stanchion frame; (d) dome roof stadium.

- *In situ* concrete slab and steel beams in a steel rigid frame. For analysis, the elastic transformed section can be used for the composite beams (Figure 2.7c). Design is based on plastic theory.
- Concrete shear wall in a rigid steel frame (Figure 2.7d). If there is compression over the whole section, the slab can be transformed into equivalent steel. If compression occurs over part of the section, use the transformed section in bending including reinforcement in tension. The shear wall is to be connected to steel sections.
- A three-dimensional steel rigid frame building with concrete core (Figure 2.7e). This can be modelled for plane frame analysis for horizontal wind loads by linking frames and replacing the shear wall by stiff vertical and horizontal members. Alternatively, the structure can be analysed as a space frame.

Great care is needed in interpreting results for design. Finite element analysis will give more accurate results.

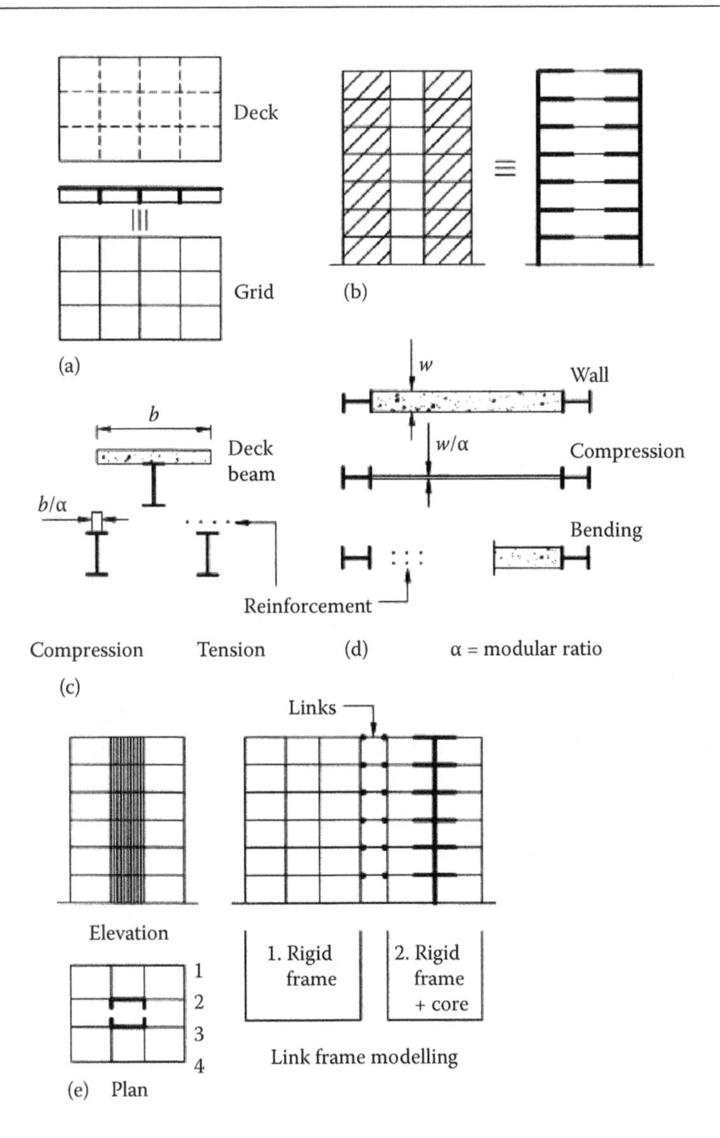

Figure 2.7 Modelling for analysis: (a) ribbed deck; (b) connected shear walls; (c) composite beam; (d) shear wall and columns; (e) rigid frame and core building.

2.9 DRAWINGS, SPECIFICATIONS AND QUANTITIES

2.9.1 Steelwork drawings

Steelwork drawings show the detail for fabrication and arrangement of the structure for erection. They are also used for taking off the materials list and preparing the bill of quantities and estimates of cost. It is essential that drawings are presented correctly and are carefully checked for accuracy.

Drawing is an essential part of the design process. The designer must ensure that the detail is such that the structure acts in the way he or she has idealized it for design. He or she must also ensure that all the detailed construction is possible, will not lead to failure and can be painted, inspected and properly maintained.

Steelwork drawings may be classified into

- General arrangement – Showing the function and arrangement of the structure;
- Making plans – Showing the location of separate numbered members for erection;
- Detail drawings – Giving details of separate members for fabrication.

Many consulting engineering practices carry out the overall analysis and design only, preparing arrangement drawings showing the member sections required. Special joint types must be carefully specified to achieve the designer's assumptions in practice. The fabricator then prepares the detail drawings for joints and shop fabrication. This enables the firm to use details and processes with which they are familiar and have the necessary equipment. Computer software is increasingly used to produce arrangement and detail drawings and take off quantities.

The present book is mainly concerned with design. Sketches showing framing arrangements, main loadbearing frames and members and details of important joints are given where appropriate. The purpose is to show the translation of the output of the analysis and design into the practical structure.

2.9.2 Specification

The specification and drawings are complementary, each providing information necessary for the execution of the work. In general terms, the specification for fabrication and erection includes:

1. General description of building, its location, access to site, etc.;
2. Description of the structural steelwork involved;
3. Types and quality of materials to be used;
4. Standard of workmanship required;
5. In some cases the order in which the work is to be carried out and the methods to be used.

Particular clauses in the steelwork specification cover:

1. Grades of steel required;
2. Workmanship and fabrication process required and the acceptance limits for dimensional accuracy, straightness, drilling, etc.;

3. Welding, methods and procedures required to eliminate defects and cracking and reduce distortion, testing to be carried out and permissible limits of defects;
4. Types and quality of bolts to be used;
5. Inspection practice and marking;
6. Erection, giving the tolerance permissible for out-of-verticality, procedures for assembly and testing for high-strength friction grip bolt joints and site-welded joints;
7. Fire protection methods to be used for the finished steel-framed building;
8. Corrosion protection for exposed steelwork where the surface preparation, protection system and testing required are described.

In all cases, the specification set out above must comply with the relevant British Standards. The designer must write clauses in the specification to cover special features in the design, fabrication or erection not set out in general clauses in codes and conditions of contract.

The aim is to ensure that the intentions in the design as to structural action, behaviour of materials, robustness and durability, etc., are met. Experience and great care are needed in writing the specification.

2.9.3 Quantities

Quantities of materials required are taken from the arrangement and detail drawings. The materials required for fabrication and erection are listed for ordering. The list comprises the separate types, sizes and quantities of hot- and cold-rolled sections, flats, plates, slabs, rounds and bolts. It is in the general form:

- Mark number
- Number off
- Description
- Weight per metre/square metre/unit length/area
- Total weight

The quantities are presented on standard sheets printed out by computer.

The bill of quantities is a schedule of the materials required and work to be carried out. It provides the basis on which tenders are to be obtained and payment made for work completed. The steelwork to be fabricated and erected is itemized under rolled and built-up beams, girders and columns, trusses and lattice girders, purlins and sheeting rails, bases, grillages, splice plates, bolts, etc. The bill requests the rate and amount for each item from which the total cost is estimated.

2.10 FABRICATION

Fabrication covers the process of making the individual elements of the steel-framed building from rolled steel sections and plates. The general process is set out briefly as follows:

1. The fabricator prepares the materials lists and drawings showing the shop details.
2. Rolled members are cut to length and drilled by numerically controlled plant.
3. Shapes of gussets, cleats, endplates, stiffeners, etc., are marked out and flame or plasma-arc cut and edges are ground. Hole locations are marked and holes drilled.
4. For built-up members, plates are flame or plasma-arc cut, followed by machining for edges and weld preparation.
5. Main components and fittings are assembled and positioned and final welding is carried out by automatic submerged-arc or gas-shielded process. Appropriate measures are taken to control distortion and cracking.
6. Members are cleaned by grit blasting, primed and given their mark number.

Careful design can reduce fabrication costs. Some points to be considered are as follows:

1. Rationalize the design so that as many similar members as possible are used. This will result in extra material being required but will reduce costs.
2. The simplest detail should be used so that welding is reduced to a minimum, sound welds can be assured and inspection and testing carried out easily.
3. Standard bolted connections should be used throughout.

The above is readily achieved in multistorey and standard factory building construction.

2.11 TRANSPORT AND ERECTION

Some brief comments are given regarding the effect of transport and erection on design:

1. The location of the site and method of transport may govern the largest size of member. Large members may be transported if special arrangements are made.

2. The method of erection and cranes to be used require careful selection. Mobile cranes can be used for single-storey buildings, while tower cranes or cranes climbing on the steelwork are required for multistorey buildings. Economies are achieved when building components are of similar size and weight, and cranes are used to capacity. Special provisions have to be made to erect large heavy members.

The above considerations affect the number and location of the site joints.

Factors considered in selecting the type of site joint are ease of assembly, appearance and cost. In general, welding is used in shop fabrication and bolts are used for site joints. Ordinary bolts in clearance holes make the cheapest joints and are used generally in all types of construction. Higher grades of ordinary bolts and friction grip fasteners are used for joints in rigid design and where strong joints are needed in simple design. Site welding is the most expensive form of jointing but gives the best appearance, though quality may be difficult to control under site conditions. Welding is essential with heavy rigid construction to achieve full strength joints. Care is needed in design to ensure that welding and inspection can be carried out easily.

Ideally, joints should be located near points of contraflexure, but such positions may not be convenient for fabrication or erection. Site joints on some types of structures are shown in Figure 2.8.

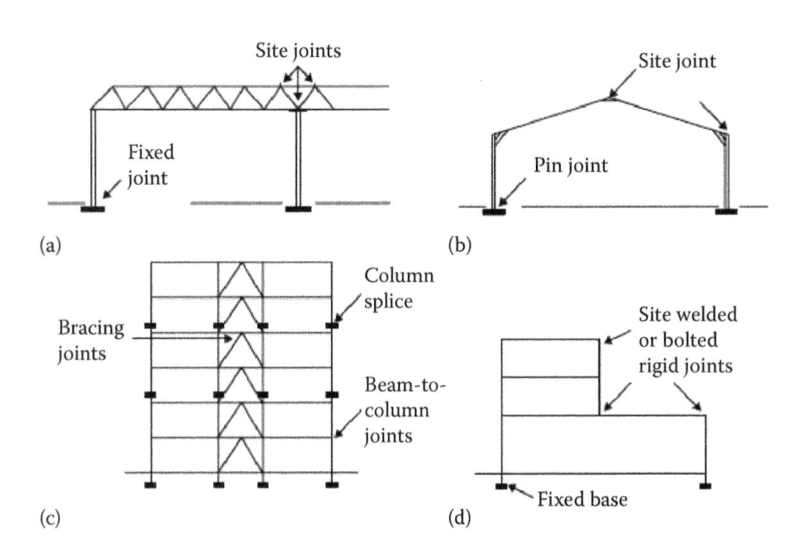

Figure 2.8 Types of joints: (a) lattice girder roof building; (b) pinned-base portal – high-grade bolts; (c) simple joints – ordinary bolts; (d) rigid frame.

Chapter 3

Structural steel design

3.1 DESIGN THEORIES

3.1.1 Development of design

The specific aim of structural design is, for a given framing arrangement, to determine the member sizes to support the structure's loads. The historical basis of design was trial and error. Then with the development of mathematics and science, the design theories – elastic, plastic and limit state – were developed, which permit accurate and economic designs to be made. The design theories are discussed; design methods given in BS 5950: Part 1 are set out briefly. Reference is also made to Eurocode 3 (EC3) and Eurocode 4 (EC4). The complete codes should be consulted.

3.1.2 Design from experience

Safe proportions for members such as depth/thickness, height/width, span/depth, etc., were determined from experience and formulated into rules. In this way, structural forms and methods of construction such as beam–column, arch–barrel vault and domes in stone, masonry and timber were developed, as well as cable structures using natural fibres. Very remarkable structures from the ancient civilizations of Egypt, Greece, Rome and the cathedrals of the middle ages survive as a tribute to the ingenuity and prowess of architects using this design basis. The results of the trial-and-error method still survive in our building practices for brick houses. An experimental design method is still included in the steel code.

3.1.3 Elastic theory

Elastic theory was the first theoretical design method to be developed. The behaviour of steel when loaded below the yield point is much closer to true elastic behaviour than that of other structural materials (Figure 3.1). All sections and the complete structure are assumed to obey Hooke's law and recover to their original state on removal of load if not loaded past yield.

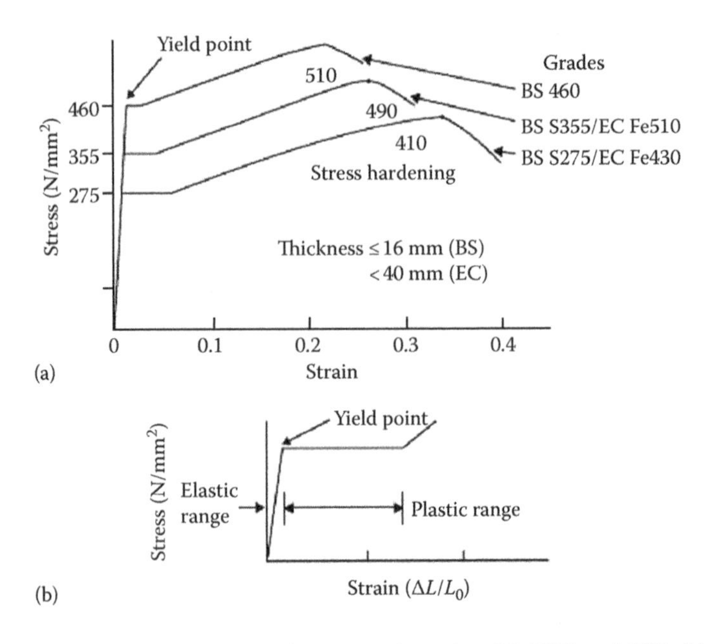

Figure 3.1 Stress–strain diagrams: (a) structural steels – BS 5950 and EC3; (b) plastic design.

Design to elastic theory was carried out in accordance with BS 449, *The Use of Structural Steel in Building.*

For design, the structure is loaded with the working loads, that is, the maximum loads to which it will be subjected during its life. Statically determinate structures are analysed using the simple theory of statics. For statically indeterminate structures, linear or first-order elastic theory is traditionally used for analysis. The various load cases can be combined by superposition to give the worst cases for design. In modern practice, second-order analysis taking account of deflections in the structure can be performed, for which computer programs and code methods are available. In addition, analysis can be performed to determine the load factor which will cause elastic instability where the influence of axial load on bending stiffness is considered. Dynamic analyses can also be carried out. Elastic analysis continues to form the main means of structural analysis.

In design to elastic theory, sections are sized to ensure permissible stresses are not exceeded at any point in the structure. Stresses are reduced where instability due to buckling such as in slender compression members, unsupported compression flanges of slender beams, deep webs, etc., can occur. Deflections under working loads can be calculated as part of the analysis and checked against code limits. The loading, deflection and elastic bending moment diagram and elastic stress distribution for a fixed base portal are shown in Figure 3.2a.

The permissible stresses are obtained by dividing the yield stress or elastic critical buckling stress where stability is a problem by a factor of safety. The one factor of safety takes account of variations in strengths of materials, inaccuracies in fabrication, possible overloads, etc. to ensure a safe design.

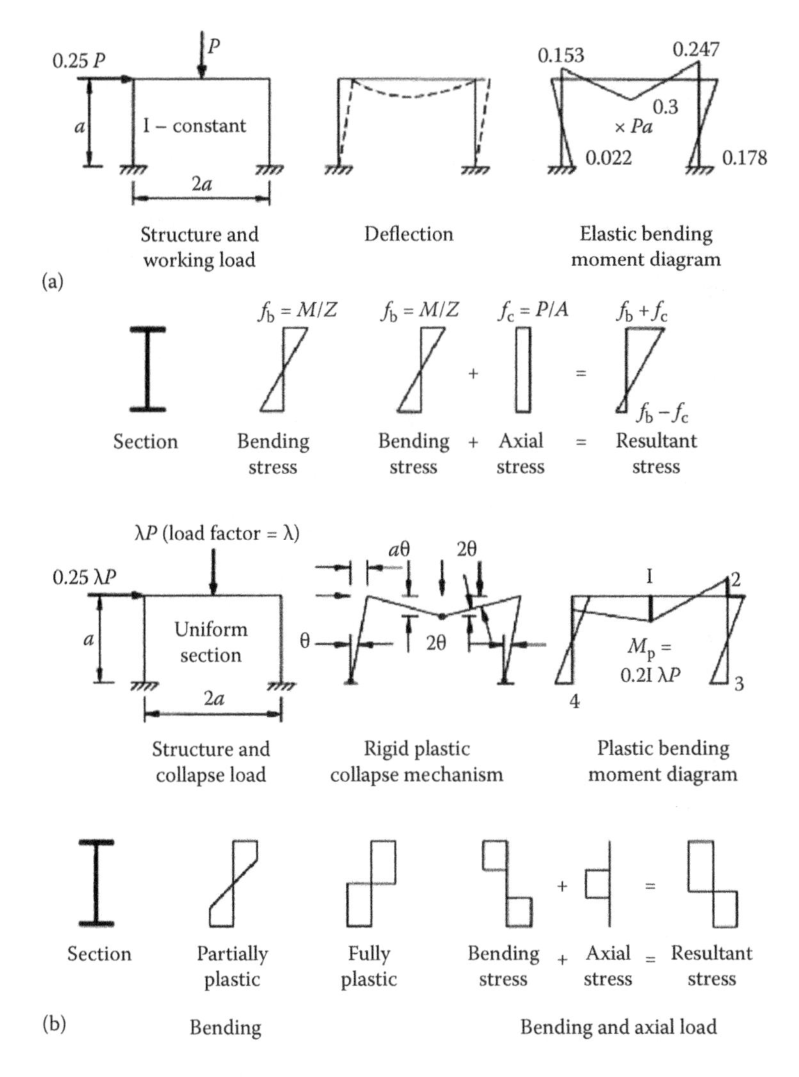

Figure 3.2 Loading, deflection, bending and stress distributions: (a) elastic analysis; (b) plastic analysis.

3.1.4 Plastic theory

Plastic theory was the next major development in design. This resulted from work at Cambridge University by the late Lord Baker, Professors Horne, Heyman, etc. The design theory is outlined.

When a steel specimen is loaded beyond the elastic limit, the stress remains constant while the strain increases, as shown in Figure 3.1b. For a beam section subjected to increasing moment, this behaviour results in the formation of a plastic hinge where a section rotates at the plastic moment capacity.

Plastic analysis is based on determining the least load that causes the structure to collapse. Collapse occurs when sufficient plastic hinges have formed to convert the structure to a mechanism. The safe load is the collapse load divided by a load factor.

In design, the structure is loaded with the collapse or factored loads, obtained by multiplying the working loads by the load factor, and analysed plastically. Methods of rigid plastic analysis have been developed for single-storey and multistorey frames where all deformation is assumed to occur in the hinges. Portals are designed almost exclusively using plastic design. Software is also available to carry out elastic–plastic analysis where the frame first acts elastically and, as the load increases, hinges form successively until the frame is converted to a mechanism. More accurate analyses take the frame deflections into account. These secondary effects are only of importance in some slender sway frames. The plastic design methods for multistorey rigid non-sway and sway frames are given in BS 5950 and EC3.

The loading, collapse mechanism and plastic bending moment diagram for a fixed-base portal are shown in Figure 3.2b. Sections are designed using plastic theory and the stress distributions for sections subjected to bending only and bending and axial load are also shown in the figure. Sections require checking to ensure that local buckling does not occur before a hinge can form. Bracing is required at the hinge and adjacent to it to prevent overall buckling.

3.1.5 Limit state theory and design codes

Limit state theory was developed by the Comitée Européen Du Béton for design of structural concrete and has now been widely accepted as the best design method for all materials. It includes principles from the elastic and plastic theories and incorporates other relevant factors to give as realistic a basis for design as possible. The following concepts are central to limit state theory:

1. Account is taken in design of all separate conditions that could cause failure or make the structure unfit for its intended use. These are the various limit states and are listed in the next section.
2. The design is based on the actual behaviour of materials in structures and performance of real structures established by tests and long-term

observations. Good practice embodied in clauses in codes and specifications must be followed in order that some limit states cannot be reached.
3. The overall intention is that design is to be based on statistical methods and probability theory. It is recognized that no design can be made completely safe; only a low probability that the structure will not reach a limit state can be achieved. However, full probabilistic design is not possible at present and the basis is mainly deterministic.
4. Separate partial factors of safety for loads and materials are specified. This permits a better assessment to be made of uncertainties in loading, variations in material strengths and the effects of initial imperfections and errors in fabrication and erection. Most importantly, the factors give a reserve of strength against failure.

The limit state codes for design of structural steel now in use are BS 5950: Part 1 (2000) and EC3 (1993). All design examples in the book are to BS 5950 and many of these design examples are to EC3 for comparison purpose. All new chapters, namely, Chapters 6, 8, 10, 11, 13, 14 and 15 are complying with EC3 and EC4 code of practices.

In limit state philosophy, the steel codes are Level 1 safety codes. This means that safety or reliability is provided on a structural element basis by specifying partial factors of safety for loads and materials. All relevant separate limit states must be checked. Level 2 is partly based on probabilistic concepts and gives a greater reliability than a Level 1 design code. A Level 3 code would entail a fully probabilistic design for the complete structure.

3.2 LIMIT STATES AND DESIGN BASIS

BS 5950 states in Clause 2.1.1.1 (and similarly the case in EC3 DDN ENV 1993-1-1 1992) that:

> The aim of structural design should be to provide with due regard to economy a structure capable of fulfilling its intended function and sustaining the specific loads for its intended life.

In Clause 2.1.3 the code states:

> Structures should be designed by considering the limit states beyond which they would become unfit for their intended use. Appropriate factors should be applied to provide adequate degree of reliability for ultimate limit state and serviceability limit state.

The limit states specified for structural steel work on the current codes are mainly in two categories:

Table 3.1 Limit states specified in BS 5950 (see Table 1 – limit states BS 5950-1:2000 and EC3 and Clause 2.3.2 of EC3)

Ultimate BS 5950	*Serviceability BS 5950*	*Ultimate EC3*
1. Strength including general yielding rupture, buckling and forming a mechanism	5. Deflection	1. Static equilibrium of the structure
2. Stability against overturning and sway stability	6. Vibration	2. Rupture or excessive deformation of a member
3. Fracture due to fatigue	7. Wind induced oscillation	3. Transformation of the structure into a mechanism
4. Brittle fracture	8. Durability	4. Instability induced by second-order effect such as lack of fit, thermal effects, sway
		5. Fatigue
		6. Accidental damage

- Ultimate limit states, which govern strength and cause failure if exceeded. Ultimate limit states concern the safety of the whole or part of the structure.
- Serviceability limit states, which cause the structure to become unfit for use but stopping short of failure or correspond to limits beyond which specified services criteria are no longer met.

The separate limit states given in Table 1 of BS 5950 and EC3 are shown in Table 3.1.

3.3 LOADS, ACTIONS AND PARTIAL SAFETY FACTORS

The main purpose of the building structure is to carry loads over or around specified spaces and deliver them to the ground. All relevant loads and realistic load combinations have to be considered in design.

3.3.1 Loads

BS 5950 and EC3 classify working loads into the following traditional types:

1. Dead loads due to the weight of the building materials (permanent action in EC3, see EC1-Action on Structures. The UK NAD advices designers to continue to use BS 6399 Part 1, CP3 Chapter V, Part 2 and BS 648). Accurate assessment is essential. Refer to BS 6399-1.
2. Imposed loads (variable actions in EC3) due to people, furniture, materials stored, snow, erection and maintenance loads. Refer to BS 6399-3.

3. Wind loads. These depend on the location, the building size and height, openings in walls, etc. Wind causes external and internal pressures and suctions on building surfaces and the phenomenon of periodic vortex shedding can cause vibration of structures. Wind loads are estimated from maximum wind speeds that can be expected in a 50-year period. They are to be estimated in accordance with BS 6399: Part 2 and CP3: Chapter V, Part 2.
4. Dynamic loads are generally caused by cranes. The separate loads are vertical impact and horizontal transverse and longitudinal surge. Wheel loads are rolling loads and must be placed in position to give the maximum moments and shears. Dynamic loads for light and moderate cranes are given in BS 2573.1.

For building, the partial factor for loads γ_f for dead loads is 1.4 and imposed loads is 1.6. The corresponding values in EC3 are γ_G (1.35) and γ_Q (1.5), respectively. Seismic loads, though very important in many areas, do not have to be considered in the United Kingdom. The most important effect is to give rise to horizontal inertia loads for which the building must be designed to resist or deform to dissipate them. Vibrations are set up, and if resonance occurs, amplitudes greatly increase and failure results. Damping devices can be introduced into the stanchions to reduce oscillation. Seismic loads are not discussed in the book.

3.3.2 Partial factors for loads/partial safety factors and design loads

Partial factors for loads for the ultimate limit state for various loads and load combinations are given in Table 2 of BS 5950, Clauses 2.3 and 2.4.3 of EC3. Part of the code table is shown in Table 3.2.

In limit state design,

Design loads = characteristic or working loads F_K × partial factor of safety γ_f

Table 3.2 BS 5950 design loads for the ultimate limit state (part of Table 2 – BS 5950 2000)

Load combination	*Design load*
Dead load	$1.4G_K$
Dead load restraining overturning	$1.0G_K$
Dead and imposed load	$1.4G_K + 1.6Q_K$
Dead, imposed and wind load	$1.2(G_K + Q_K + W_K)$

Note: G_K = dead load; Q_K = imposed load; W_K = wind load.

Table 3.3 BS 5950 design strength p_y and EC3 nominal value of yield strength f_y

Steel grade		*Thickness (mm)*	*BS 5950 p_y (N/mm²)*	*EC3 f_y (N/mm²)*
BS 5950	EC3			
S275	Fe430 (S275)	≤16	275	275
		≤40	265	275
S355	Fe 510 (S355)	≤16	355	355
		≤40	345	355

Note: For hot rolled structural steel, for more information, see Table 9-BS 5950 and EN 10025.2, Table 3.1-EC3.

3.4 STRUCTURAL STEELS – PARTIAL SAFETY FACTORS FOR MATERIALS

Some of the design strengths, p_y (BS 5950) and f_y (EC3), of structural steels used in the book, taken from Table 9 in BS 5950, are shown in Table 3.3. Also refer to Table 3.1 of EC3.

Design strength is given by

$$\text{Design strength} = \frac{\text{Yield or characteristic strength}}{\text{Partial factor of safety } \gamma_m}$$

In BS 5950, the partial safety factor for materials $\gamma_m = 1.0$. In EC3, the partial safety factors for resistance are given in Section 5.1.1. The value for member design is normally 1.1.

3.5 DESIGN METHODS FROM CODES – ULTIMATE LIMIT STATE

3.5.1 Design methods from BS 5950

The design of steel structures may be made to any of the following methods set out in Clause 2.1.2 of BS 5950:

- Simple design;
- Continuous design or rigid design;
- Semi-continuous design or semirigid design;
- Experimental verification.

The clause states:

> the details of the connections should be such as to fulfil the assumptions made in the relevant design method without adversely affecting any other parts of the structure.

(a) *Simple design*

The connections are assumed not to develop moments that adversely affect the member or structure. The structure is analysed, assuming that it is statically determinate with pinned joints. In a multistorey beam–column frame, bracing or shear walls acting with floor slabs are necessary to provide stability and resistance to horizontal loading.

(b) *Continuous design or rigid design*

The connections are assumed to be capable of developing actions arising from a fully rigid analysis, that is, the rotation is the same for the ends of all members meeting at a joint.

The analysis of rigid structures may be made using either elastic or plastic methods. In Section 5 of the code BS 5950, methods are given to classify rigid frames into non-sway, that is, braced or stiff rigid construction, and sway, that is, flexible structures. The non-sway frame can be analysed using first-order linear elastic methods including subframe analysis. For sway frames, second-order elastic analysis or methods given in the code (BS 5950) (extended simple design or the amplified sway method) must be used. Methods of plastic analysis for non-sway and sway frames are also given.

(c) *Semicontinuous design or semirigid design*

The code BS 5950 states that in this method some degree of joint stiffness short of that necessary to develop full continuity at joints is assumed. The relative stiffness of some common bolted joints is shown in the behaviour curves in Figure 3.3. Economies in design can be achieved if partial fixity is taken into account. The difficulty with the method lies in designing a joint to give a predetermined stiffness and strength.

The code BS 5950 further states that the moment and rotation capacity of the joint should be based on experimental evidence which may permit some limited plasticity. However, the ultimate tensile capacity of the fastener is not to be the failure criterion. Computer software where the semirigid joint is modelled by an elastic spring is available to carry out the analyses. The spring constant is taken from the initial linear part of the behaviour curve. Plastic analysis based on joint strength can also be used.

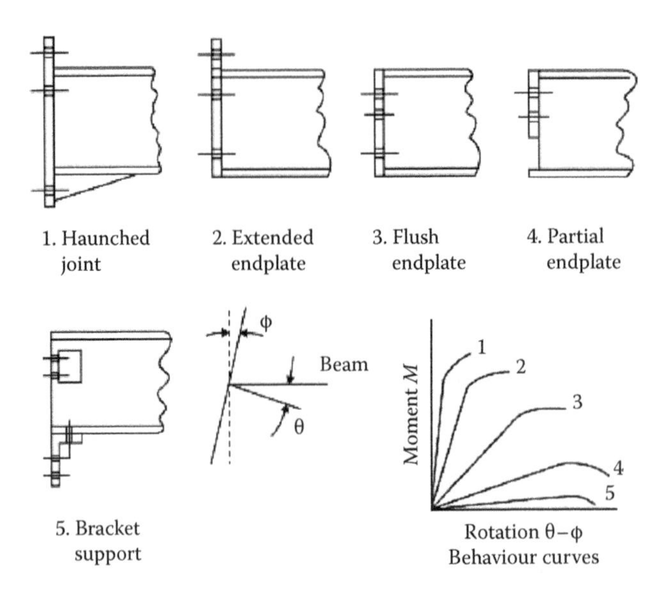

Figure 3.3 Beam–column joint behaviour curves.

The code BS 5950 also gives an empirical design method. This permits an allowance to be made in simple beam–column structures for the inter-restraint of connections by an end moment not exceeding 10% of the free moment. Various conditions that have to be met are set out in the clause. Two of the conditions are as follows:

- The frame is to be braced in both directions.
- The beam-to-column connections are to be designed to transmit the appropriate restraint moment in addition to the moment from eccentricity of the end reactions, assuming that the beams are simply supported. Further information on details design procedures are given in the SCI publications P-263 and P-183.

(d) *Experimental verification*

The code BS 5950 states that where the design of a structure or element by the above methods is not practicable, the strength, stability and stiffness may be confirmed by loading tests as set out in Section 7 of the code.

3.5.2 Analysis of structures – EC3

The methods of calculating forces and moments in structures given in EC3, Section 5.2 are set out briefly:

1. Simple framing statically determinate structures – use statics. All joints are pinned and the structure should be laterally restrained in-plane and out-of-plane to provide sway stability and resist horizontal forces (actions).
2. Statically indeterminate structures. All joints are rigid (for elastic design) or full strength (for plastic analysis). Joints can transmit all moments and forces (actions) – elastic global analysis may be used in all cases. Plastic global analysis may be used where specific requirements are met.
3. Elastic analysis – Linear behaviour may be assumed for first- and second-order analysis where sections are designed to plastic theory. Elastic moments may be redistributed.
4. Effects of deformation – Elastic first-order analyses are permitted for braced and non-sway frames. Second-order theory taking account of deformation can be used in all cases.
5. Plastic analysis – Either rigid plastic or elastic–plastic methods can be used. Assumptions and stress–strain relationships are set out. Lateral restraints are required at hinge locations.

3.5.3 Member and joint design

Provisions for member design from BS 5950 and EC3 and are set out briefly:

1. Classification of cross-sections – In both codes, member cross-sections are classified into class 1 cross-section, class 2 cross-section, class 3 cross-section and class 4 cross-section in EC3 correspond to plastic, compact, semicompact and slender, respectively. Only the plastic cross-section can be used in plastic analysis (Figure 3.2b).
2. Tension members – Design is based on the net section. The area of unconnected angle legs is reduced.
3. Compression members:

   - Short members – Design is based on the squash resistance;
   - Slender members – Design is based on the flexural buckling resistance.

4. Beams – Bending resistances for various cross-section types are

   - Plastic and compact – design for plastic resistance;
   - Semicompact – design for elastic resistance;
   - Slender – buckling must be considered;
   - Biaxial bending – use an interaction expression;
   - Bending with unrestrained compression flange – design for lateral torsional buckling.

   Shear resistance and shear buckling of slender webs to be checked. Tension field method of design is given in both codes.
   Combined bending and shear must be checked in beams where shear force is high.

Webs checks – Check web crushing and buckling in both codes. A flange-induced buckling check and crippling resistance of the web are given in Eurocode No. 3 (EC3).

5. Members with combined tension and moment – Checks cover single axis and biaxial bending:

   - Interaction expression for use with all cross-sections;
   - More exact expression for use with plastic and compact cross-sections where the moment resistance is reduced for axial load.

6. Members with combined compression and moment – Checks cover single axis and biaxial bending:

   - Local capacity check – Interaction expression for use with all cross-sections; more exact expression for use with plastic and compact cross-sections where the moment resistance is reduced for axial load;
   - Overall buckling check – Simplified and more exact interaction checks are given which take account of flexural and lateral torsional buckling.

7. Members subjected to bending shear and axial force – Design methods are given for members subjected to combined actions.
8. Connections – Procedures are given for design of joints made with ordinary bolts, friction grip bolts, pins and welds.

3.6 STABILITY LIMIT STATE

Design for the ultimate limit state of stability is of the utmost importance. Horizontal loading is due to wind, dynamic and seismic loads and can cause overturning and failure in a sway mode. Frame imperfections give rise to sway from vertical loads.

BS 5950 states that the designer should consider stability against overturning and sway stability in design.

1. Stability against overturning – To ensure stability against overturning, the worst combination of factored loads should not cause the structure or any part of it (including the foundations) to slide, overturn or lift off its seating. Checks are required during construction.
2. Sway stability – The structure must be adequately stiff against sway. The structure is to be designed for the applied horizontal loads and in addition a separate check is to be made for notional horizontal loads. The notional loads take account of imperfections such as lack

of verticality. The loads applied horizontally at roof and floor level are taken as the greater of:

- 1% of the factored dead loads;
- 0.5% of the factored dead plus imposed load.

Provisions governing their application are given.

3.7 DESIGN FOR ACCIDENTAL DAMAGE

3.7.1 Progressive collapse and robustness

In 1968, a gas explosion near the top of a 22-storey precast concrete building blew out side panels, causing building units from above to fall onto the floor of the incident. This overloaded units below and led to collapse of the entire corner of the building. A new mode of failure termed 'progressive collapse' was identified where the effects from a local failure spread and the final damage is completely out of proportion to the initial cause. New provisions were included in the Building Regulations at that time to ensure that all buildings of five stories and over in height were of sufficiently robust construction to resist progressive collapse as a result of misuse or accident.

3.7.2 *Building Regulations 1991* (also see Approved Document A – Structure [2004 Edition] incorporating 2010 and 2013 amendments)

Part A, *Structure of the Building Regulations*, Section 5, A3/A4 deals with disproportionate collapse. The Regulations state that all buildings must be so constructed as to reduce the sensitivity to disproportionate collapse due to an accident. The main provisions are summarized as follows:

1. If effective ties complying with the code are provided, no other action is needed (Section 2.7.3(a)).
2. If ties are not provided, then a check is to be made to see if load bearing members can be removed one at a time without causing more than a specified amount of damage.
3. If in (2), it is not possible in any instance to limit the damage, the member concerned is to be designed as a 'protected' or 'key' member. It must be capable of withstanding 34 kN/m^2 from any direction.
4. Further provisions limit damage caused by roof collapse.

The Building Regulations should be consulted.

3.7.3 BS 5950 requirements for structural integrity

Clause 2.4.5 of BS 5950 ensures that design of steel structures complies with the Building Regulations. The main provisions are summarized. The complete clause should be studied. Also, see Clause 5.2.2 of EC3 for structural stability and frames.

(a) *All buildings*

Every frame must be effectively tied at roof and floors and columns must be restrained in two directions at these levels. Beam or slab reinforcement may act as ties, which must be capable of resisting a force of 75 kN at floor level and 40 kN at roof level.

(b) *Certain multistorey buildings*

To ensure accidental damage is localized, the following recommendations should be met.

- Sway resistance – No substantial part of a building should rely solely on a single plane of bracing in each direction.
- Tying – Ties are to be arranged in continuous lines in two directions at each floor and roof. Design forces for ties are specified. Ties anchoring columns at the periphery should be capable of resisting 1% of the vertical load at that level.
- All horizontal ties and all other horizontal members should be capable of resisting a factored tensile load of not less than 75 kN (and should not be considered as additive to other loads). Also see Clause 2.4.5.3 of BS 5950.
- Columns – Column splices should be capable of resisting a tensile force of two-thirds of the factored vertical load.
- Integrity – Any beam carrying a column should be checked for localization of damage ([c] below).
- Floor units should be effectively anchored to their supports.

(c) *Localization of damage*

The code states that a building should be checked to see if any single column or beam carrying a column could be removed without causing collapse of more than a limited portion of the building. If the failure would exceed the specified limit, the element should be designed as a key element. In the design check, the loads to be taken are normally dead load plus one-third wind load plus one-third imposed load; the load factor is 1.05. The extent of damage is to be limited.

(d) *Key element*

A key element is to be designed for the loads specified in (c) above plus the load from accidental causes of 34 kN/m^2 acting in any direction. Also see Clause 2.4.5.4 key elements, BS 5950: 2000.

3.8 SERVICEABILITY LIMIT STATES

The serviceability limit states are listed in BS 5950 as deflection, vibration and oscillation, repairable damage due to fatigue, durability and corrosion. Deflection limits and vibration will be discussed in this section. Fatigue and corrosion are treated in Section 3.9.

Breaching serviceability limit states renders the structure visually unacceptable, uncomfortable for occupants or unfit for use without causing failure. Exceeding limits can also result in damage to glazing and finishes. Also see Clause 7 of the EC3 for vertical deflection, horizontal deflection and dynamic effects.

3.8.1 Deflection limits

Deflection is checked for the most adverse realistic combination of serviceability loads. The structure is assumed to be elastic. Deflection limits are given in Table 8 of BS 5950 and Table 4.1 of EC3. Some values are

- Beams carrying plaster or other brittle finishes – span/360 in BS 5950 and span/350 in EC3;
- Horizontal deflection of columns in each storey of a multistorey building – storey height/300 in BS 5950.

EC3 recommends two limiting values for vertical deflections δ_2 and δ_{max},

δ_2 = deflection due to variable loads

δ_{max} = final deflection (see Figure 3.4)

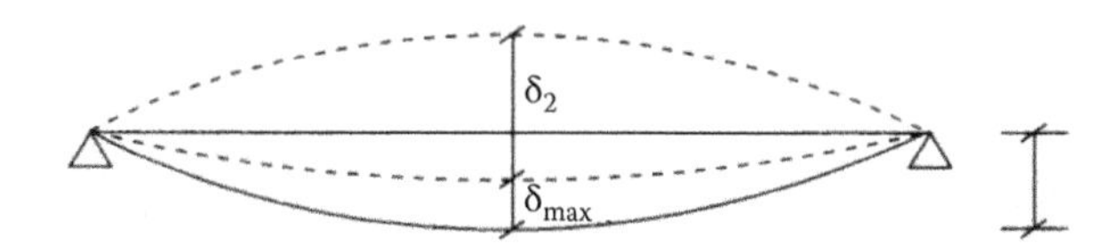

Figure 3.4 Final vertical deflection δ_{max}.

3.8.2 Vibration

Design for vibration is outside the scope of the book. Some brief notes are given.

Vibration is caused by wind-induced oscillation, machinery and seismic loads. People walking on slender floors can also cause vibration. Resonance occurs when the period of the imposed force coincides with the natural period of the structure when the amplitude increases and failure results. Damping devices can be installed to reduce the amplitude.

Wind loading can cause flexible tall buildings and stacks to vibrate at right angles to the wind direction due to vortex shedding. Gust-induced vibration occurs in the wind direction. Designs are made to ensure that resonance does not occur or devices to break up vortices are installed. Long-span flexible roofs such as cable, girder or net are prone to aerodynamic excitation.

3.9 DESIGN CONSIDERATIONS

Other considerations listed as limit states in BS 5950 and EC3 are discussed.

3.9.1 Fatigue

Fatigue failure is an ultimate limit state, but, if repairable, a serviceability limit state failure occurs in members subjected to variable tensile stress, which may be at values well below failure stress. There is virtually no plastic deformation. Bridges, crane girders, conveyor gantries, etc., are subject to fatigue.

Fatigue tests are carried out to determine endurance limits for various joints and members. Failure usually occurs at welded joints. Tests subjecting joints to pulsating loads are made. Some types of welded joints tested are shown in Figure 3.5.

The joint giving the least disturbance to stress flow gives the best results. Some comments on welded construction are as follows:

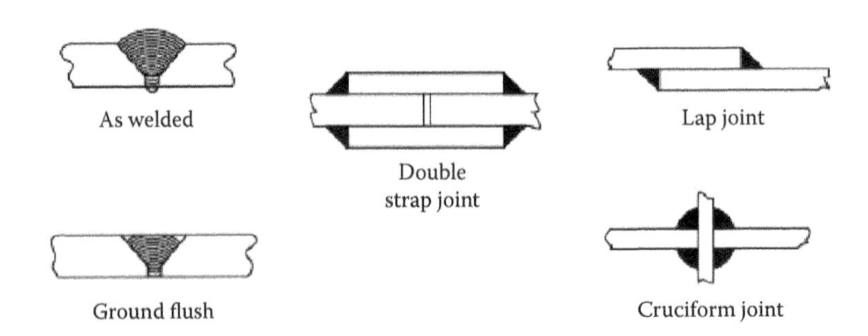

Figure 3.5 Load-carrying welds.

1. Butt welds give the best performance. Fatigue strength is increased by grinding welds flush.
2. Fillet-welded joints do not perform well. The joints in order of performance are shown in Figure 3.5. The effect of non-load bearing fillet welds is also significant.
3. For plate girders the best results are given when the web flange weld is continuous and made by an automatic process. Intermittent welds give low fatigue strength.

Structures are usually subjected to widely varying or random stress cycles. Methods are available for estimating the cumulative damage for a given load spectrum. The life expectancy of the structure can be estimated if the maximum stress is specified.

Some 'good practice' provisions in design are as follows:

1. Avoid stress concentrations in regions of tensile stress, for example, taper thicker plates to meet thinner plates at joints, locate splices away from points of maximum tensile stress, and use continuous automatic welding in preference to intermittent or manual welding.
2. Reduce working stresses depending on the value of the maximum tensile stress, the ratio of maximum to minimum stress, the number of cycles and the detail.
3. No reduction in ability to resist fatigue is needed for bolted joints. Sound, full penetration butt welds give the best performance.
4. Cumulative damage must be considered.

BS 5950 states that fatigue need only be considered when the structure or element is subjected to numerous stress fluctuations. Wind-induced oscillation should be considered. Fatigue must be taken into account in some heavy crane structures.

3.9.2 Brittle fracture

Brittle fracture starts in tensile stress areas at low temperatures and occurs suddenly with little or no prior deformation. It can occur at stresses as low as one-quarter of the yield stress and it propagates at high speed with the energy for crack advance coming from stored elastic energy.

The mode of failure is by cleavage, giving a rough surface with chevron markings pointing to the origin of the crack. A ductile fracture occurs by a shear mechanism. The change from shear to cleavage occurs through a transition zone at 0°C or lower. The Charpy V-notch test, where a small beam specimen with a specified notch is broken by a striker, is the standard test. The energy to cause fracture is measured at various temperatures. The fracture becomes more brittle and energy lower at lower temperatures.

Steels are graded A–E in order of increasing notch ductility or resistance to brittle fracture.

The likelihood of brittle fracture occurring is difficult to predict with certainty. Important factors involved are as follows:

- Temperature – Fractures occur more frequently at low temperatures.
- Stress – Fractures occur in regions of tensile stress.
- Plate thickness – Thick plates, which are more likely to contain defects, have a higher risk of failure.
- Materials – Alloys included and production processes produce small grain size in steels which improves fracture toughness at low temperatures.

Precautions necessary to reduce the likelihood of brittle fracture occurring are as follows:

- Materials – Select steels with high Charpy value and use thin plate.
- Design detail – Some obvious 'don'ts' are
  - Avoid abrupt changes of section; taper a thick plate to meet a thin one;
  - Do not locate welds in tension in high stress areas;
  - Fillet welds should not be made across tension flanges;
  - Avoid intermittent welds.

Detailing should be such that inspection and weld testing can be readily carried out.

- Fabrication – Flame cut edges should be ground off. Welding practice should be of highest quality and adequate testing carried out.
- Erection – No bad practices such as burning holes or tack-welding temporary fixings should be permitted.

Brittle fracture is an ultimate limit state in BS 5950 and EC3. Maximum thickness of sections and plates for various grades of steels are given in Table 9 and Table 3.1 in the BS 5950 and EC3, respectively.

3.9.3 Corrosion protection

Corrosion and durability is a listed serviceability limit state in BS 5950 and EC3, respectively. For durability requirements, see relevant sections of EC3. Steels are particularly susceptible to corrosion, an electrochemical process where iron is oxidized in the presence of air, water and other pollutants. Corrosion is progressive and leads to loss of serviceability and eventual failure. It is very necessary to provide steelwork with suitable protection against corrosion. The choice of system depends on the type and degree of pollution and length of life required. A long maintenance-free life is assured if the correct system is used and applied correctly (BS 5493, 1977).

Surface preparation is the most important single factor in achieving successful protection against corrosion. All hot-rolled steel products are covered with a thin layer of iron oxide termed mill scale. If this is not removed, it will break off under flexure or abrasion and expose the steel to rusting. All mill scale, rust and slag spatter from welding must be removed before paint is applied. Methods of surface preparation are

- Manual cleaning using scrapers and brushes, etc.;
- Flame cleaning using a torch to loosen the scale, which may be removed by brushing;
- Pickling in a tank of acid, used as a preparation for galvanizing;
- Blast cleaning, where iron grit or sand is projected against the steel surface; manual and automatic processes are used.

BS 5493 defines cleanliness quality and the desirable surface profile for the cleaned steel.

Two types of protective coatings used are

- Metallic – Metal spraying and galvanizing;
- Nonmetallic – Paint systems.

Metal spraying is carried out by atomizing metal wire or powder by oxy-acetylene flame in a gun and projecting the molten droplets onto the surface of the steel part. Zinc and aluminium coatings are used.

In hot-dip galvanizing, rust and mill scale are removed by pickling and the cleaned metal is immersed in a bath of molten zinc. The thickness of coating depends on the time of immersion and speed of withdrawals.

A common paint system consists of a primer of zinc chromate or phosphate and under coat and finish coat of micaceous iron oxide paint. Many other paint types are used (chlorinated rubber, epoxy, urethane, bituminous paints, etc.).

The design and details can greatly influence the life of a paint coating. Some important points are as follows:

- Use detail that sheds water, and avoid detail that provides places where water can be trapped, such as in upturned channel sections. If such sections are unavoidable, they should have drain holes.
- Box sections should be sealed.
- Access for maintenance should be provided.

BS 5493 gives recommendations for design and detail.

Weathering steels containing copper, nickel and chromium can also be used. The alloying elements form a dense self-healing rust film which protects the steel from further corrosion. These steels are used in exposed skeletal structures.

3.9.4 Fire protection

(a) *General considerations*

Fire causes injury and loss of life, damage to and destruction of finishes, furnishings and fittings and damage to and failure of the structure itself. Design must aim at the prevention or minimization of all of the above effects. Injury and loss of life are caused by toxic gases generated by combustion of furnishings, etc., as well as by heat. Destruction of property and structural damage and failure are caused by heat and burning of combustible material.

The means of prevention and control of damage may be classified as:

- Early detection by smoke and heat detectors or manual sighting followed by extinction of the fire by automatic sprinklers, manual application of water, foams, etc.;
- Containment by dividing the building into fireproof compartments to prevent fire spread and smoke travels, and provision of fireproof escape routes;
- Fire protection of load bearing structural member ties and connections to ensure collapse does not occur before people can escape or the fire be extinguished and that the building can be subsequently repaired.

The last two control methods form an essential part of the design considerations for steel structures. All multistorey commercial and residential buildings require fire protection of structural members, but single-storey and some other industrial buildings do not need protection.

(b) *Fire and structural steelwork, fire resistance*

Structural steelwork performs badly in fire. Temperatures commonly reach 1200°C at the seat of the fire, while the critical temperature for steel is about 550°C. At this temperature, the yield stress of steel has fallen to about 0.7 of its value at ambient temperatures, that is, to the stress level in steel at working loads.

The Fire Research Station carries out tests to determine the behaviour of steel members in fire and the efficiency of protective measures. The ability of a structural element to continue to support load is termed the fire resistance and this is stated in terms of time (½, 1, 2, 4 h). Fire resistance is determined by testing elements and protection systems in furnaces heated to various temperatures up to 1100°C. The resistance required depends on the type and height of buildings, contents and the type and location of the structural member and whether a fire extinguishing system is provided. This is given in the Building Regulations.

(c) *Type of fire protection*

Examples of fire protection for columns and floor beams in steel-framed buildings are shown in Figure 3.6. Fire resistance periods for various types of protection have been established by tests on loaded structural members (details in BS 476, also see relevant sections in EC3). Some notes on the various types of protection used are given below:

- Solid protection for columns, where the concrete assists in carrying the load, is not much used in modern construction. Beams can also be cased in concrete. A concrete thickness of 50 mm will give 2 h protection.
- Brick-clad steel-framed buildings, where brick provides the walling and fire protection, are a popular building system.
- Hollow casing can be applied in the form of prefabricated casing units or vermiculite gypsum plaster placed on metal lathing.

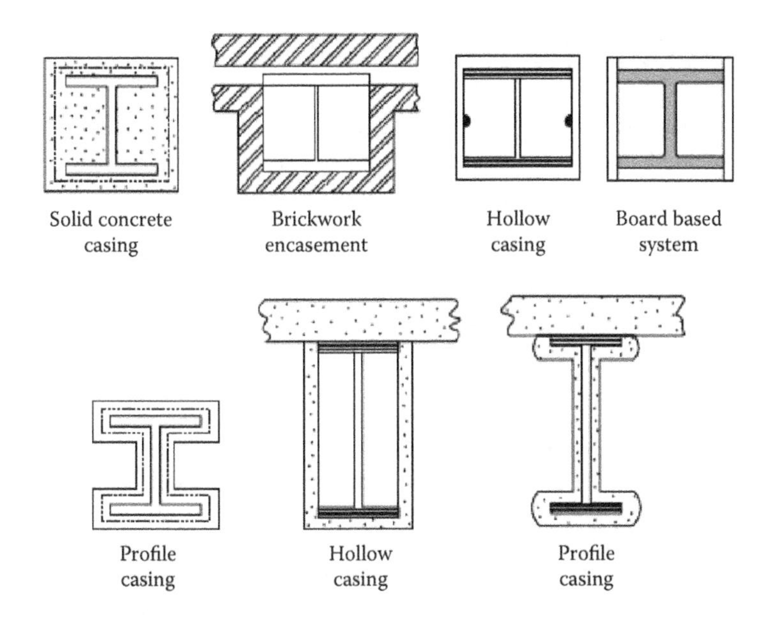

Figure 3.6 Types of fire protection for columns and beams.

- Profile casing, where vermiculite cement is sprayed on to the surface of the steel member, is the best system to use for large plate and lattice girders and is the cheapest protection method. A thickness of 38 mm of cement lime plaster will give 2 h protection.
- Intumescent coatings inflate into foam under the action of heat to form the protective layer.
- Fire resistant ceilings are used to protect floor steel.
- Board systems are used to form rectangular encasements around steel members, such as internal beams and columns. The type and thickness of the boards used influence the level of fire resistance that can be achieved.

Another system of fire protection that can be used with frames of box sections is to fill these with water, which is circulated through the members. This method has been used in Europe and the United States. The aim of the method is to ensure the steel temperature does not reach a critical level.

Concrete-filled hollow sections also have increased fire resistance.

(d) *Fire engineering*

A scientific study of fires and behaviour of structures in fire has led to the development of fire engineering. Most of the work has been carried out in Sweden. The aims of the method are to determine fire load and predict the maximum temperatures in the steel frame and its resistance to collapse. The method can be used to justify leaving steelwork unprotected in certain types of buildings.

(e) *Building Regulations*

The statutory requirements for fire protection are set out in the Building Regulations, Part B, *Fire Safety*. A brief summary of the main provisions is as follows:

- Buildings are classified according to use, which takes account of the risk, severity of possible fire or danger to occupants. Assembly and recreational, industrial and storage buildings carry the highest risk.
- Large buildings must be divided into fireproof compartments to limit fire spread. The compartment size depends on the use, fire load, ease of evacuation, height and availability of sprinklers.
- Minimum periods of fire resistance are specified for all buildings. These depend on the purpose group, building height and whether a sprinkler system is installed.
- Every load bearing element, that is frame, beam, column or wall, must be so constructed as to have the fire resistance period specified.

The complete Building Regulations should be consulted.

(f) *BS 5950 Part 8: Fire protection and EC3 EN1993-1-2 structural fire design*

The codes set out data and procedures for checking fire resistance and designing protection for members in steel-framed buildings. Some of the main provisions in the code are summarized below:

- Strength reduction factors for steel members at elevated temperatures are given.
- Methods for determining the strength of members at high temperatures which depend on the limiting temperatures of protected and unprotected members are given. These can be used to establish if load bearing steelwork can be left unprotected.
- Procedures to be followed to determine the thickness of fire protection required by testing, or from calculations using specified material properties, are given.

The code also has provisions for portal frames, slabs, walls, roofs, composite floors, concrete-filled hollow sections, water-filled hollow sections, etc. The code should be consulted.

Chapter 4

Preliminary design

4.1 GENERAL CONSIDERATIONS

Preliminary design may be defined as a rapid approximate manual method of designing a structure as opposed to carrying out rigorous analysis and detailed design. The overall aim for the design of a given structure (from a structural engineering point of view) is to identify critical loads, estimate design actions and select sections. The problem is bound up with conceptual design, alternative systems, idealization, identification of critical members and rationalization. The process depends greatly on the designer's experience and use of appropriate design aids. The term is also often applied to manual design.

4.2 NEED FOR AND SCOPE OF PRELIMINARY DESIGN METHODS

Preliminary design is needed for the following reasons:

- To obtain sections and weights for cost estimation;
- To compare alternative proposals;
- To obtain initial sections for computer analysis;
- To check a completed design.

The need for approximate manual methods is more important than ever because a 'black box' (computer analysis and design process) design era is taking over. It is necessary to know if output is right, wrong or complete nonsense. Preliminary design must not replace normal rigorous design and certified checking must still be carried out.

Methods of preliminary analysis not dependent on member sizes are set out for both elastic and plastic theories. Redundant structures are treated as statically determinate by approximately locating points of contraflexure by use of subframes or by assigning values of actions at critical positions. Handbooks give solutions to commonly used members and frames such as continuous beams and portals.

Design aids for sizing members are given in many handbooks in tabular and chart form (*Steel Designers Manual* 1986, 1994, 2002, 2012; Steelwork Design Guide to BS 5950-1 2000, 2001; Steel Construction Institute, 1987). These give load capacities for a range of sections taking account of buckling where appropriate. Also see Steelwork Design Guide to Eurocode 3, Part 1.1 – Introducing Eurocode 3: A Comparison of EC3, Part 1.1 with BS 5950, Part 1 and the following design guides/documents:

SCI P355 Design of Composite Beams with Large Web Openings, 2011
SCI P358 Joints in Steel Construction: Simple Joints to Eurocode 3, 2014
SCI P359 Composite Design of Steel Framed Buildings, 2011
SCI P360 Stability of Steel Beams and Columns, 2011
SCI P361 Steel Building Design: Introduction to the Eurocodes, 2009
SCI P362 Steel Building Design: Concise Eurocodes, 2009
SCI P363 Steel Building Design: Design Data, 2013
SCI P364 Steel Building Design: Worked Examples – Open Sections, 2009
SCI P365 Steel Building Design: Medium Rise Braced Frames, 2009
SCI P374 Steel Building Design: Worked Examples – Hollow Sections, 2008
SCI P375 Fire Resistance Design of Steel Framed Buildings, 2012
SCI P385 Design of Steel Beams in Torsion, 2011
SCI P391 Structural Robustness of Steel Framed Buildings, 2011
SCI P394 Wind Actions to BS EN 1991-1-4, SCI, 2013
SCI P397 Elastic Design of Single-Span Steel Portal Frame Buildings to Eurocode 3, 2013
SCI P398 Joints in Steel Construction: Moment-Resisting Joints to Eurocode 3, 2013
SCI P399 Design of Steel Portal Frame Buildings to Eurocode 3, 2015
SCI P400 Interim Report: Design of Portal Frames to Eurocode 3: An Overview for UK Designers, 2013

4.3 DESIGN CONCEPT, MODELLING AND LOAD ESTIMATION

4.3.1 Design concept

Alternative designs should be considered. Some implications are set out. Two examples are

- Single-storey industrial buildings (Figure 2.4):
  - Lattice girders and cantilever columns;
  - Portal frames;

- Circular multistorey buildings (Figure 2.2):
  - Core and perimeter columns;
  - Core, umbrella girder and perimeter hangers.

In both examples, all systems have been used. It is not clear which gives the most economical solution. Often steel weights are nearly the same for each system. In multistorey buildings, the steel frame is only about 15% of the total cost, so a small weight saving indicated by a preliminary design is of doubtful value. Decisions are often made on grounds other than cost.

The design concept should express clear force paths. It should include, where possible, repetition of members, floor panel sizes, column stacks, frames and bracing so that only a small number of separate members or frames need to be designed.

4.3.2 Modelling

Some points relevant to modelling for preliminary design are set out.

- In a new structure, the designer can model it such that it is statically determinate. The sections from this design could be used as input for a redundant structure. The simple structure will give indicative sizes for costing.
- In checking a finished design or evaluating an existing structure, the design concept and model must be established and loadbearing elements and frames identified. The approximate design check can then proceed.

4.3.3 Load estimation

Typical dead load values for various types of construction are required. These are available from handbooks. Imposed loads are given in the codes. Some comments on load estimation are as follows:

- Most loads are distributed. Beam and column reactions are point loads. Floor loading is expressed as equivalent uniform loads.
- Loads are assessed on the tributary floor area supported by the member.
- Loads are cumulative from the roof down. Imposed loads are reduced depending on the number of floors involved.
- Wind loads generally act horizontally, but uplift due to suction is important in some cases.
- The structure is taken to be pin jointed for load estimation.

4.4 ANALYSIS

The purpose of analysis is to determine the critical actions for design. Some methods that can be used in preliminary analysis are given.

4.4.1 Statically determinate structures

Figure 4.1 shows some common types of statically determinate construction.

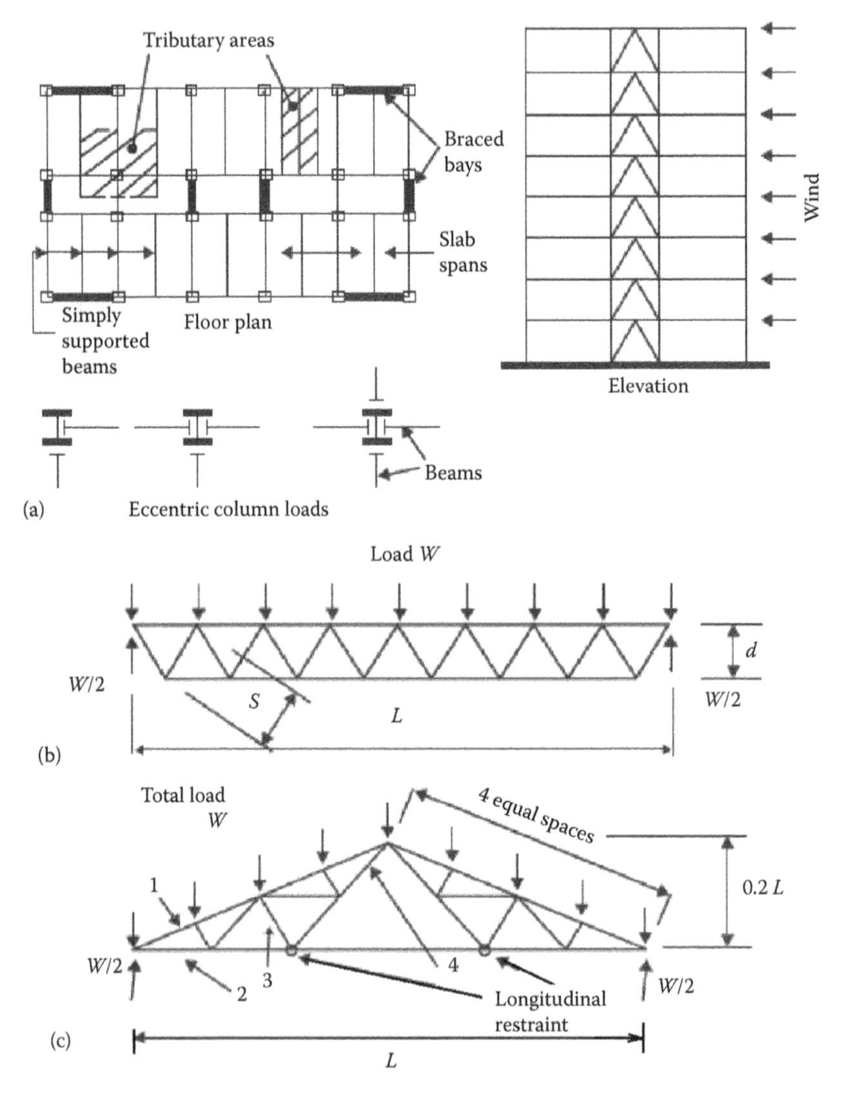

Figure 4.1 Statically determinate structures: (a) multistorey building; (b) lattice girder; (c) roof truss.

(a) *Floor systems*

The floor system (Figure 4.1a) consists of various types of simply supported beams. For uniform loads, the maximum moment is $WL^2/8$ where W is the design load and L is the span commonly in kN/m and in m, respectively. This gives a safe design whether the beams are continuous or some two-way action in the slab is considered.

(b) *Columns in multistorey buildings*

Columns are designed for axial load and moment due to eccentric floor beams, Figure 4.1a. Loads are estimated as set out in Section 4.3.3. Beam reactions act at a nominal 100 mm from the column face or at the centre of the length of stiff bearing, whichever gives the greatest eccentricity. The resulting moment on the column may be divided equally between its upper and lower lengths at that particular level where the column carries the beams, as an approximation.

(c) *Lattice girders, trusses and bracing*

For lattice girders (Figure 4.1b), the critical actions are

Chord at mid-span = $WL/8d$

Web member at support = $WS/2d$

where W is the design load; d is the depth of girder; and S is the length of web member.

For a pitched roof truss (Figure 4.1c), force coefficients can be determined for a given truss such as the Fink truss shown. The axial forces are

Top chord, member 1: $F = -1.18W_{max}$

Bottom chord, member 2: $F = 1.09W_{max}$

Web members 3, 4: $F = -0.23W, + 0.47W$

where W is the total load. Tension is negative, compression positive.

(d) *Three-pinned portal*

For the three-pinned portal (Figure 4.2),

Horizontal reaction $H = WL/8h$

Eaves moment $M_X = Ha$

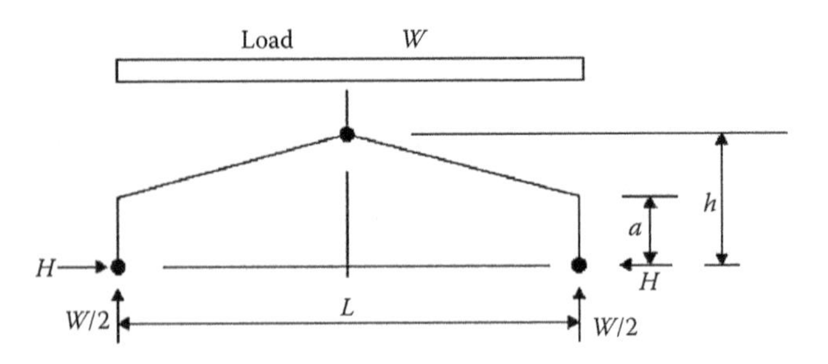

Figure 4.2 Three-pinned portal.

where W is the vertical load; L is the span; h is the height to crown; and a is the height of columns.

4.4.2 Statically indeterminate structures

(a) *General comments*

For elastic analysis, if the deflected shape under load is drawn, the point of contraflexure may be located approximately and the structure analysed by statics. The method is generally not too accurate.

For plastic analysis, if sufficient hinges are introduced to convert the structure to a mechanism, it is analysed by equating external work done by the loads to internal work in the hinge rotations. The plastic moment must not be exceeded outside the hinges.

Design aids in the form of formulae, moment and shear coefficients, tables and charts are given in handbooks (*Steel Designers Manual*, 1986, 1994, 2002, 2012). Some selected solutions are given.

(b) *Continuous beams*

The moment coefficients for elastic and plastic analysis for a continuous beam of three equal spans are shown in Figure 4.3. For other cases, see *Steel Design Manual* (2012).

(c) *Pitched roof pinned-base portal*

Portal design is usually based wholly on plastic theory (Figure 4.4). As a design aid, solutions are given in chart form for a range of spans. A similar chart could be constructed for elastic design (Chapter 5 gives detailed designs).

It is more economical to use a lighter section for the rafter than for the column, rather than a uniform section throughout. The rafter is haunched at the eaves. This permits use of a bolted joint at the eaves and ensures that the hinge there forms in the column.

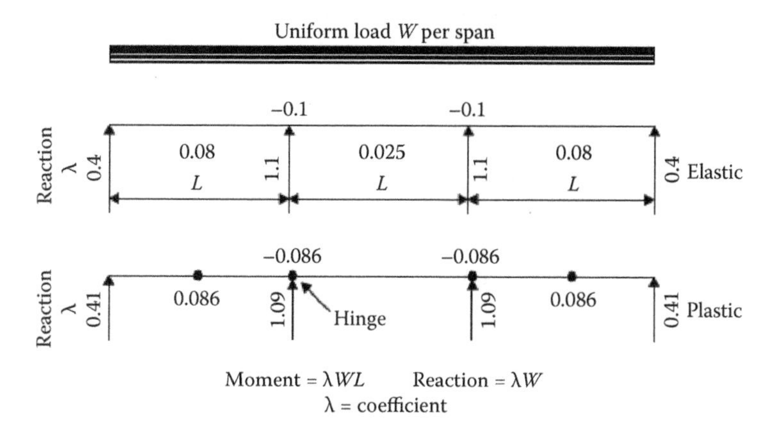

Figure 4.3 Continuous beam.

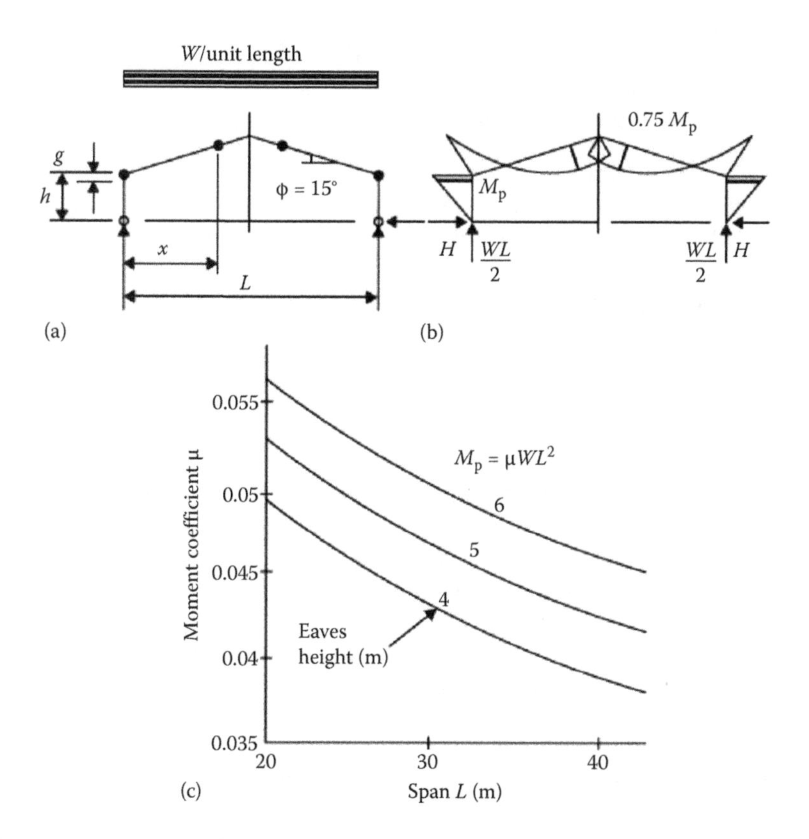

Figure 4.4 Plastic analysis: (a) frame load and hinges; (b) plastic bending moments; (c) analysis chart.

Let M_p be the plastic moment of resistance of the column and qM_p be the plastic moment of resistance of the rafter, where $q = 0.75$ for chart. Then

Column hinge $M_p = H(h - g)$

Rafter hinge $qM_p = qH(h - g)$

$$= \frac{WLx}{2} - Wx^2 - H(h + x\tan\phi)$$

where

W = roof load per unit length;
L = span;
H = horizontal reaction;
h = eaves height;
g = depth of column hinge below intersection of column and rafter centrelines (0.3–0.5 m for chart);
ϕ = roof slope (15° for chart);
x = distance of rafter hinge from support.

Equate $dH/dx = 0$, solve for x and obtain H and M_p.

A chart is given in Figure 4.4c to show values of the column plastic moment M_p for various values of span L and eaves height h.

(d) *Multistorey frames subjected to vertical loads*

Elastic and plastic methods are given.

- Elastic analysis – The subframes given in Clause 5.6.4.1 and Figure 11 of BS 5950 (1998) can be used to determine actions in particular beams and columns (Figure 4.5a). The code also enables beam moments to be determined by analysing the beam as continuous over simple supports. EC3 BS EN 1993-1-1 2005, Clauses 5.1–5.6 provide similar important information on structural analysis.
- Plastic analysis – The plastic moments for the beams are $\pm WL/16$. The column moments balance the moment at the beam end as shown in Figure 4.5b and hinges do not form there. This analysis applies to a braced frame.

(e) *Multistorey frames subjected to horizontal loads*

The following two methods were used in the past for analysis of multistorey buildings.

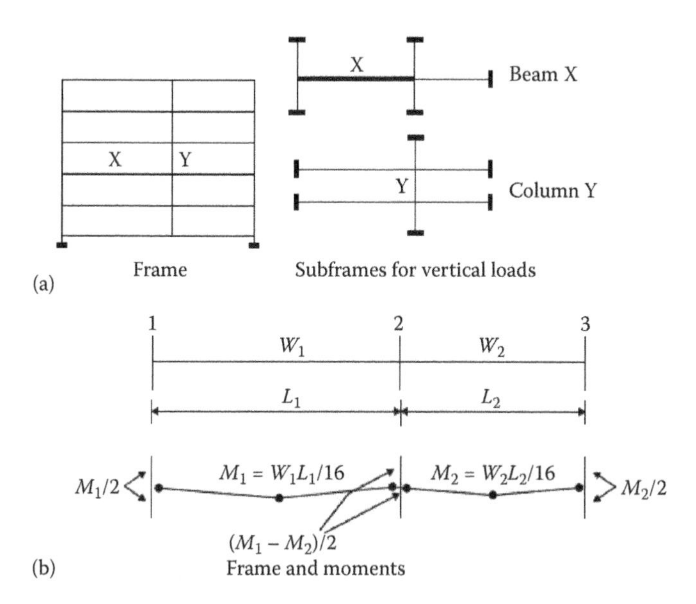

Figure 4.5 Multistorey frames: (a) elastic analysis; (b) plastic analysis.

(i) PORTAL METHOD

The portal method is based on two assumptions:

- The points of contraflexure are located at the centres of beams and columns.
- The shear in each storey is divided between the bays in proportion to their spans. The shear in each bay is then divided equally between the columns.

The column end moments are given by the product of the shear by one-half the storey height. Beam moments balance the column moments. External columns only resist axial force, which is given by dividing the overturning moment at the level by the building width. The method is shown in Figure 4.6.

(ii) CANTILEVER METHOD

In the cantilever method, two assumptions are also made:

- The axial forces in the columns are assumed to be proportional to the distance from the centre of gravity of the frame. The columns are to be taken to be of equal area.
- The points of contraflexure occur at the centres of the beams and columns.

The method is shown in Figure 4.7.

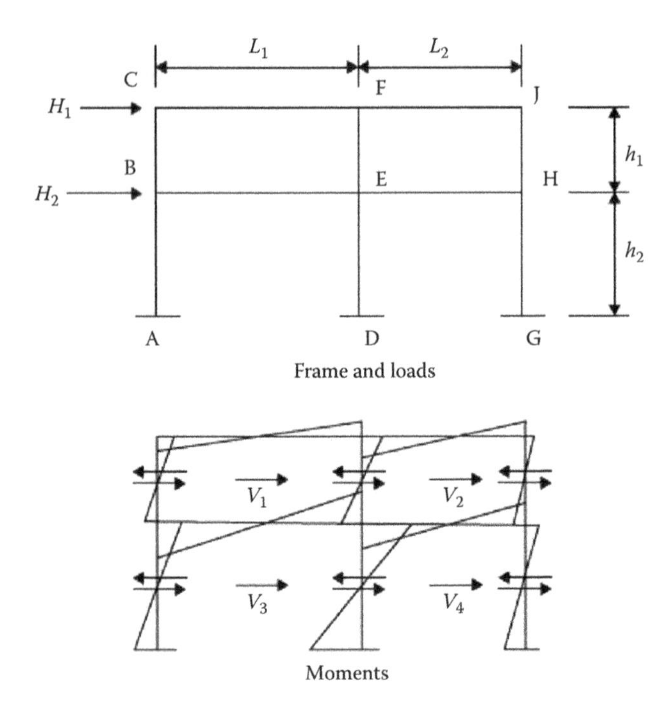

Top column and beam moments
$V_1 = H_1L_1/(L_1 + L_2)$ $\quad$ $V_2 = H_1 - V_1$
$M_{CB} = M_{BC} = V_1h_1/4$
$M_{JH} = M_{HJ} = V_2h_1/4$
$M_{EF} = M_{FE} = (V_1 + V_2)h_1/4$
$V_3 = (H_1 + H_2)L_1/(L_1 + L_2)$ $\quad$ $V_4 = H_1 + H_2 - V_3$
Find moments in lower columns and centre beam

Figure 4.6 Portal method.

4.5 ELEMENT DESIGN

4.5.1 General comments

Element design is the process of sizing sections to resist actions obtained from analysis. Member design is required for members subjected to:

- Axial load – ties, struts;
- Bending and shear – beams with fully supported and unsupported compression flanges (universal beams, built-up sections, lattice girders and composite sections);
- Axial load and bending – beam-columns to be designed for local capacity and overall buckling.

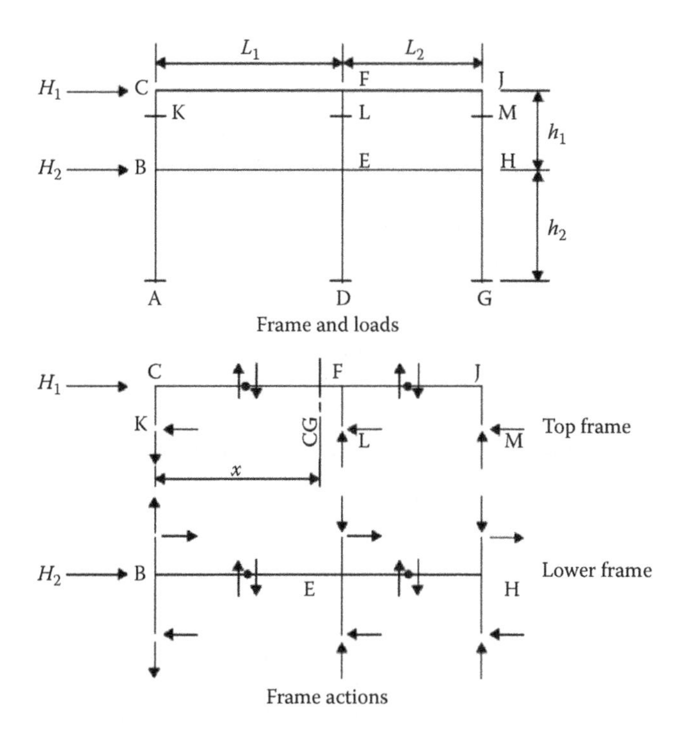

Actions in top frame
Calculate centre of gravity of columns x
Express column forces in terms of R, for example, at K = xR
Take moments about K, ΣM = O, solve for R
Obtain column forces and beam shears
Take moments about beam centres, obtain column shears
Calculate top frame moments

Similarly find actions in lower frame

Figure 4.7 Cantilever method.

4.5.2 Ties and struts

(a) *Ties*

Section size is based on the effective area allowing for bolt holes, defined in Clause 3.4 of BS 5950. Certain commonly used sections may be selected from load capacity tables (Steelwork Design Guide to BS 5950-1, 2000; Steel Construction Institute, 2001; Steelwork Design Guide to Eurocode 3, Part 1.1 – Introducing Eurocode 3: A Comparison of EC3, Part 1.1 with BS 5950, Part 1).

(b) *Struts and columns*

The load capacity depends on the cross-sectional area, effective length (in BS 5950-1, 2000), buckling length (in EC3, see Clause 5.5.1.5 and Annex E) and least radius of gyration. Certain commonly used sections may be selected directly from capacity tables in the handbook cited above.

Estimation of effective length of a member is important. This depends on whether ends are held in position or sway can occur and how effectively the ends are restrained in direction (BS 5950, Table 22 and Appendices D and E. Also see Clauses 6.3.1.1 and 16.5 column buckling resistance of EC3 BS EN 1993-1-1. Figure 16.4 of EC3 shows column buckling curves).

If sway is prevented, the effective length (l) is equal to or less than the actual length (L). If sway occurs, it is greater. The effective lengths for members in some common situations are shown in Figures 4.8 and 4.9.

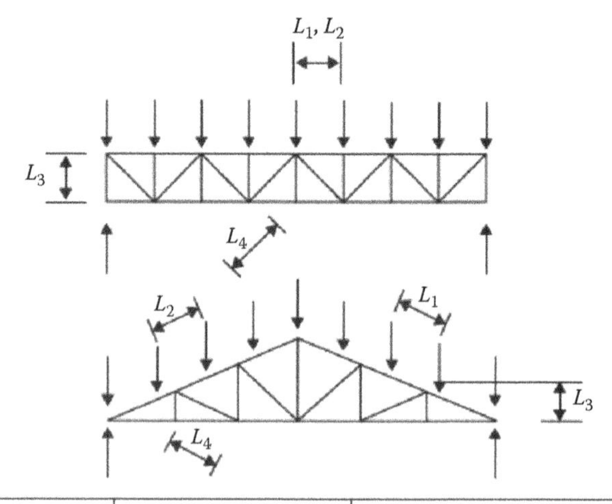

Member	Section	Effective length l
Top chord L_1, L_2	X, Y	$L_{1X} = 0.85\ L_1$ $L_{2Y} = L_2$
Internal members L_3, L_4	X, Y, V	$L_{3V} = 0.85\ L_3$ $L_{3X} = L_{3Y} = 0.85\ L_3$ L_4 = same (Table 25, BS 5950)

Figure 4.8 Effective length for lattice girder and truss members.

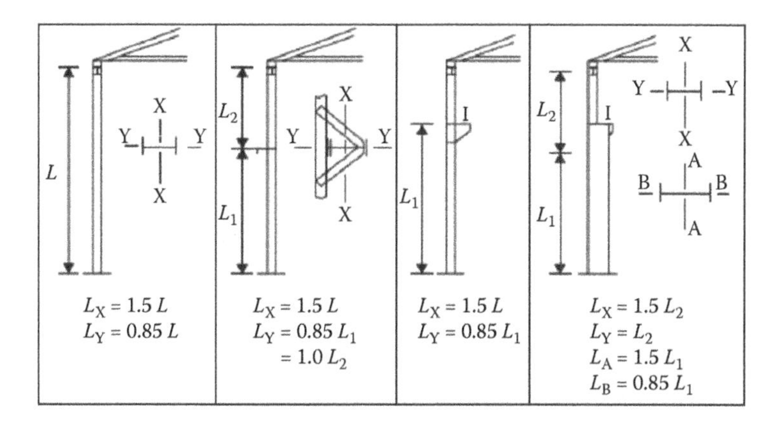

Figure 4.9 Effective lengths for side stanchions in single-storey buildings.

The following applies for a rectangular multistorey building:

- Frame braced in both directions, simple design $l = 0.85L$;
- Frame braced in both directions – rigid in transverse direction, simple in longitudinal direction $l_X = 0.7L$, $l_Y = 0.85L$;
- Frame unbraced in transverse direction, braced in longitudinal direction $l_X > L$, $l_Y = 0.85L$.

Here l_X is the transverse effective length (X–X axis buckling); l_Y is the longitudinal effective length (Y–Y axis buckling); and L is the column length.

The capacities in axial load for some commonly used universal column sections are shown in Figure 4.10. Capacities for sections at upper and lower limits of serial sizes only are shown.

The buckling length in EC3 is identical to the effective length in BS 5950.

4.5.3 Beams and girders

(a) *Universal beams*

If the compression flange of the beam is fully restrained or the unsupported length is less than $30r_y$ where r_y is the radius of gyration about the minor axis, the beam will reach its full plastic capacity. Lateral torsional buckling reduces the capacity for longer unsupported compression flange lengths.

For a beam with unsupported compression flange, the effective length depends on

- Spacing of effective lateral restraints;
- End conditions, whether the end is torsionally restrained or the compression flange is laterally restrained or if it is free to rotate in plan;
- Whether the load is destabilizing; if so, the effective length is increased by 20%.

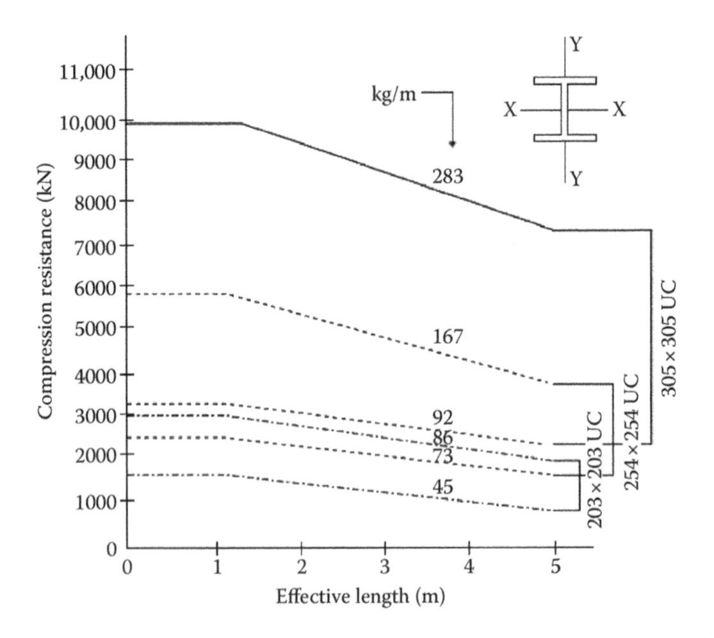

Figure 4.10 Compression resistance – Y–Y axis buckling. In the EC3 and all the new Eurocodes, X–X axis is used as equipment to Y–Y axis of the BS 5950 and Y–Y axis (in BS 5950) is replaced by Z–Z axis in the new European codes.

Refer to BS 5950, Section 4.3 and EC3 Section 6.3.1.1.

The *Steel Designers Manual* (1986, 1994, 2012) gives tables of buckling resistance moments for beams for various effective lengths. A design chart giving buckling resistance moments for some sections is shown in Figure 4.11.

(b) *Plate girders*

The simplified design method given in BS 5950, Section 4.4.4.2 is used where the flanges resist moment and the web shear. For a restrained compression flange (Figure 4.12):

Flange area $bT = M/Dp_y$

Web thickness $t = V/q_{cr}d$

where

M, V = applied moment, shear;
D, d = overall depth, web depth;
T, t = flange thickness, web thickness;
b = flange width;

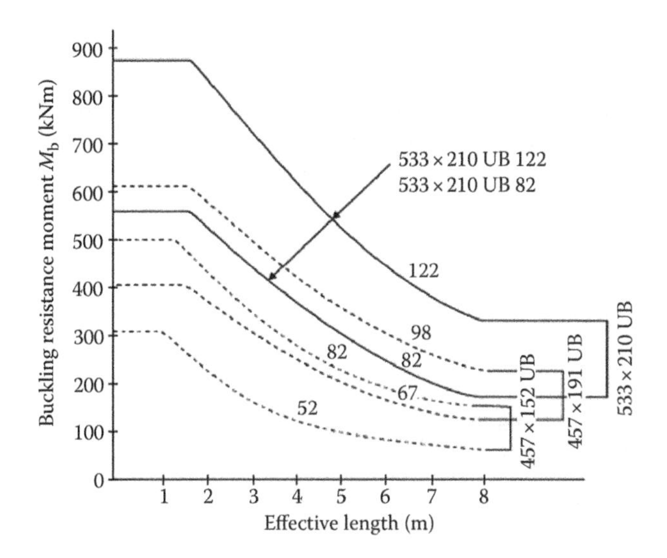

Figure 4.11 Buckling resistance moment – conservative approach.

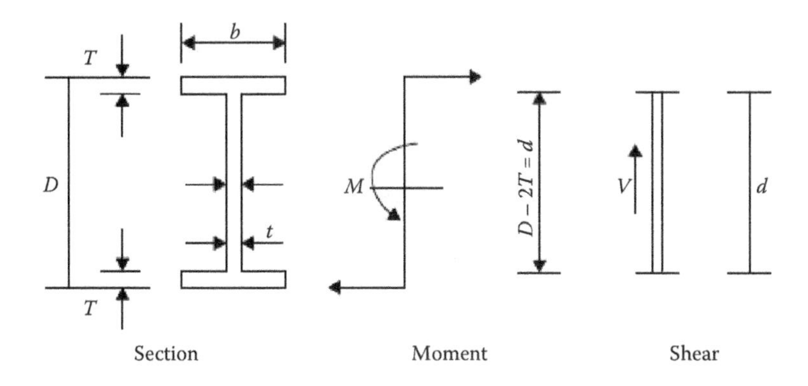

Figure 4.12 Plate girder.

p_y = design strength;
q_{cr} = critical shear strength (depends on d/t and stiffener spacing – see Clause 4.4.5.4 and Annex H.2 of BS 5950).

(For plate girder design to EC3, please see Chapters 14 and 15.)

(c) *Lattice girders*

Suitable members can be selected from load capacity tables (Steel Construction Institute, 2001, 2013). Lattice girders are analysed in Section 4.4.1(c) (Figure 4.1b).

4.5.4 Beam–columns

The beam–column in a multistorey rigid frame building is normally a universal column subjected to thrust (axial load) and moment about the *major axis*. A section is checked in accordance with BS 5950, Section 4.8.3, using the simplified approach. See Chapter 8 for the design of EC3.

- Cross-section capacity or local capacity at support:

$$\frac{F_c}{A_g p_y} + \frac{M_X}{M_{cx}} \leq 1$$

- Member buckling resistance or overall buckling:
  It is to be checked that the following relationships are both satisfied.

$$\frac{F_c}{P_c} + \frac{m_x M_X}{p_y Z_X} \leq 1 \text{ (buckling generally)}$$

$$\frac{F_c}{P_{cy}} + \frac{m_{LT} M_{LT}}{M_b} \leq 1 \text{ (buckling about minor axis)}$$

where

F_c, M_X = applied load, applied moment in the segment, L_X, governing P_{cx};
A_g = gross area;
p_y, P_c = design strength, compressive resistance (the smaller of P_{cx} or P_{cy}, see Clause 4.7.4 of BS 5950);
M_{cx} = plastic moment capacity;
M_b = buckling resistance moment, see Clause 4.3 and I.4 of BS 5950;
m_x = equivalent uniform moment factor for major axis flexural buckling;
m_{LT} = equivalent uniform factor, see Clause 4.8.3.3.4 of BS 5950;
M_{LT} = maximum major axis moment in the segment length L, governing M_b.

Resistance and capacity tables are given in Steel Construction Institute (2001, 2012, 2013). Interaction charts for local capacity and overall buckling can be constructed for various sections for particular values of effective length (Figure 4.13).

4.5.5 Members in portal frames

Members in portal frames designed to plastic theory are discussed. The column is sized for the plastic hinge moment and axial load at the eaves.

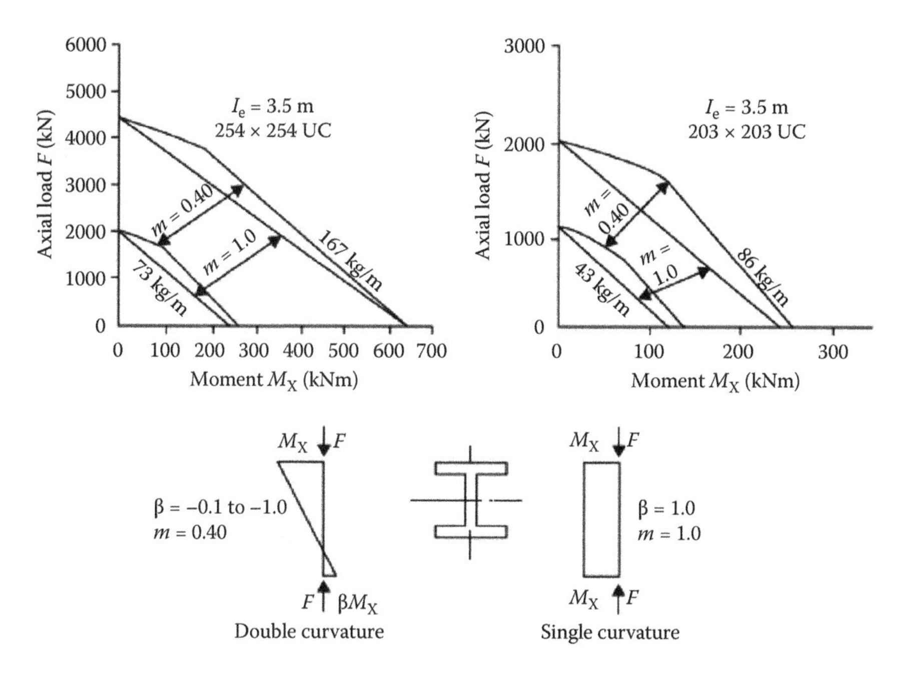

Figure 4.13 Beam–column interaction charts for local capacity/overall buckling.

About 90% of capacity is required to resist moment. Lateral restraints must be provided at the hinge and within a specified distance to ensure that the hinge can form.

In the rafter, a haunch at the eaves ensures that that part remains elastic. The hinge forms near the ridge and restraints to the top flange are provided by the purlins.

Provisions for plastic design of portals are shown in Figure 4.14 and a detailed design is given in Section 5.2.

4.6 EXAMPLES

4.6.1 Ribbed dome structure

(a) *Specification*

A kiosk required for a park is to be hexagonal in plan on a 16 m diameter base, 3 m high at the eaves and 5 m at the crown (Figure 4.15). There are to be three braced bays with brick walls. The roof is felt on timber on purlins with ceiling.

Deal load = 1.0 kN/m^2 on slope

Imposed load = 0.75 kN/m^2 on plan

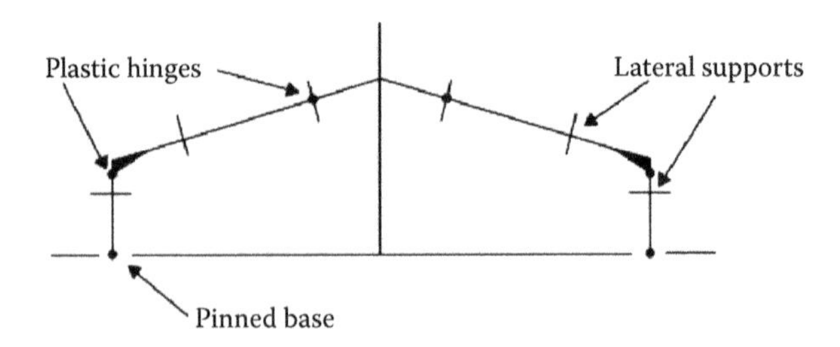

Figure 4.14 Plastic-designed portal.

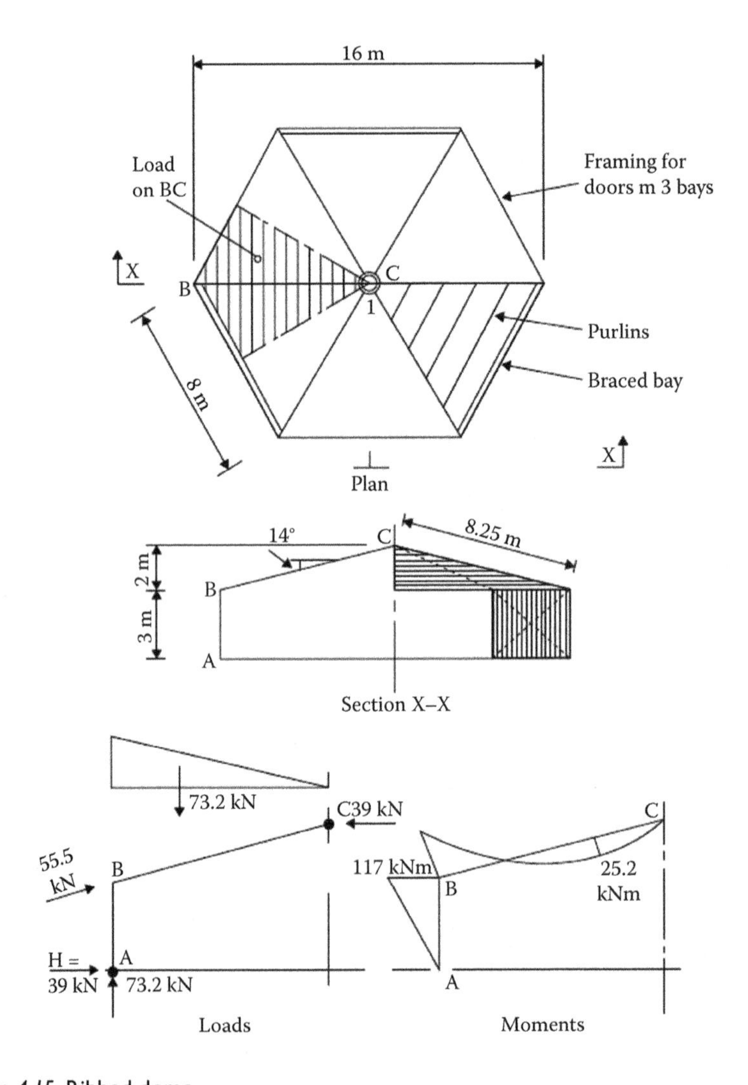

Figure 4.15 Ribbed dome.

Design using 3 No. three-pinned arches as shown in the figure, steel grade S355. For thickness <16 mm, p_y = 355 N/mm^2.

(b) *Analysis and design*

Design load = 1.4/cos 14° + 1.6 × 0.75 = 2.64 kN/m^2

Roof load on half portal = 8^2(cos 30°)2.64/2 = 73.2 kN

Reaction H = 73.2 × 8/3 × 5 = 39 kN

Moment M_B = 39 × 3 = 117 kNm

Axial load in BC = 73.2 sin 14° + 39 cos 14° = 55.5 kN

The maximum sagging moment is shown.
Assume eaves moment to be 85% of section capacity S_x:

S_x required = 117 × 10^3/(355 × 0.85) = 387 cm^3

Select 250 × 150 × 6.3 RHS, for which S_x = 402 cm^3.
A rigorous check can be made using Lee et al. (1981). This gives the out-of-plane effective lengths for the column as 2.6LBA and for the rafter as 1.4LBC. The moment is 81% of section capacity and the axial load in the rafter is 9%. The section must be adequately supported laterally.

4.6.2 Two-pinned portal – plastic design

(a) *Specification* (Figure 4.16a)

Span 30 m, eaves height 5 m, roof slope 15°, spacing 6 m. Dead load 0.5 kN/m^2, imposed load 0.75 kN/m^2. Steel grade S275. Determine sections for the portal.

(b) *Analysis and design*

Design load = (1.4 × 0.5 + 1.6 × 0.75)6 = 11.4 kN/m

The portal rafter is to have 75% of the moment of the columns. From Figure 4.4,

Column: moment M_{pc} = 0.047 × 11.4 × 30^2 = 482.2 kNm

axial load F = 11.4 × 15 = 171 kN

Rafter: moment M_{pR} = 0.75 × 482.2 = 361.7 kNm

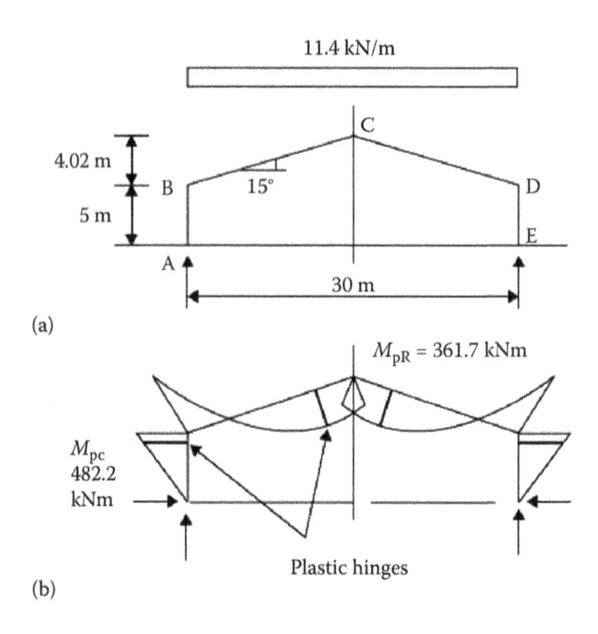

Figure 4.16 Two-pinned portal: (a) elevation; (b) plastic moment diagram.

For moment resistance of 90% of moment capacity,

Section resistance = 482.2/0.9 = 536.1 kNm

Try 533 × 210 UB 82, where A = 105 cm², S_x = 2060 cm³. Check:

$$\frac{171 \times 10}{275 \times 105} + \frac{482.2 \times 10^3}{275 \times 2060} = 0.91$$

Rafter:

$S_x = 361.7 \times 10^3/275 = 1315$ cm³

For

457 × 191 UB 67, S_x = 1470 cm³

The lateral support system must be designed. Section 5.2 gives a complete design.

Chapter 5

Single-storey, one-way-spanning buildings

5.1 TYPES OF STRUCTURES

Some of the most important steel-framed buildings fall in the classification of single-storey, one-way-spanning structures. Types shown in Figure 5.1 include:

- Truss or lattice girder and stanchion frames;
- Portals in various types of construction:
  - o Universal beams;
  - o Built-up tapered sections – lattice;
- Arches in single section or lattice construction.

The first type with pitched roof truss and cantilever columns included the historical mill building. It is still favoured for flat-roofed buildings using lattice girders. Portals now form the most popular building type for single-storey factories and warehouses. The pinned-base portal designed using plastic theory is almost exclusively adopted in the United Kingdom. In the United States, the built-up tapered section portal is commonly used. Arches are an architecturally pleasing and structurally efficient form of construction with a wide use including exhibition buildings and sports halls, warehouses, etc.

Detailed designs are given for a pinned-base portal using plastic theory and a two-pinned arch using elastic theory. The portal is designed using universal beam sections. Two designs are given for the arch – one using rectangular hollow sections and one lattice construction.

The design of a portal constructed from built-up tapered sections is outlined briefly.

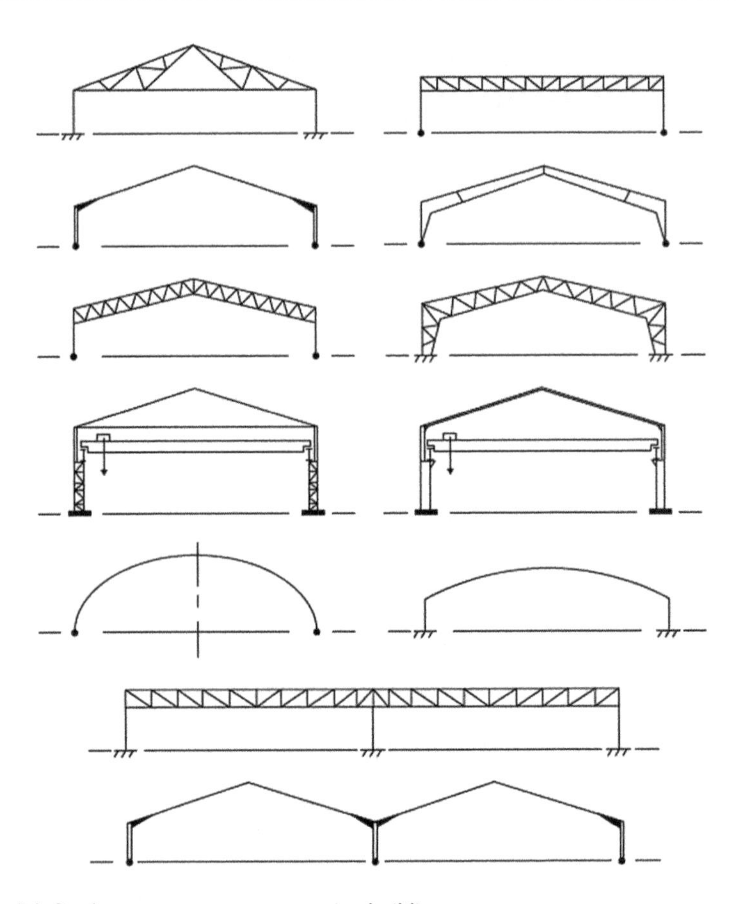

Figure 5.1 Single-storey, one-way-spanning buildings.

5.2 PINNED-BASE PORTAL – PLASTIC DESIGN

See Chapter 6 for one-way spanning pinned – base portal frame-plastic design to EC3.

5.2.1 Specification and framing plans

The portal has a span of 40 m, height at eaves 5 m and roof slope 15°. The portal spacing is 6 m and the building length is 60 m. The location is an industrial estate on the outskirts of a city in the northeast of the United Kingdom. The framing plans are shown in Figure 5.2. The material is Grade S275 steel.

Standard universal beam portal construction is adopted. The haunched front at the eaves is shown in the figure. The frame is of simple design longitudinally with braced bays at each end. Purlins and sheeting rails are cold-rolled sections.

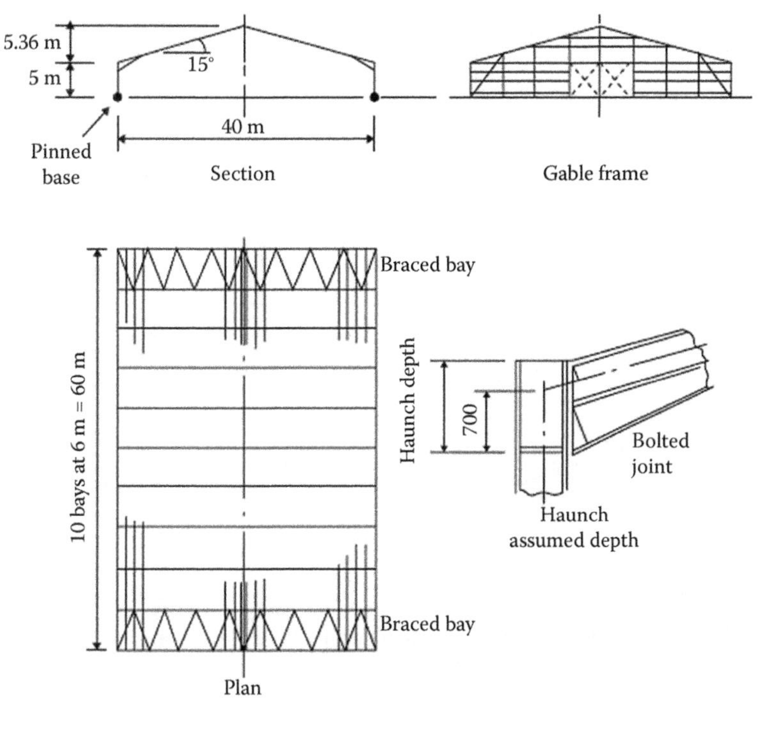

Figure 5.2 Framing plans.

5.2.2 Dead and imposed loads

(a) *Dead load*

$$
\begin{aligned}
\text{Sheeting} &= 0.1\ \text{kN/m}^2 \\
\text{Insulation} &= 0.1 \\
\text{Purlins} &= 0.05 \\
\text{Rafter} &= 0.15 \\
\hline
\text{Total} &= 0.4\ \text{kN/m}^2 \text{ on slope} \\
&= 0.41\ \text{kN/m}^2 \text{ on plan}
\end{aligned}
$$

Therefore,

Load on roof = 0.41 × 6 = 2.46 kN/m

Walls = 0.4 × 5 × 6 = 12 kN

The dead load is shown in Figure 5.3a.

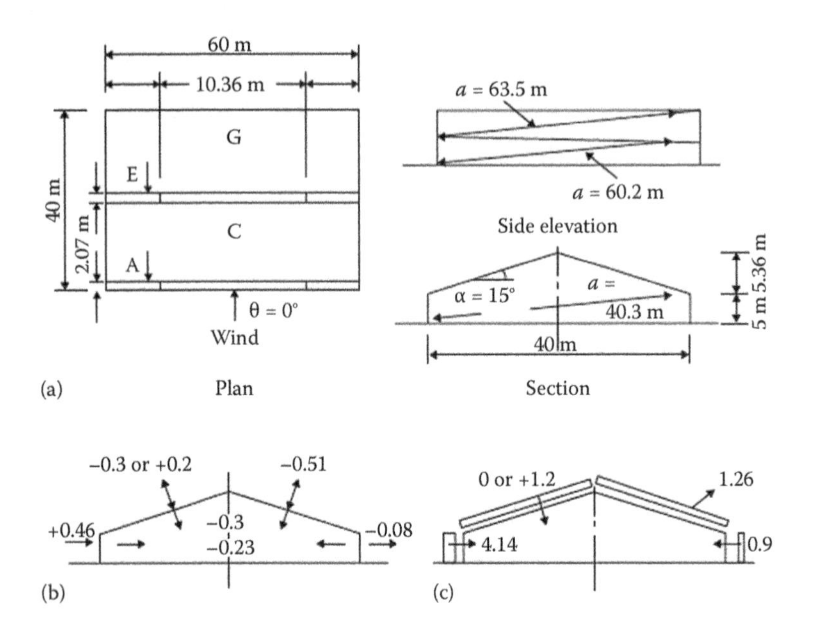

Figure 5.3 Transverse wind loads: (a) dimensions; (b) external and internal pressures (kN/m^2); (c) net member loads (kN/m).

(b) *Imposed load*

Imposed load from BS 6399: Part 1 = 0.75 kN/m^2 = 4.5 kN/m on roof. This is shown in Figure 5.3b.

Note that the purlin load is 0.95 kN/m^2. For purlin centres 1.5 m, select Wara Multibeam A170/170 purlins with a safe load of 1.19 kN/m^2.

5.2.3 Wind loads

The wind load is in accordance with BS 6399: Part 2.

(a) *Location of buildings*

Outskirts of city in northeast England on an industrial estate with clear surroundings, 10 km to sea and 75 m above sea level.

(b) *Building dimensions*

Plan 60 m × 40 m; height to eaves 5 m, to rooftop 10.36 m (Figure 5.2).

(c) *Building data – Sections 1.6 and 1.7 of code*

Building type factor: for portal sheet, $k_b = 2$ (Table 1 of BS 6399: Part 2).

Reference height $H_r = 10.36$ m $= H$

Dynamic augmentation factor $C_r = 0.05$ (Figure 3)

(d) *Wind speed – Section 2.2 of code*

Basic wind speed $V_b = 25$ m/s (Figure 6)

Site wind speed $V_s = V_b S_a S_d S_s S_p$

where

Sea-altitude factor = 1.001, $\Delta_s = 1.075$, $\Delta_s = 75$ m
$S_d = 1.0$ (Table 3)
$S_s = S_p = 1.0$ = Seasonal and probability factors

Therefore,

$V_s = 1.075 \times 25 = 26.9$ m/s

Effective wind speed $V_e = V_s S_b$

where

S_b = terrian and building factor (Table 4)
= 1.45 for walls, 1.7 for roof

$V_e =$ 39 m/s for walls, 45.7 m/s for roof

(e) *Transverse wind load – Section 2.1 of code*

Dynamic pressure $q_s = 0.613 V_e^2 / 10^3$ kN/m^2

Walls: $q_s = 0.93$ kN/m^2; roof: $q_s = 1.25$ kN/m^2

External pressure coefficients C_{pe} are as follows:

- Wall (code, Section 2.4, Table 5), $D/H = 40/874$. For windward wall, $C_{pe} = +0.6$; leeward wall, $C_{pe} = -0.1$.
- Roof (code, Section 2.5). From Figure 5.3a, $\theta = 0°$, $\alpha = +15°$, $b_L = 2H =$ 20.72 m. From code, Table 10 for duopitch roofs, C_{pe} values for A: -1.302+0.2; C: –0.3 or +0.2; E: –1.1; G: –0.5. Use uplift value for C (negative) and G in analysis.

The internal pressure coefficient C_{pi} (code, Section 2.6), for four walls equally permeable, is −0.3 (Table 16).

Size effect factors C_a (Clause 2.1.3.4, Figure 4) for $H > 10$ –15 m, B = 10 km to sea are (Figure 5.3a):

- Walls: $a = 60.2$ m, $C_a = 0.82$ (Curve B);
- Roof: $a = 63.5$ m, $C_a = 0.81$.

The internal and external pressures are shown in Figure 5.3b:

External surface pressure $p_e = q_s C_{pe} C_a$

Internal surface pressure $p_i = q_s C_{pi} C_a$

Net pressure $p = p_e - p_i$

The net member loads (= $6p$ kN/m) are shown in Figure 5.3c.

Note that in accordance with Clause 2.1.3.6, the overall horizontal loads could be reduced by the factor $0.85(1 + C_r)$ where C_r is the dynamic augmentation factor. This reduction will not be applied in this case.

(f) *Longitudinal wind loads* (Figure 5.4)

The high values of the external pressure coefficients in the edge strips on the walls and roof will be taken into account. The windward end is taken to be permeable to give internal pressure. The portal frame X next to the windward end subject to the highest uplift load is analysed.

$a = 40.3$ m, $C_a = 0.85$ (Figure 4, Curve B)

External pressure coefficients C_{pe} are as follows:

- Walls, $b = B$ or $2H = 20.72$ m. The various zones are taken from Figure 12. For A, $C_{pe} = -1.3$; for B, $C_{pe} = -0.8$ (Table 5).
- Roof, $b_w = W$ or $2H = 20.72$ m. The various zones are taken from Figure 20. For A, $C_{pe} = -1.6$; for B, $C_{pe} = -1.5$; for C, $C_{pe} = -0.6$ (from Table 10 for roof angle $\alpha = 15°$).

Internal pressure coefficient $C_{pi} = +0.2$ (Table 16).

$$\text{Total pressure} = p_e + p_i$$
$$= q_s C_a (C_{pe} + C_{pi})$$

$$\text{Wall load for frame X } (q_s = 0.93 \text{ kN/m}^2)$$
$$= 0.93 \times 1.5 \times 4.14 \times 0.85 \times 2.07/6$$
$$+ 0.93 \times 1.0 \times 0.85 \times 1.86 \times 5.07/6 + 0.93 \times 1.0 \times 0.85 \times 3$$
$$= 1.69 + 1.24 + 2.37 = 5.3 \text{ kN/m}$$

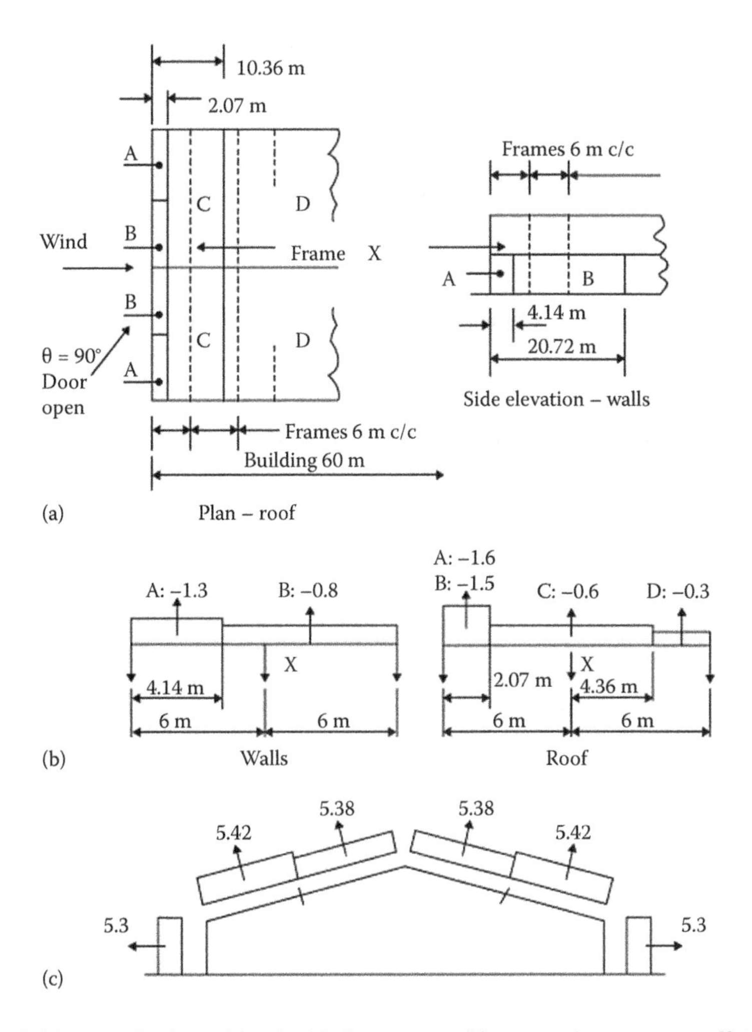

Figure 5.4 Longitudinal wind loads: (a) dimensions; (b) external pressure coefficients; (c) member loads (kN/m).

Roof load for frame X (q_s = 1.25 kN/m^2):

$$\begin{aligned}\text{Load for section A} &= (1.25\times1.8\times2.07\times0.85\times1.04/6)\\ &+(1.25\times0.8\times3.93\times0.85\times4.04/6)\\ &+(1.25\times0.8\times4.36\times0.85\times3.82/6)\\ &+(1.25\times0.5\times1.64\times0.85\times0.82/6)\\ &=0.69+2.25+2.36+0.12=5.42\text{ kN/m}\end{aligned}$$

$$\text{Load for section B} = (0.69 \times 1.7/1.8) + 2.25 + 2.36 + 0.12$$
$$= 5.38 \text{ kN/m}$$

The portal loads are shown in Figure 5.4. The analysis will be performed for a uniform wind load on the roof of 5.42 kN/m.

5.2.4 Design load cases

(a) *Dead and imposed loads*

Design load = (1.4 × dead) + (1.6 × imposed)

Roof: (1.4 × 2.46) + (1.6 × 4.5) = 10.64 kN/m

Walls: 1.4 × 12 = 16.8 kN

The design loads are shown in Figure 5.5a.

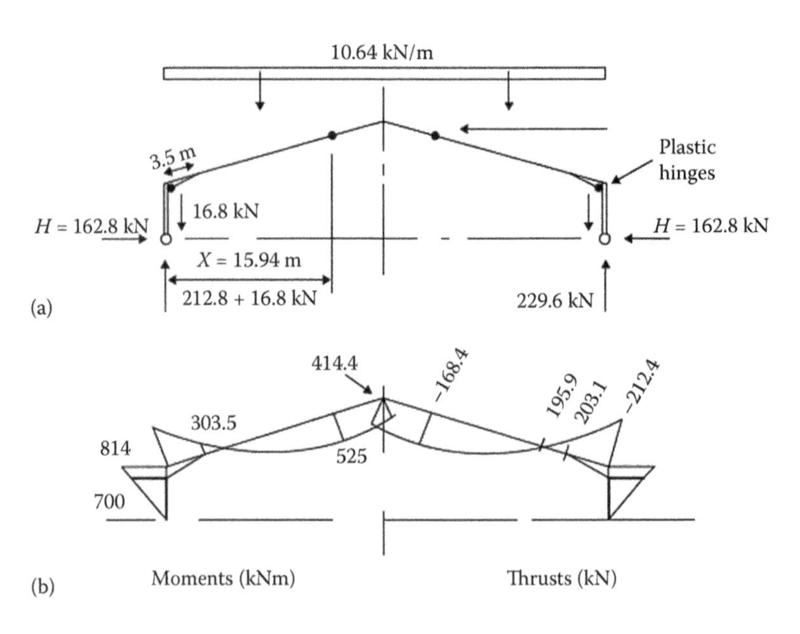

Figure 5.5 Dead and imposed load – plastic analysis: (a) loads, reactions and hinges; (b) moments and thrusts.

(b) *Dead and wind loads*

Wind uplift is important in checking roof girder stability. Both cases, wind transverse ($\theta = 0°$) with internal suction and wind longitudinal ($\theta = 90°$) with internal pressure, are examined:

Design load = 1.4 × wind − 1.0 × dead resisting uplift

The characteristic wind loads for the two cases are shown in Figures 5.3 and 5.4. The design loads are shown in Figure 5.6.

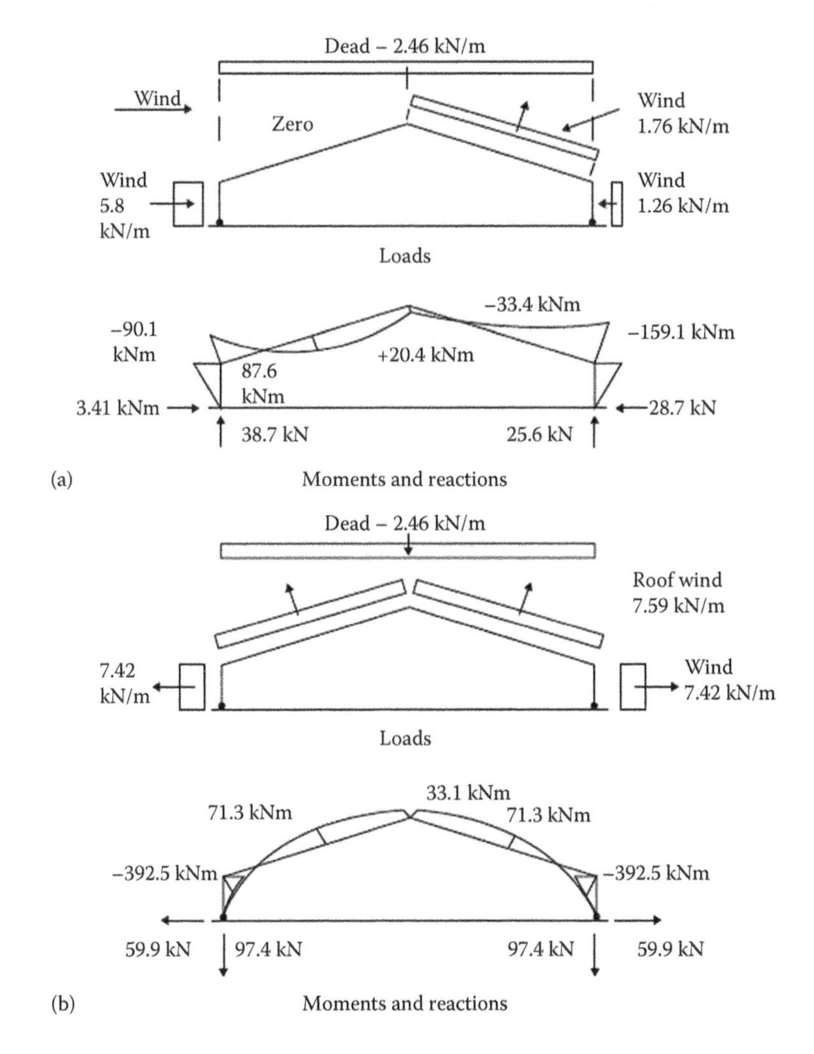

Figure 5.6 **Design wind loads: (a) transverse; (b) longitudinal.**

5.2.5 Plastic analysis and design

Plastic design is adopted for dead and imposed loads.

(a) *Uniform portal*

Collapse occurs when hinges form in the column at the bottom of the haunches and either side of the ridge. The hinges ensure that no hinges form in the rafter near the eaves (Figure 5.5a).

If the rafter hinge forms at x from the column, then the following two equations can be formed to give the value of the plastic moment:

Column: $M_p = 4.3H$

Rafter: $M_p = 212.8x - 5.32x^2 - (5 + 0.27x)H$

$$H = \frac{212.8x - 5.32x^2}{9.3 + 0.27x}$$

Put $dH/dx = 0$ to give equation

$$x^2 + 68.7x - 1375 = 0$$

Solve to give

$$x = 16.2 \text{ m}$$

From which

$$H = 150 \text{ kN}$$

Plastic moment $M_p = 645$ kNm

Plastic modulus $S_x = 645 \times 10^3/275 = 2346$ cm^3

Select 610 × 229 UB 101 or 610 × 229 UB 113, $S_x = 2880$ cm^3, $T =$ 14.8 mm, $p_y = 275$ N/mm^2. This allows for axial load in column.

This section would require further checks. The basic frame weight neglecting the haunch is 5/93 kg.

(b) *Nonuniform portal*

Assume that the plastic bending capacity of the rafter is 75% that of the column. The equations for the plastic moments at the hinges can be rewritten to give

Column: $M_p = 4.3H$

Rafter: $0.75M_p = 212.8x - 5.32x^2 - (5 + 0.27x)H$

Then

$$H = \frac{212.8x - 5.32x^2}{8.23 + 0.27x}$$

Put $dH/dx = 0$ and solve to give $x = 15.9$ m, $H = 162.8$ kN.

Column: $M_p = 700$ kNm

$S_x = 2545$ cm^3

Select 610 × 229 UB 101, $S_x = 2880$ cm^3, $T = 14.8$ mm, $p_y = 275$ N/mm^2.

Rafter: $M_p = 0.75 \times 700 = 525$ kNm

$S_x = 1909$ cm^3

Select 533 × 210 UB 82, $S_x = 2060$ cm^3.

Weight = 4406 kg (neglecting haunches)

Figure 5.5b shows moments and thrusts at critical sections. Further checks are carried out below on the nonuniform frame.

5.2.6 Dead and wind loads

The stability of the rafter must be checked for possible uplift loads due to wind. The two wind load cases – wind transverse and wind longitudinal – acting with the dead loads are considered. Elastic analyses based on the sections obtained from the plastic design are carried out. The frame loads and bending moment diagrams are shown in Figure 5.6. For transverse wind, the zero wind load case on the windward rafter only is shown.

The stability check on the leeward girder for the wind transverse case is carried out in Section 5.2.8.

5.2.7 Plastic design – checks

(a) *Sway stability*

Clause 5.5.4.2.1 is satisfied as $L \leq 5h$, $L = 40$ m, $5h = 5 \times 5 = 25$ m. $h_r \leq 0.25L$, $h_r = 5.36$ m, $0.25L = 0.25 \times 40 = 10$ m. Sway stability of the portal is checked using the procedure given in Clause 5.5.4.2.2 of BS 5950 for gravity loads, where the following condition must be satisfied for Grade S275 steel:

$$\frac{L_b}{D} \le \frac{44L}{\Omega h}\left(\frac{\rho}{4+(\rho L_r/L)}\right)\left(\frac{275}{P_{yr}}\right) \text{ then } \lambda_r = 1$$

where

$$\rho = (2I_c/I_r)(L/h) \text{ (for single bay frame)}$$

in which I_c, I_r are moments of inertia of the column and rafter – 75,800 cm³ and 47,500 cm³, L is the span – 40 m and h is the column height – 5 m. Thus

$$\rho = (2 \times 75{,}800 \times 40)/(47{,}500 \times 5) = 25.5$$

Also,

Haunch depth = 1011.93 mm (Figure 5.2)

< 2 × rafter depth D = 528.3 mm,

$$L_h = (\text{say})\frac{\text{span}}{10} = \frac{40}{10} = 4 \text{ m}$$

$$P_{yr} = 275 \text{ N/mm}^2,$$

$$L_b = 40 - \left(\frac{2D_h}{D_s + D_h}\right) \times 4,\ D_h = D_s$$

$$= 36.00$$

where W_r is the factored vertical load on the rafter (Figure 5.5a) and W_0 is the load causing plastic collapse of rafter treated as fixed-ended beam of span 40 m.

$$\Omega = W_r / W_0$$

$$W_r = 40 \times 10.64$$

$$W_0 = \frac{275 \times 2060 \times 16}{10^3 \times 40} = 226.6\,\text{kN}$$

Thus,

$$\Omega = 40 \times 10.64/226.6 = 1.88$$

Finally,

L_r = developed length of rafter = 41.4 m

Substituting into the code expression gives

$$\frac{36.00}{0.528} = 68.2 < \frac{44 \times 40}{1.88 \times 5}\left[\frac{25.5}{4 + 25.5 \times 41.4/40}\right]\left(\frac{275}{275}\right)$$

$$< 157.1 \text{ then } \lambda_r = 1$$

and the portal is satisfactory with respect to sway stability, under this load condition.

For horizontal loads (Clause 5.5.4.2.3 of BS 5950, 2000)

$$\lambda_r = \lambda_{sc}/(\lambda_{sc} - 1),$$

$$\lambda_{sc} = \frac{220\,DL}{\Omega_h L}\left(\frac{\rho}{4 + \rho L_r/L}\right)\left(\frac{275}{P_{yr}}\right)$$

$$= \left(\frac{220 \times 528.3 \times 40}{1.88 \times 5 \times 36}\right)$$

$$\times \left(\frac{25.5}{4 + 25.5 \times 41.4/40}\right)\left(\frac{275}{275}\right)$$

$$= 11.256$$

$$\therefore \lambda_r = \frac{11.526}{11.256 - 1} = 1.09$$

$$\therefore \lambda_p > 1 \therefore \text{satisfactory}$$

(b) *Column*

(i) CHECK CAPACITY AT HINGE (FIGURE 5.5a)

$$M_p = 700 \text{ kNm}$$

$$F = 212.8 + (0.97 \times 16.8/5) = 216.1 \text{ kN}$$

The column uses 610 × 229 UB 101, with A = 129 cm², S_x = 2880 cm³, r_x = 24.2 cm, r_y = 4.75 cm, u = 0.863, × = 43.1.

From Steel Construction Institute (2000, Vol. 1, note 3.2.6) the axial load ratio is given by

$$n = (216.1 \times 10)/(129 \times 275) = 0.061$$

$$< \text{change value } 0.468 \quad \frac{(D-2T)t}{A}$$

$$\text{Reduced } S_r = 2880 - 3962 \times 0.061^2 = 2865\,\text{cm}^3$$

$$> S_x \text{ required } 2545\,\text{cm}^3 \quad \frac{(700 \times 1000)}{275}$$

Column section is satisfactory.

(ii) COLUMN RESTRAINTS AND STABILITY

A torsional restraint is provided by stays from a sheeting rail at the plastic hinge in the column as shown in Figure 5.7. The distance to the adjacent restraint using the conservative method given in Clause 5.3.3(a) of BS 5950 L_m may be taken = L_u given by

$$L_u = \frac{38_{ry}}{\left[(F_c/130) + (x/36)^2 (P_y/275)^2\right]^{1/2}}$$

where

$$F_c = 216.1 \times 10/129 = 16.8 \text{ N/mm}^2$$

Thus,

$$L_u = \frac{38 \times 47.5}{\left[(16.8/130) + (43.1/36)^2\right]^{1/2}} = 1444\,\text{mm}$$

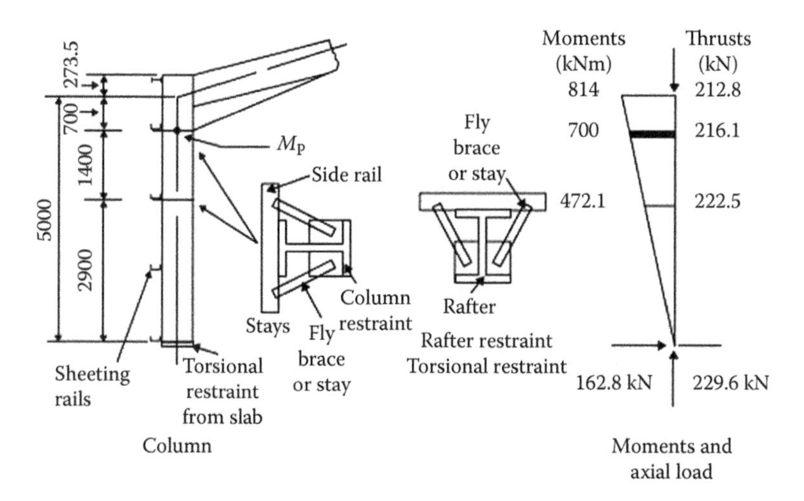

Figure 5.7 Torsional column restraint.

Place a sheeting rail and stays at 1400 mm below the hinge as shown in Figure 5.7.

The column is now checked between the second restraint and the base over a length of 2.9 m. Actions at the restraint shown on Figure 5.7 are

$F = 222.5$ kN

$M = 472.1$ kNm

In-plane:

$l_{ex} = 5.0$ m*

$\lambda_X = 5000/242 = 20.6$

Out-of-plane:

$\lambda_X = 2900/47.5 = 61.1$

$p_c = 200.8$ N/mm^2 (Table 24 for strut curve c)

$\lambda/x = 61.1/43 = 1.42$

$V = 0.97$ (Table 19)

$\lambda_{LT} = uv\lambda\sqrt{\beta_W}, \beta_w = 1$

* Steel Construction Institute (1987, Vol. 2, portal design example).

See Clause 4.3.6.9 of BS 5950-1-2000

$\lambda_{LT} = 0.863 \times 0.97 \times 61.1 = 51.2$

$p_b = 235.1$ N/mm^2

$M_b = 235.1 \times 288/10^3 = 677.1$ kNm

$\beta = 0.0$, $m = 0.57$ (Table 18)

Combined–local capacity has been checked above. Overall buckling:

$$\frac{F_c}{P_{cy}} + \frac{m_{LT}M_{LT}}{M_b} + \frac{m_y M_y}{P_y Z_y} \leq 1$$

$$\frac{222.5 \times 10}{200.8 \times 129} + \frac{0.6 \times 472.1}{677.1} + 0.0 = 0.49$$

The column section is satisfactory – adopt 610 × 229 UB 101 or 610 × 229 UB 113.

(c) *Rafter*

(i) CHECK CAPACITY AT HINGE (FIGURE 5.5a)

$M_p = 525$ kNm

$F = 146$ kN

The rafter uses 533 × 210 UB 82, with $A = 105$ cm^2, $S_x = 2060$ cm^3, $r_y = 4.38$ cm, $u = 0.863 \times = 41.6$.

$n = (146 \times 10)/(105 \times 275) = 0.051$

$< \text{change value, } 0.458$

$S_r = 2060 - (2871 \times 0.051^2) = 2053$ cm^3

$> S_x$ required, 1909 cm^3

The rafter section is satisfactory.

(ii) CHECK STRESSES IN HAUNCH

The haunch length is normally made about one-tenth of the span. The proposed arrangement is shown in Figure 5.8, with a haunch 3.5 m long. This is checked first to ensure that the stresses remain elastic. The actions are shown in the figure.

At the beginning of the haunch the actions are

$$M = 763.8 \text{ kNm}$$

$$F = 211.5 \text{ kN}$$

(Section AA in Figure 5.8). The flange of the UB is neglected but it supports the web. The properties are listed in Table 5.1. From the table, A = 150.3 cm², Z_x = 4193 cm³. Thus, maximum stress is given by

$$\frac{211.5\times10}{150.3}+\frac{763.8\times10^3}{4193}=196.3\text{ N/mm}^2$$

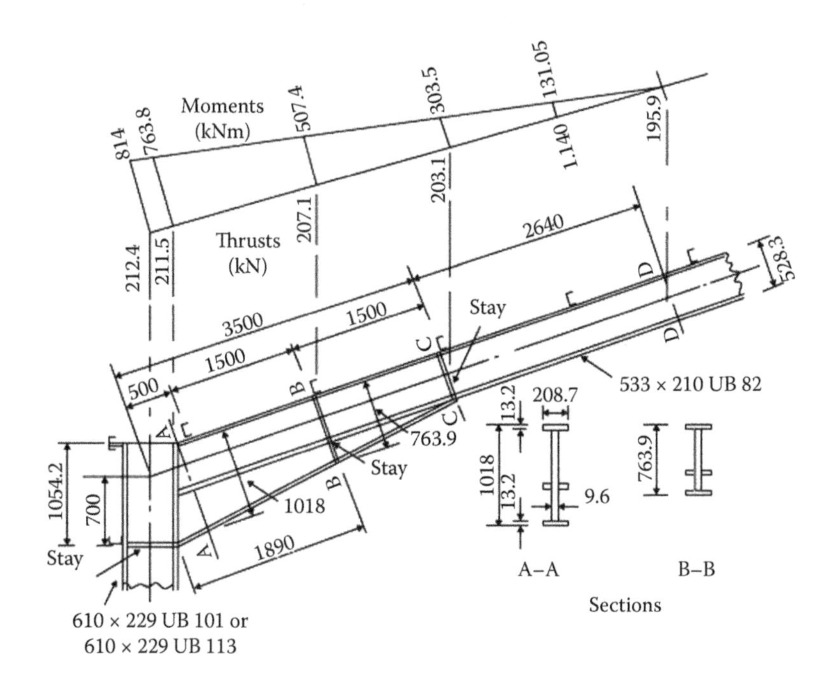

Figure 5.8 Haunch arrangement – moments and thrusts.

At the end of the haunch the actions are

$$M = 303.5 \text{ kNm}$$

$$F = 203.1 \text{ kN}$$

Maximum stress is given by

$$\frac{203.1 \times 10}{104} + \frac{303.5 \times 10^3}{1800} = 188.1 \text{ N/mm}^2$$

The stresses are in the elastic region.

(iii) HAUNCH RESTRAINTS AND STABILITY

The spacing of restraints to the compression flange of the haunch is designed to comply with Clause 5.5.3.5.2 of BS 5950.

The properties of the haunch sections at AA and BB (Figure 5.8), calculated using the formula from Appendix B.2.5 of BS 5950, are listed in Table 5.1. The UB properties and moments and thrusts are also shown.

Check haunch between sections AA and BB:

Effective length, $L_E = 1500$ mm

$$\lambda = 1500/36.5 = 41.1$$

$$p_c = 236.4 \text{ N/mm}^2 \text{ (Table 24(c))}$$

$$\lambda/x = 41.1/64.9 = 0.63$$

Slenderness factor $\nu = 0.99$ (Table 19),

$$\lambda_{LT} = u\nu\lambda\sqrt{\beta_W}, = 0.822 \times 0.99 \times 41.1 = 33.4$$

$$p_b = 273 \text{ N/mm}^2$$

Equivalent uniform moment factor $m_x = 0.86$, $\beta_w = 1$ (Clause 4.3.6.9 BS 5950).

Table 5.1 Properties of rafter haunch sections

Section	A *(cm²)*	I_X *(cm⁴)*	I_Y *(cm⁴)*	Z_X *(cm³)*	S_X *(cm³)*	Y_Y *(cm)*	u	X	M *(kNm)*	F *(kN)*
AA	150.3	213,446	2007	4193	5128	3.65	0.822	89.1	763.8	211.5
BB	125.5	109,723	2005	–	3374	3.99	0.842	64.9	507.4	207.1
CC	104	47,500	2010	1800	2060	4.38	0.865	41.6	303.5	203.1

For section AA

$$\frac{F_c}{P_c} + \frac{m_x M_X}{P_y Z_X} = \frac{211.5 \times 10}{236.4 \times 150.3} + \frac{0.86 \times 763.8 \times 10^3}{275 \times 4193} = 0.60$$

$$\frac{F_c}{P_{cy}} + \frac{m_{LT} M_{LT}}{M_b} \leq 1$$

$$\therefore \frac{211.5 \times 10}{236.4 \times 150.3} + \frac{0.86 \times 763.8 \times 10^3}{273 \times 5128} = 0.52$$

Thus, the section is satisfactory.

A similar check is carried out on the haunch section between sections BB and CC. This gives a maximum combined criterion of 0.62.

The rafter and haunch meet conditions specified in BS 5950. The limiting spacing L_s for the compression flange restraints is given for Grade S275 steel by the following equation:

Haunch depth/rafter depth = 1054.2/528.3 = 2.0

$$K_1 = 1.4$$

$$L_s = \frac{620 \times 4.38 \times 10}{1.4[72 - (100/41.6)^2]^{0.5}} = 2384.0 \text{ mm}$$

Put restraints to the compression flange at sections BB and CC, the end of the haunch.

Check rafter between end of haunch CC and point of contraflexure DD:

$\lambda = 2640/43.8 = 60.3$

$p_c = 220$ N/mm^2 (Table 24(b))

$\lambda/x = 60.3/41.6 = 1.45$

$\nu = 0.97$ (Table 19)

$\lambda_{LT} = u\nu\lambda\sqrt{\beta_W}, = 0.865 \times 0.97 \times 60.3 = 50.6$

$p_b = 236.5$ N/mm^2 (Table 16)

Combined criterion:

$$\frac{F_c}{P_{cy}} + \frac{m_{LT}M_{LT}}{M_b} \leq 1; \quad \frac{203.1\times 10}{104\times 220} + \frac{303.5\times 10^3}{236.5\times 2060} = 0.7$$

(iv) RAFTER NEAR RIDGE – PLASTIC ANALYSIS

Under dead and imposed load, a hinge forms near the ridge as shown in Figure 5.5a. Purlins are spaced at not greater than L_m adjacent to the hinge. These restrain the compression flange (Clause 5.3.3 of code). At the hinge,

$$F_c = 146 \times 10/15 = 14 \text{ N/mm}^2$$

$$L_u = \frac{38\times 43.8}{[(14/130)+(41.6/36)]^{1/2}} = 1385 \text{ mm}$$

$$D/B = 2.5, F_c < 80/\text{mm}^2$$

$$\therefore L_m = \phi L_u, \ \beta_u = 0.44 + \frac{41.6}{270} + \frac{14}{200} = 0.66$$

$$\beta = 0.6 \quad 0 < \beta < \beta_u$$

$$\therefore \phi = 1.17$$

$$\therefore L_m = 1.17 \times 1385 = 1620 \text{ mm}$$

Put purlins at 1050 mm near the hinge.

Stays to the bottom flange are not required at this hinge, which is the last to form. However, under load reversal due to wind, a stay is required and will be located at the hinge position. The purlin arrangement is shown in Figure 5.9.

5.2.8 Rafter under wind uplift

The bending moment diagram for the case of dead and transverse wind load is shown in Figure 5.9 for the leeward portal rafter. The bottom flange is in compression over the unrestrained length between stays at the eaves and the plastic moment near the ridge of 12.9 m.

The stability is checked in accordance with Section 4.3 of BS 5950. The member is loaded between the lateral restraints, so the slenderness correction factor n is determined.

The rafter uses 533 × 210 UB 82, with r_Y = 4.38 cm, u = 0.865, x = 41.6, S_x = 2060 cm^3, A = 105 cm^2.

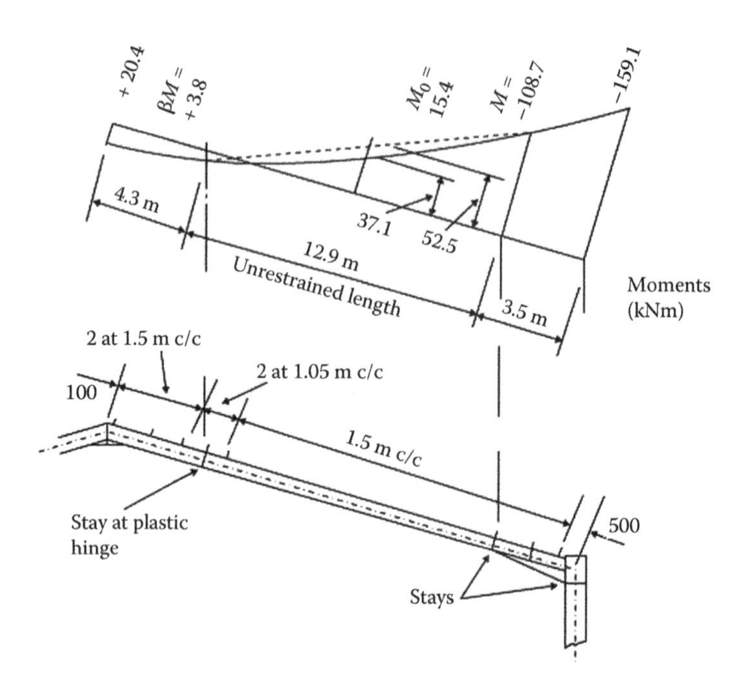

Figure 5.9 Wind uplift moments – rafter stays: dead and transverse wind loads.

$\lambda = 12{,}900/43.8 = 295$

$\lambda/x = 295/41.6 = 7.09$

$v = 0.728$ (Table 19)

From the moment diagram, $\beta = -3.8/108.7 = -0.03$, $m_{LT} = 0.59$.

$\lambda_{LT} = 0.865 \times 0.728 \times 295 = 185.76$

$p_b = 46$ N/mm^2

$M_b = 94.76$ kNm

(Figures 5.6a and 5.9). The thrust in the rafter at the location of moment $M = 108.7$ kNm = 39.5 kN (compression).

$\lambda = 295$

$p_c = 21.5$ N/mm^2 (Table 24(b))

$p_c = 21.5 \times 105/10 = 225.8$ kN

Combined:

$$(39.5/225.8) + (0.59 \times 108.7/94.76) = 0.84$$

This is satisfactory.

The elastic stability could also have been checked taking account of restraint to the tension flange (BS 5950, Appendix G) (Figure 5.6b). The case of maximum uplift can be checked in a similar manner. The rafter again is satisfactory.

5.2.9 Portal joints

(a) *Eaves joint bolts – dead and imposed load*

The joint arrangement is shown in Figure 5.10a. Assume the top four rows of bolts resist moment. Joint actions are

$$M = 814 \text{ kNm}$$

$$F = 212.8 \text{ kN}$$

$$V = 162.8 \text{ kN}$$

With bolts at lever arms 684 mm, 784 mm, 884 mm and 984 mm, the bolt group modulus is given by

$$\frac{\sum y^2}{y_{max}} = \frac{2(684^2 + 784^2 + 884^2 + 984^2)}{984} = 5757 \text{ mm}$$

and the maximum bolt tension is

$$T = \frac{814 - (162.8 \times 0.78)}{5.757} = 119.3 \text{ kN}$$

Use 22 mm diameter Grade 8.8 bolts with capacity of $0.8P_tA_t = 136$ kN.

$$\text{Shear per bolt} = 212.8/14 = 15.2 \text{ kN}$$

$$\text{Shear capacity} = P_s = P_sA_s = 114 \text{ kN}$$

Combined shear and tension (Clause 6.3.4.4):

$$\frac{F_s}{P_s} + \frac{F_t}{P_{nom}} = \frac{15.2}{114} + \frac{119.4}{136} = 1.01 < 1.4$$

which is satisfactory.

The joint must also be checked for the reverse wind moment of 392.5 kN.

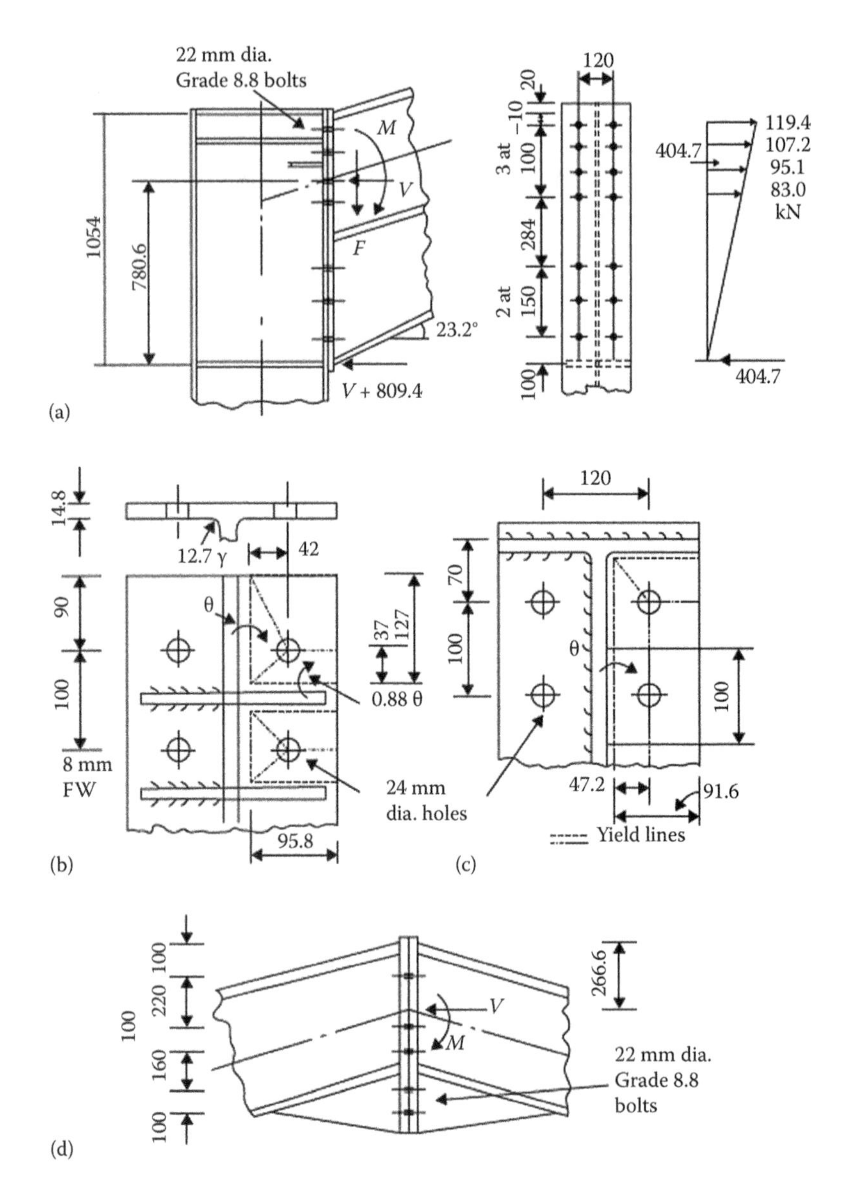

Figure 5.10 Portal joints: (a) joint arrangement and bolt loads; (b) UB flange; (c) endplate; (d) ridge joint.

(b) *Column UB flange*

The yield line pattern is shown in Figure 5.10b. For the top bolt the work equation is

$$(127+103)M\theta+(95.8+30+41.8)0.88M\theta$$
$$=119.4\times 42\theta\times 10^3$$
$$M=13.3\times 10^3 \text{ N mm per mm}$$
$$< MR=275\times 14.8^2/4=15.1\times 10^3 \text{ N mm per mm}$$

The flange is satisfactory.
A further stiffener is required between the second and third bolts.

(c) *Rafter endplate*

For the weld, try 8 mm fillet – strength 1.2 kN/mm; 100 mm of weld resists 120 kN. The yield line pattern is shown in Figure 5.10c.
The second bolt is critical in determining endplate thickness t. The work equation is

$$(100+76)M\theta=107.2\times 47.2\theta\times 10^3$$
$$M=28.7\times 10^3 \text{ N mm per mm}$$
$$=265\times t^2/4$$
$$t=20.8 \text{ mm}$$

Provide 22 mm plate, p_y = 265 N/mm^2.

(d) *Check column web shear*

Shear = (404.7 × 2) + 162.8 = 967.5 kN

Shear capacity = 0.6 × 275 × 602.2 × 10.5/10^3 = 1053.2 kN

The web is satisfactory.

(e) *Column stiffeners*

Top and bottom stiffener loads = 967.5 kN

Try two 20 × 100 stiffeners:

Capacity = 20 × 100 × 2 × 265/100 = 1060 kN

Use weld with 8 mm fillet.

(f) *Haunch flange* (Figure 5.10a)

Load = 967.5 sec 23.2° = 1052.6 kN

Flange thickness required to carry load

$$= (1052.6 \times 10^3)/(265 \times 208.7) = 19\,\text{mm}$$

The haunch could be cut from 533 × 210 UB 122, where flange thickness is 21.3 mm. Alternatively, a small length of web of the 533 × 210 UB 82 can be counted on to carry part of the load. This reduces the bolt lever arm slightly.

(g) *Base plate and HD bolts*

Provide 20 mm base plate and 4 No. 22 mm diameter HD bolts.

(h) *Ridge joint* (Figure 5.10d)

Joint actions are

M = 414.4 kNm

V = 162.8 kN

$\sum y^2 = 2(680^2 + 580^2 + 420^2) = 1.95 \times 10^6\ \text{mm}^2$

Maximum bolt tension is given by

$$T = \left(\frac{414.4 - (162.8 \times 0.27)}{1.95 \times 10^3} \right) \times 680 = 129.2\,\text{kN}$$

Provide 22 mm diameter Grade 8.8 bolts.

5.2.10 Serviceability check

The outward deflection of the columns at the eaves is calculated using a classical method set out in British Constructional Steelwork Association Publication 19 (1963).

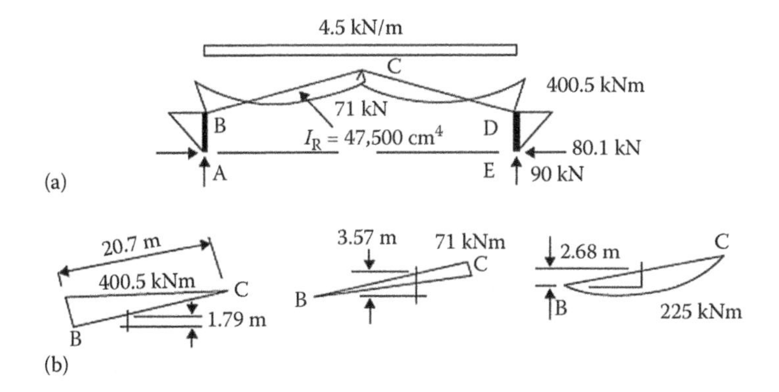

Figure 5.11 Portal bending moments: (a) overall; (b) separate elements of rafter moments.

The portal bending moment diagram due to the unfactored imposed load is shown in Figure 5.11. The separate elements forming the rafter moments are shown. The uniform load causes a moment $Wl^2/8$ = 225 kNm.

The outward deflection at the eaves is given by

$$\delta_B = \left(\sum \text{Areas of bending momment diagram on BC}\right) \times (\text{Level arm to level BD})/EI_R$$

$$\begin{aligned}\delta_B &= [-(400.5\times 20.7\times 1.79/2)+(71\times 20.7\times 3.75/2) \\ &\quad +(2\times 225\times 20.7\times 2.68/3)]\times 10^5/(205\times 47{,}500) \\ &= 36.2\,\text{mm}\ (>h/300=16.7\,\text{mm})\end{aligned}$$

The metal sheeting can accommodate this deflection.

A drawing of the portal is shown in Figure 5.12 or see Clause 5.5.4.2.2 BS 5950: 2000 for gravity and Clause 5.5.4.2.3 for horizontal loads.

5.3 BUILT-UP TAPERED MEMBER PORTAL

5.3.1 General comments

The design process for the tapered welded I-section member portals commonly used in the United States is reviewed briefly (Figure 5.13).

The design is made using elastic theory. The portal members are made deeper at the eaves. The columns taper to the base and the rafters may be single or double tapered as shown in the figure. This construction is theoretically the most efficient with the deepest sections at points of maximum moments.

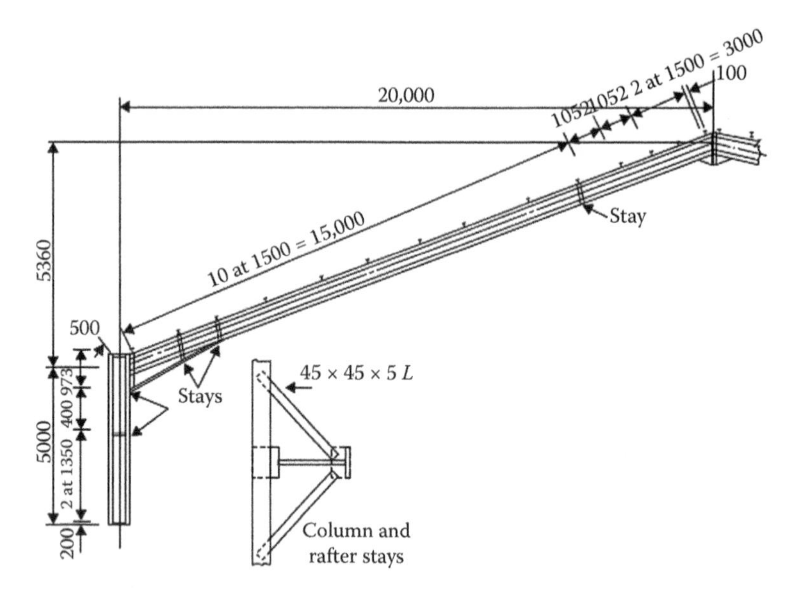

Figure 5.12 Plastic-designed portal.

5.3.2 Design process

The complete design method is given in Lee et al. (1981). The analysis may be carried out by dividing the frame into sections, over which the mid-point properties are assumed to be constant, and using a frame program.

A series of charts are given to determine the in-plane effective lengths of the portal members. The tapered members are converted into equivalent prismatic members. Charts are given for single- and double-tapered members. Another series of charts gives the effective length factors for sway-prevented and permitted cases. The in-plane effective lengths depend on the spacing of the restraint. The design can then proceed to the required code. The textbook by Lee et al. should be consulted for the complete treatment.

5.4 TWO-PINNED ARCH

5.4.1 General considerations

The arched roof is most commonly constructed in the form of a circular arc. It has been used extensively for sports arenas, bus and rail terminals, warehouses, etc. Many variations in arched roof construction are possible including circular and parabolic shapes, three-pinned, two-pinned and fixed types, multiarched roofs and barrel vaults in three-way grids.

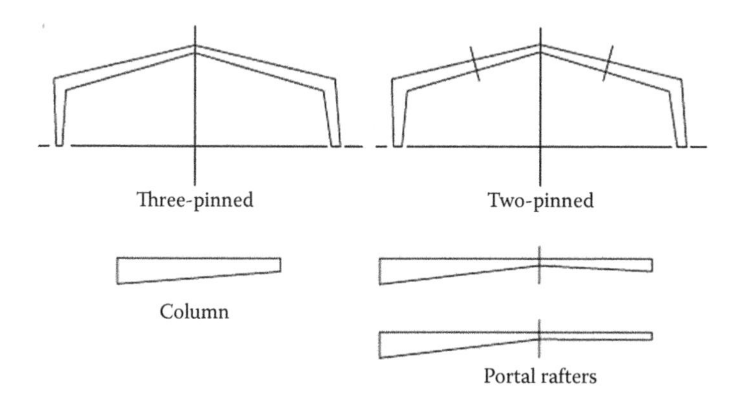

Figure 5.13 Tapered member portals.

A design is made for a rib in an arched roof building to the same general specification as that for the portals considered earlier in the chapter. This enables a comparison to be made with these structures. This is redesigned as a lattice arch.

The arch rib is sized for dead and imposed load. The maximum conditions occur when the imposed load covers about two-thirds of the span. The wind load causes uplift and tension in the arch.

Arch stability is the critical feature in design. The two-hinged arch buckles in-plane into an anti-symmetrical shape with a point of contraflexure at the crown. The expression for effective length is

$$l = (1.02 \text{ to } 1.25) \times (\text{Arch length}/2)$$

The higher factors apply to high-rise arches (Johnson, 1976; Timoshenko and Gere, 1961). Lateral supports are also required.

5.4.2 Specification

The two-pinned arch has span 48 m, rise 10 m, spacing 6 m. Dead load is 0.4 kN/m^2 and imposed load 0.75 kN/m^2 on plan. A section through the building is shown in Figure 5.14. It has a clear span of 40 m between side walls and the steel ribs extend outside. The arch rib is to be a rectangular hollow section in Grade S275 steel.

5.4.3 Loading

(a) *Dead load*

Roof design load = 1.4 × 0.4 × 6 = 3.36 kN/m

Arch rib1 kN/m = 1.4 × 5.25 = 7.4 kN

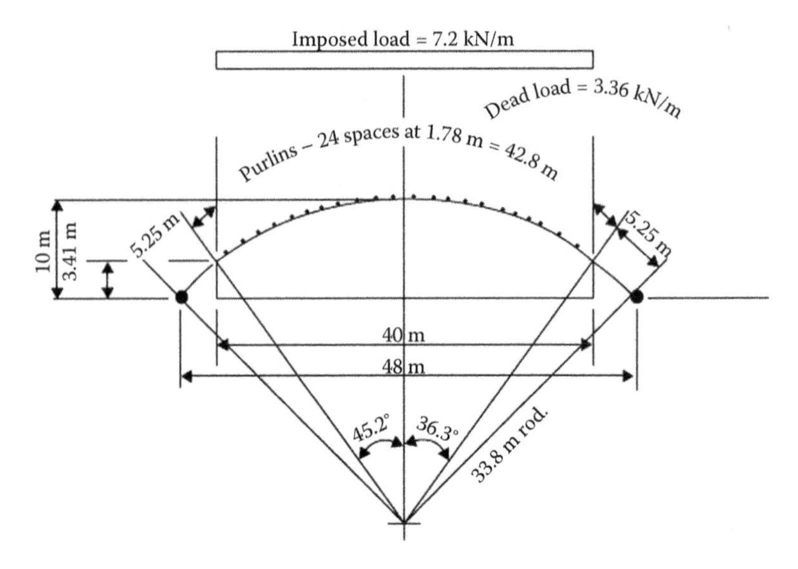

Figure 5.14 Section through building.

The loads are applied at 13 points on the roof, as shown in Figure 5.15a. The joint coordinates are shown in Figure 5.15b.

(b) *Imposed load*

Roof design load = 1.6 × 0.75 × 6 = 7.2 kN/m on plan

The loads for two cases are shown in Figure 5.15c and d:

- Imposed load over the whole span;
- Imposed load over 63% of span.

The second case is found to give maximum moments.

(c) *Wind load*

Basic wind speed is 45 m/s, ground roughness 3, building size C, height 10 m.

Design wind speed = 0.69 × 45 = 31.1 m/s

Dynamic pressure = $0.613 \times 31.1^2/10^3 = 0.59$ kN/m^2

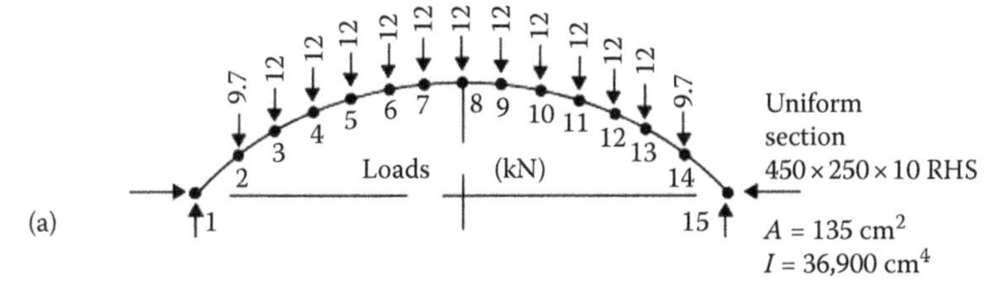

(a)

No.	x	y	No.	x	y	No.	x	y
1	0.0	0.0	6	16.91	9.25	11	34.53	8.32
2	4.0	3.44	7	20.44	9.51	12	37.86	7.03
3	6.97	5.39	8	24.0	10.0	13	41.03	5.39
4	10.14	7.03	9	27.56	9.51	14	44.0	3.44
5	13.47	8.32	10	31.09	9.25	15	48.0	0.0

(b)

10.7 20.1 23.4 24.4 25.1 25.5 25.6 25.5 25.1 24.4 23.4 22.1 10.7

Loads (kN)

(c)

10.7 20.1 23.4 24.4 25.1 25.5 25.6 25.5 25.1

Loads (kN)

(d)

4.4 8.8 8.8 8.8 10.1 11.3 11.3 11.3 10.1 8.8 6.9 5.0 2.5

α = 0 (internal pressure)

Loads (kN)

(e)

Figure 5.15 Arched roof loading: (a) design dead loads – joints; (b) joint coordinates (m); (c) design imposed load – whole span; (d) design imposed load – 63% of span; (e) characteristic wind load.

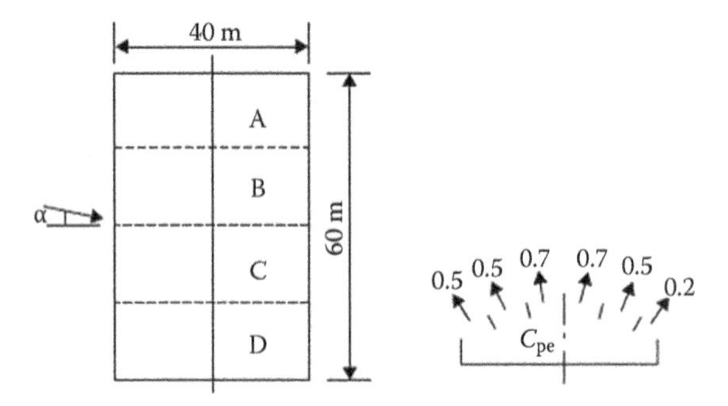

Figure 5.16 External pressure coefficients for wind angle $\alpha = 0$.

External pressure coefficients C_{pe}, taken from Newberry and Eaton (1974) for a rectangular building with arched roof, are shown in Figure 5.16. For $\alpha = 90°$; $C_{pe} = -0.8$ in division A, causing maximum uplift.

Wind load puts the arch in tension if the internal suction C_{pi} is taken as -0.3 for the case where the four walls and roof are equally permeable.

5.4.4 Analysis

Analyses are carried out for:

1. Dead load;
2. Imposed load over whole roof;
3. Imposed load on 63% of roof.

The joint coordinates are shown in Figure 5.15b.

Successive trials are needed to establish that case 3 gives the maximum moment in the arch rib. The wind load causes tension in the arch, and the analysis is not carried out for this case.

The bending moment diagrams for the three load cases are shown in Figure 5.17. The arch thrusts at critical points are noted in the diagrams.

5.4.5 Design

(a) *Maximum design conditions*

For dead and imposed loads over whole span, at joint 14

$$F = 117.8 + 234.5 = 352.3 \text{ kN}$$

$$M = 32.7 + 98.6 = 131.3 \text{ kNm}$$

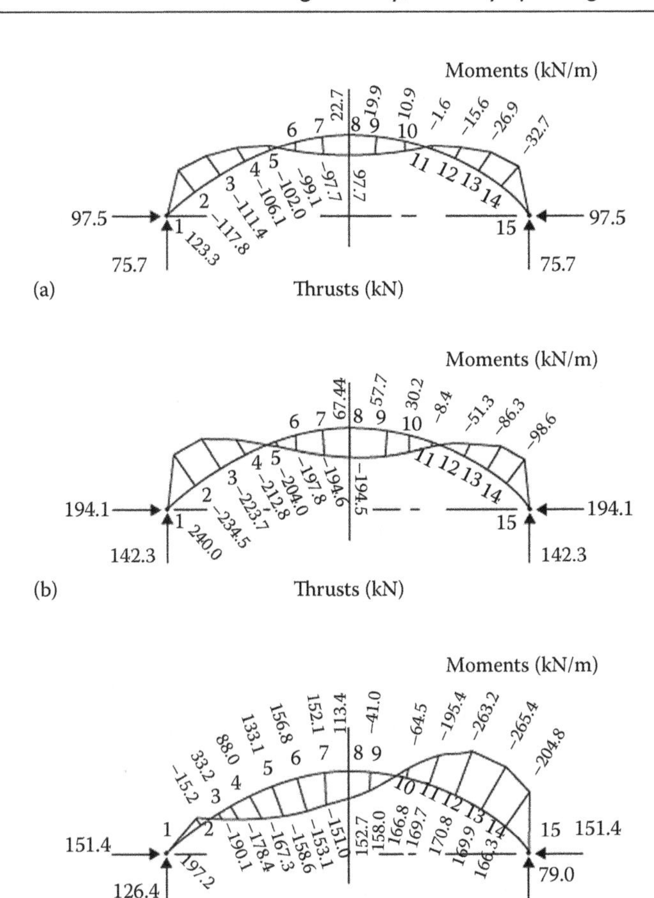

Figure 5.17 Bending moments: (a) dead loads; (b) imposed load – whole span; (c) imposed load – 63% of span.

For dead and imposed loads over 63% of span, at joint 13

$$F = 111.4 + 169.9 = 281.3 \text{ kN}$$

$$M = 26.9 + 265.4 = 292.3 \text{ kNm}$$

(b) *Trial section*

This is to be 450 × 250 × 10 RHS × 106 kg/m, for which A = 135 cm², S_x = 2000 cm³, r_X = 16.5 cm, r_Y = 10.5 cm.

(c) *Arch stability*

The in-plane buckled shape of the arch is shown in Figure 5.18b. The arch is checked as a pin-ended column of length S equal to one-half the length of the arch. The effective length factor is 1.1.

$$l/r_X = 1.1 \times 26{,}700/165 = 178 < 180$$

$$p_c = 58.8 \text{ N/mm}^2 \text{ (Table 24(a))}$$

Lateral stability is provided by supports at the quarter points as shown in Figure 5.18a.

$$l/r_Y = 13{,}350/105 = 127.1$$

Lateral torsional buckling (Appendix B.2.6 of code):

$$\lambda = 127.1,\ D/B = 1.8$$

Limiting λ (for $D/B = 1.8$ for $p_y = 275$ N/mm²) = 395

$$p_b = 275 \text{ N/mm}^2$$

$$M_b = 275 \times 2000/10^3 = 550.0 \text{ kNm}$$

(d) *Member buckling resistance check*

For joint 14:

$$\frac{352.3 \times 10}{58.8 \times 135} + \frac{0.95 \times 131.3}{550} = 0.68$$

For joint 13:

$$\frac{281.3 \times 10}{58.8 \times 135} + \frac{0.95 \times 292.3}{550} = 0.88$$

The arch rib selected is satisfactory.

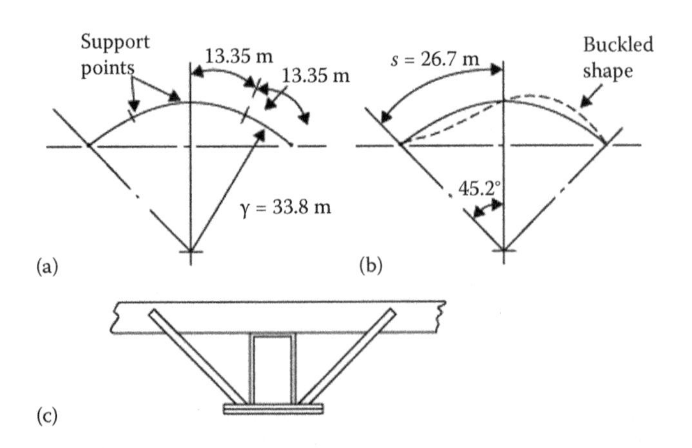

Figure 5.18 Arch stability: (a) lateral support; (b) in-plane buckling; (c) support.

Note that if the imposed load is extended to joint 11, that is, 70% of span, F = 299.5 kN, M = 280.6 kNm and capacity check gives 0.88.

5.4.6 Construction

The arch is fabricated in three sections 18.8 m long. The rib is bent to radius after heating. The sections are joined by full strength welds on site with the arch lying flat. It is then lifted into its upright position. The arch site welds and springing detail are shown in Figure 5.19.

5.4.7 Lattice arch

(a) *Specification*

An alternative design is made for a lattice arch with structural hollow section chords 1 m apart and mean radius 33.8 m, the same as the single section rib designed above.

The arrangement of the arch, the lateral restraints and a section through the rib are shown in Figure 5.20. The coordinates of the joints and the loads applied at the purlin points are listed in Table 5.2. The critical load case consisting of dead load plus imposed load covering 63% of the span is shown.

A preliminary design is made using results from the single rib design above. This is compared with the design made for an accurate analysis of the lattice arch.

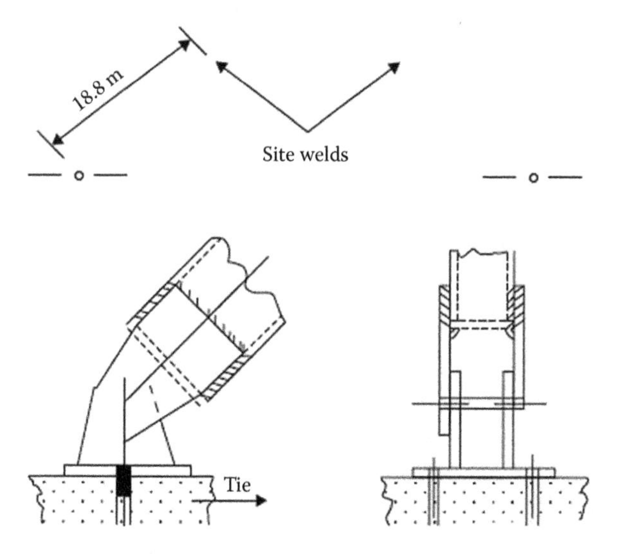

Figure 5.19 Construction details.

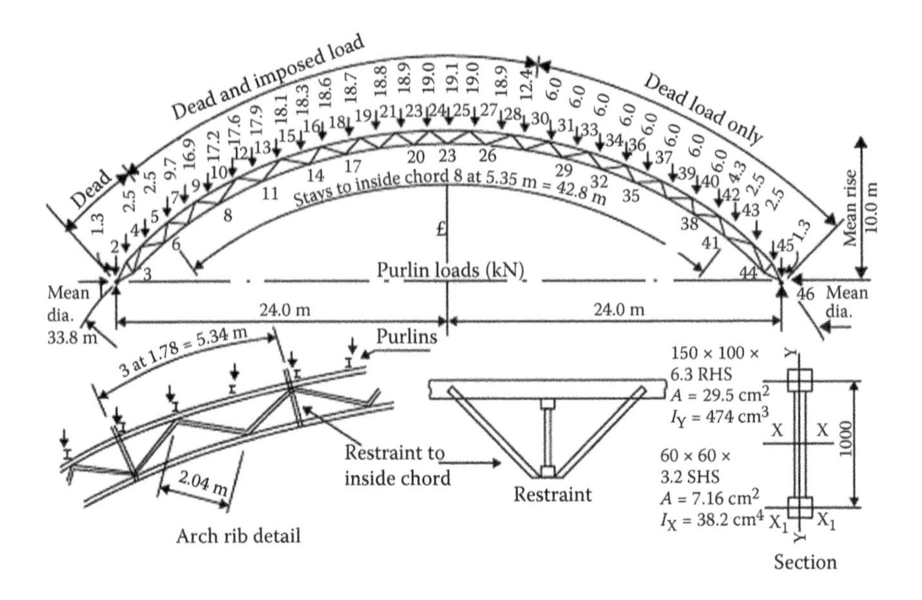

Figure 5.20 Lattice arch.

(b) *Preliminary design*

(i) TRIAL SECTION (FIGURE 5.20)

The chords are 150 × 100 × 6.3 RHS, with A = 29.5 cm², I_Y = 474 cm³, r_x = 5.52 cm, r_Y = 4.01 cm, S_x = 147 cm³.

The web is 60 × 60 × 3.2 SHS, with A = 7.16 cm², r = 2.31 cm, S = 15.2 cm.

(ii) MAXIMUM DESIGN CONDITIONS

For joint 13, dead and imposed load on 63% of span:

F = 281.3 kN

M = 292.3 kNm (tension in top chord)

Purlin load = 6 kN (dead load on top chord)

For joint 7, dead and imposed load on 63% of span:

F = 248.7 kN

M = 172.0 kNm (compression in top chord)

Purlin load = 18.7 kN (dead and imposed load on top chord)

Table 5.2 Lattice arch-coordinates and loads

No.	x (m)	y (m)	Load (kN)	No.	x (m)	y (m)	Load (kN)	No.	x (m)	y (m)	Load (kN)
1	0	0	1.3	17	17.01	8.76	–	33	34.68	8.79	6.0
2	0.93	1.59	2.5	18	16.80	9.74	18.7	34	36.39	8.18	6.0
3	1.61	0.85	–	19	18.58	10.07	18.8	35	37.65	6.57	–
4	2.28	2.75	2.5	20	20.47	9.31	–	36	38.06	7.49	6.0
5	4.29	3.04	–	21	20.37	10.31	18.9	37	39.69	6.70	6.0
6	3.69	3.85	9.7	22	22.17	10.45	19.0	38	40.77	4.97	–
7	5.19	4.88	16.9	23	24.0	9.5	–	39	41.27	5.83	6.0
8	7.23	4.97	–	24	24.0	10.5	19.1	40	42.81	4.88	6.0
9	6.73	5.83	17.2	25	25.83	10.45	19.0	41	43.71	3.04	–
10	8.31	6.70	17.6	26	27.53	9.31	–	42	44.31	3.85	4.3
11	10.35	6.57	–	27	27.63	10.31	18.9	43	45.72	2.75	2.5
12	9.94	7.49	17.9	28	29.42	10.07	12.4	44	46.39	0.85	–
13	11.61	8.18	18.1	29	30.99	8.76	–	45	47.07	1.59	2.5
14	13.62	7.84	–	30	31.20	9.74	6.0	46	48.0	0	1.3
15	13.31	8.79	18.3	31	32.96	9.31	6.0				
16	15.04	9.31	18.6	32	34.38	7.84	–				

(iii) MEMBER CAPACITIES

Compression

For the whole arch buckling in plane:

$$r_X = \{[(29.5 \times 2 \times 500^2) + (2 \times 474)]/(2 \times 29.5)\}^{1/2} = 500 \text{ cm}$$

$$l/r_X = 1.1 \times 26\,700/500 = 58.7$$

The slenderness ratios are based on the mean diameter of the arch. For arch chord in plane of arch, $l/r_{X1} = 0.85 \times 3560/40.1 = 75.3$.

For arch chord out of plane (lateral restraints are provided at 5.34 m centres).

$$l/r_Y = 5340/55.2 = 96.6$$

$$p_c = 164.8 \text{ N/mm}^2 \text{ (Table 24(a))}$$

$$P_c = 164.8 \times 29.5/10 = 486.2 \text{ kN}$$

Bending

$$M_c = 275 \times 147/10^3 = 40.4 \text{ kNm}$$

(iv) DESIGN CHECK

For joint 13, check compression in bottom chord:

Load = (281.3/2) + (292.3/1) = 432.9 kN

Capacity = 486.2 kN

For joint 7, check compression and bending in the top chord. Secondary bending due to the purlin load is

$$M = WL/12 = 18.7 \times 3.56/12 = 5.55 \text{ kNm}$$

and

Compression force = (248.7/2) + (170.2/1) = 296.4 kN

Cross-section capacity check:

$$\frac{296.4}{486.2} + \frac{5.55}{40.4} = 0.74$$

The section selected is satisfactory.

(v) WEB MEMBERS

The maximum shear in the arch at the support is 45.0 kN, so that

Web member force = 45 × 2.04 = 91.8 kN

Select from capacity, tables in Steel Construction Institute (2000): 60 × 60 × 3.2 SHS with capacity 135 kN for an effective length of 2 m. Make all the internal members the same section.

(c) *Accurate design*

A computer analysis is carried out using the arch dimensions, coordinates, section properties and loads shown in Figure 5.20 and Table 5.2.

The maximum design conditions from the analysis are given below:

- Top chord –

Member 16–18, $F = -326.9$ kN, $M = 1.06$ kNm

Member 36–37, $F = 167.8$ kN, $M = -8.71$ kNm

- Bottom chord –

Member 35–38, $F = -400.0$ kN, $M = -1.84$ kNm

- Web members –

Member 28–29, $F = -99.9$ kN, $M = -0.23$ kNm

(i) CHECK CHORD MEMBERS (150 × 100 × 6.3 RHS)

Capacities (Section 5.4.7(b)) are

Tension $P_t = 29.5 \times 275/10 = 811.2$ kN

Compression $P_c = 486.2$ kN

Bending $M_c = 40.4$ kNm

For member 16–18:

$(326.9/486.2) + (1.06/40.4) = 0.7$

For member 36–37:

$(167.8/811.2) + (8.71/30.5) = 0.5$

For member 35–38:

$$(400/486.2) + (1.84/30.5) = 0.88$$

The section chosen is satisfactory.

(ii) CHECK WEB MEMBER (60 × 60 × 3.2 SHS)

This member could be reduced to 60 × 60 × 3 SHS with capacity in compression of 128 kN for 2.0 m effective length. Make all web members the same section – 60 × 60 × 3 SHS × 5.29 kg/m.

Chapter 6

Single-storey, one-way-spanning pinned-base portal-plastic design to EC3

6.1 TYPE OF STRUCTURE

The following analysis and check were carried out for the same pinned-base portal-frame with the specification and framing plans detailed in Section 5.2.1.

6.2 SWAY STABILITY

Sway stability of the portal is checked using the procedure given in Appendix B.4.1 of EC3, where the following condition must be satisfied for grade S275 steel.

6.2.1 For dead and imposed load

(a) *For truly pinned bases*

$$V_{sd}/V_{cr} = [N_r/N_{r,cr} + (4.0 + 3.3R_p)(N_c/N_{cr})]$$

N_r = Axial compression in the rafter at shallow end of haunch
= 203.1 kN

$$N_{r,cr} = \pi^2 EI_r/s^2$$
$$= \pi^2 \times 210 \times 47,500 \times 10^4/[(20/\cos\ 15)^2 \times 10^6]$$
$$= 2296.36\ \text{kN}$$

$$R_p = I_c s / I_r h$$
$$= 87{,}300 \times 20.706/(47{,}500 \times 5) = 7.61$$

$$N_c = \text{Axial compression in column at column mid height}$$
$$= 212.8 + 16.8 = 229.6 \text{ kN}$$

$$N_{c,cr} = \pi^2 E I_c / h^2$$
$$= \pi^2 \times 210 \times 87{,}300 \times 10^4 / 5.0^2 \times 10^6 = 72{,}375.78 \text{ kN}$$

$$V_{sd}/V_{cr} = [203.1/2296.36 + (4.0 + 3.3 \times 7.61) \times (229.6/72{,}375.78)]$$
$$= 0.18 < 0.2 \text{ (ENV 1993-1-1, Clause 5.2.6.3)}$$

Structure classified as a sway case, second-order effects therefore must be accounted for as done by Merchant-Ranking (ENV 5.2.5.2, 1993-1-1 Appendix K.6).

(b) *For nominally pinned bases*

$$V_{sd}/V_{cr} = [N_r/N_{r,cr} + (2.9 + 2.7R_p)(N_c/N_{cr})] \times [1/(1 + 0.1\,R_p)]$$
$$V_{sd}/V_{cr} = [203.1/2296.36 + (2.9 + 2.7 \times 7.61)(229.6/72{,}375.78)]$$
$$\times [1/(1 + 0.1 \times 7.61)]$$
$$= 0.088 < 0.1$$

There is no need to allow for second-order frame effects. This indicates a more economical solution. Other load combinations need to be checked too.

6.3 ARCHING STABILITY CHECK-RAFTER, SNAP THROUGH

$$V_{sd}/V_{cr} = [L/D][(\Omega - 1)/(55.7(4 + L/h))][I_r/(I_c + I_r)] \times [f_{yr}/275][1/\tan 2\theta_r]$$

Ω = factored vertical load/maximum load to cause failure of the rafter as a fixed ended beam

Factored vertical load = 2.46 × 1.35 + 4.5 × 1.5 = 10.07 kN

Maximum load to cause failure of the rafter as a fixed ended beam

$$= Mp \times 16/L = [2060 \times 10^3 \times 275 \times 16 / (40 - 6.76) \times 10^6]$$
$$= 272.68 \text{ kN}$$

$\Omega = 10.071 \times (40 - 6.76)/272.68 = 1.22$

$$V_{sd}/V_{cr} = [40/528.3][(1.22 - 1)/(55.7(4 + 40/5))]$$
$$\times [47{,}500/(75{,}800 + 47{,}500)][275/275][1/\tan 2 \times 15]$$
$$= 0.019 < 0.12$$

Very small value, and $1/(1 - V_{sd}/V_{cr}) = 1/(1 - 0.019) = 1.019 \cong 1.00$, no need to modify the partial factor of safety and allow for second order effects.

6.4 CHECK THE COLUMN

The loading details are

M_{sd} = 821.8 kNm, V_{sd} = 162.8 kN, N_{sd} = 229.6 kN

6.4.1 Section classification

Check if the column section is plastic to accommodate plastic hinge formation (Figure 6.1). 610 × 229 UB 101 fails the EC check, therefore 610 × 229 UB 113 is selected instead. Section properties 610 × 229 UB 113

h = 607.6 mm, b = 228.2 mm, t_w = 11.1 mm, t_f = 17.3 mm,

$d = 547.6$ mm, $\lambda_1 = 93.9$, $\varepsilon = 86.6$, $i_{LT} = 55.5 = \left(I_z I_w / W_{pl,y}^2\right)^{1/2}$

i_z = 48.8 mm, i_y = 246 mm, $\alpha = 1630 = (I_w/I_t)^{0.5}$,

$W_{pl,y} = 3280 \times 10^3$ mm³, A = 14,400 mm², $I_w = I_z h^2 s/4$,

$h_s = h - t_f$, $I_{zz} = 3430 \times 10^4$ mm⁴, $I_t = 111 \times 10^4$ mm⁴

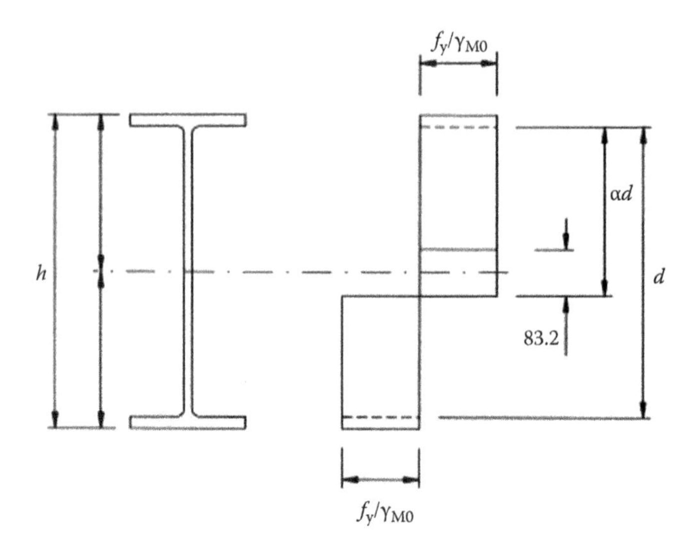

Figure 6.1 Plastic stress distribution rafter.

As the web is under combined axial and bending moment, we need to calculate α as follows:

$$d_c = N_{sd}/(f_y \times t_w/\gamma_{M0}) = 229.6 \times 1000/(275 \times 11.2/1.1) = 82 \text{ mm}$$

$$\alpha d = d/2 + d_c/2$$

$$\alpha = 0.576$$

$$d/t_w < 396\varepsilon/(13\alpha - 1.0)$$

$$\varepsilon = 0.92$$

$$d/t_w = 547.3/11.1 = 49.3 < 396\varepsilon/(13\alpha - 1.0) = 56.2$$

Therefore, the web is plastic.

$$c/t_f = 114.1/17.3 = 6.6 < 10\varepsilon = 9.2$$

Therefore, the flange is plastic. Therefore, the whole section is plastic.

6.4.2 Moment of resistance

It is to be checked if the moment resistance is not reduced by

1. The coincident of shear force, or
2. The coincident of axial force.

(a) *Coincident of shear force*

It is to be checked that $V_{sd} < 0.5V_{pl,Rd}$

$$V_{pl,Rd} = A_v \times (f_y \sqrt{3})/\gamma_{M0}$$

$$A_v = 1.04\,h \times t_w = 7011\,\text{mm}^2$$

$$V_{pl,Rd} = 7011(275/\sqrt{3})/1.1 = 1012\,\text{kN}$$

$$\therefore V_{sd} = 162.8 < 0.5 \times 1012 = 506\,\text{kN (satisfactory)}$$

(b) *Coincident of axial force*

The following have to be checked:

(a) $N_{sd} < 0.5 \times A_{web} \times f_y/\gamma_{M0}$

$$A_{web} = A_{gross} - 2A_{flange} = 6504\ \text{mm}^2$$

$$N_{sd}(229.6\ \text{kN}) \leq 0.5 \times 6504 \times 275/1.1 = 813\ \text{kN}$$

$$= \text{plastic tensile resistance of the web (satisfactory)}$$

(b) $N_{sd} \leq 0.25 \times$ plastic tensile resistance of the section

$$N_{sd} \leq 0.25 \times A_{gross} \times f_y/\gamma_{M0}$$

$$N_{sd}(229.6\ \text{kN}) \leq 0.25 \times 14{,}400 \times 275/1.1 = 900\ \text{kN (satisfactory)}$$

Therefore, the effect of shear and axial forces on the plastic resistance moment can be neglected, and the $M_{pl,Rd}$ need not be reduced.

6.4.3 Column buckling between intermediate restraints

For the bending moment shown in Figure 6.2, a plastic hinge may occur at the underside of the hunch. Therefore, find the stable length with a plastic hinge as follows:

Maximum spacing between restraint at plastic hinge and stays = $0.4(C_p)^{0.5} \lambda_1 i_{LT}$ where $C_p = 1/[M_{hinge}/(C_1 W_{pl,y} f_y) + (4N_{hinge}/Af_y)(i_{LT}/i_z)^2]$.

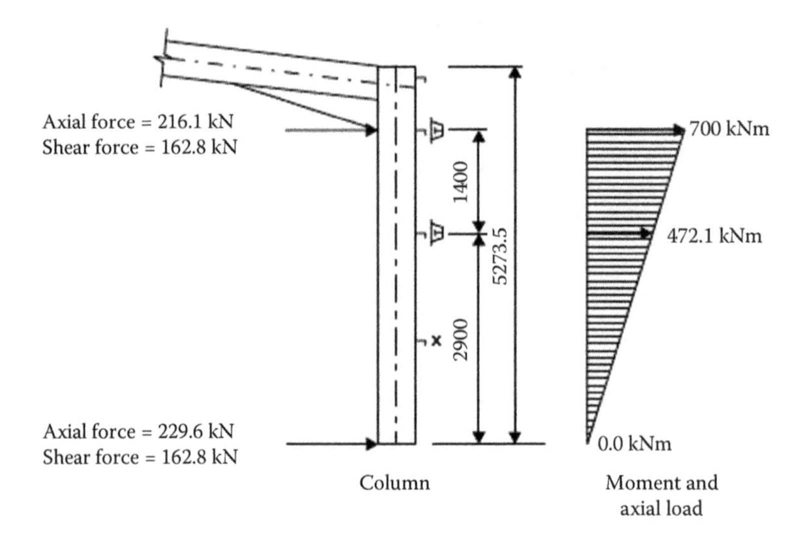

Figure 6.2 Torsional column restraint.

For $C_p = 1$, first approximation

$$\text{spacing} = 0.4(C_p)^{0.5}\lambda_1 i_{LT} = 0.4 \times 186.8 \times 55.5$$
$$= 1927\text{ mm} > 1400\text{ mm (satisfactory)}$$

To allow for the coincident axial force try 1400 mm centre-to-centre tensional restraint from stay.

Moment at first stay beneath hunch = 472.1 kNm

$$C_p = 1/[M_{hinge}/(C_1 W_{pl,y} f_y) + (4N_{hinge}/Af_y)(i_{LT}/i_z)^2]$$

for $k = 1.0$, $C_1 = 1.88 - 1.4\psi + \psi^2$

$\psi = 472.1/700 = 0.674$

$C_1 = 1.173 < 2.7$ (satisfactory)

(see ENV 1993-1-1 Annex F, Tables F.1.1 and F.1.2).

For i_{LT} equal approximately $i_z/0.9$

$$C_p = 1/[M_{hinge}/(C_1 W_{pl,y} f_y) + (4N_{hinge}/Af_y)(i_{LT}/i_z)^2]$$
$$C_p = 1/[700 \times 10^6/(1.173 \times 3280 \times 10^3 \times 275)$$
$$+ 4 \times 216.1 \times 10^3/(14,400 \times 275)(1/0.9)^2]$$
$$C_p = 1/(0.66 + 0.269) = 1.076$$

$L_{cmax} = 0.4(C_p)^{0.5} \lambda_1 i_{LT} = 0.4 \times (1.076)0.5 \times 86.8 \times 55.5 = 2000\ \text{mm} > 1400\ \text{mm}$. Therefore, a spacing or 1400 mm is satisfactory.

Between the stay at 2900 mm from the base and the rail lateral restraint is provided from rail and bracing. Therfore, a spacing of 1350 mm is satisfactory for rails below 2900 mm height up column, as the moment is lower.

$$\psi = 0.0, \quad k = 1.0, \quad C_1 = 1.88, \quad i_{LT} = i_z/0.9,$$

$$M = 472.1\ \text{kNm}\ N = 222.5\ \text{kN}$$

$$\therefore C_p = 1.8 \text{ and}$$

$$L_{cmax} = 0.4(C_p)^{0.5} \lambda_1 i_{LT} = 2585\ \text{mm} > 1350\ \text{mm}$$

$\therefore$ satisfactory.

6.4.4 Column buckling between torsional restraints

Buckling resistance to combined axial and bending have to be checked.

$$N_{sd}/N_{b,Rd,z} + k_{LT} M_{y,sd}/M_{b,rd,y} \leq 1.0$$

$$N_{sd} = 229.6\ \text{kN}$$

$$M_{y,sd} = 472.1\ \text{kNm}$$

To calculate $N_{b,Rd,z}$, k_{LT} and $M_{b,rd,y}$, we need to determine λ and λ_{LT} first (see Appendix F.3.4).

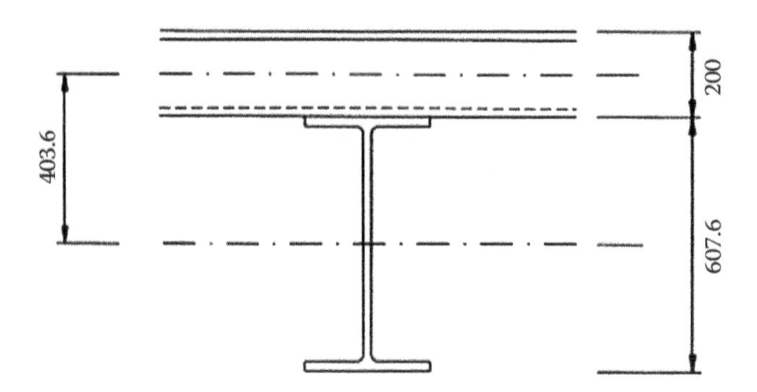

Figure 6.3 Column and side rail.

From Figure 6.3 assume the side rail depth = 200 mm.

a = 607.3/2 + 200/2 = 403.8 mm

$$i_s^2 = i_y^2 + i_z^2 + a^2$$
$$i_s^2 = 246^2 + 48.8^2 + 403.8^2$$
$$= 225{,}952 \text{ mm}^2$$

$$h_s = h - t_f/2 = 607.3 - 17.3 = 590.0 \text{ mm}$$
= distance between shear centres of flanges

$\alpha = [a^2 + (I_w/I_z)/I_z]/i_s^2$ for doubly symmetrical I sections,

$I_w = I_z(h_s/2)^2$

$$\alpha = [a^2 + (h_s/2)^2]/\,i_s^2 = [403.8^2 + (590.0/2)^2]/225{,}952 = 1.106$$

$$\lambda = (L_t/i_z)/\left[\alpha + I_t L_t^2/\left(2.6\pi^2 I_z i_s^2\right)\right]^{1/2}$$
$$= (2900/48.8)/[(1.106 + 111 \times 2900^2)/$$
$$(2.6 \times 3.14^2 \times 3430 \times 225{,}952)]^{1/2}$$
$$= 53.6$$

$\lambda_{LT} = \left(m_t^{0.5} c\right)\left[(W_{pl,y}/A)\left(2a/i_s^2\right)\right]^{0.5} \lambda$ (see Appendix F.3.4)

$\psi_t = 0.0/472.1 = 0.0$

$y = \lambda/(L_t/i_z) = 53.6/(2900/48.8) = 0.9$

$m_t = 0.53$

$c = 1.0$ (see Appendix F.3.3, for load combinations including lateral loads, m_t should be obtained from Clause F.3.3.1.2)

$$\lambda_{LT} = (0.53^{0.5} \times 1)[(3280 \times 10^3/14,400)(2 \times 403.8/225,952)]^{0.5} \times 53.6$$
$$= 35.2$$

$N_{b,Rd} = \chi A f_y/\gamma_{M1}$

$\chi_{min} = 1/[\phi + (\phi^2 - \bar{\lambda}^2)^{0.5}]$

$\phi = 0.5[1 + \alpha(\bar{\lambda} - 0.2) + \bar{\lambda}^2]$

$h/b = 607.3/228.2 = 2.66$

From ENV Table 5.5.1 and ENV 5.5.1.2 use curve b for hot rolled sections.

$\alpha = 0.34$

$\bar{\lambda} = \lambda/\lambda_1 = 53.6/86.8 = 0.62$

$\phi = 0.5[1 + 0.34(0.62 - 0.2) + 0.62^2] = 0.76$

$\chi_z = 1/[0.76 + (0.76^2 - 0.62)^{0.5}] = 1.14$

$$N_{b,Rd} = \chi A f_y / \gamma_{M1}$$
$$= 1.14 \times 14{,}400 \times 275 / 1.1 = 4104 \text{ kN}$$

$M_{b,Rd,y} = \chi_{LT} W_{pl,y} f_y / \gamma_{M1}$ (ENV 5.5.2, Appendix D.3.2)

$\bar{\lambda}_{LT} = 35.5/86.8 = 0.4$

$$\chi_{LT} = 1/\left[\phi_{LT} + \left(\phi_{LT}^2 - \bar{\lambda}_{LT}^2\right)^{0.5}\right]$$

$$\phi_{LT} = 0.5\left[1 + \alpha_{LT}(\bar{\lambda}_{LT} - 0.2) + \bar{\lambda}_{LT}^2\right]$$
$$= 0.5[1 + 0.21(0.4 - 0.2) + 0.4^2] = 0.9540$$

$$M_{b,Rd,y} = \chi_{LT} W_{pl,y} f_y / \gamma_{M1} = 0.9540 \times 3280 \times 1000 \times 275/1.1$$
$$= 783.9 \text{ kNm}$$

Calculate k_{LT}

$\psi = 0.0$ (ENV 5.5.3)

$\beta_{M,LT} = 1.8 - 0.7\psi = 1.8$ (ENV 5.5.4)

$$\mu_{LT} = 0.15\bar{\lambda}_z \beta_{M,LT} - 0.15, \ \mu_{LT} \le 0.9$$
$$= 0.15 \times 0.62 \times 1.8 - 0.15 = 0.0174 \le 0.9$$

$$k_{LT} = 1 - [\mu_{LT} N_{sd} / (\chi_z A f_y)], \ k_{LT} \le 1.0$$
$$= 1 - (0.0174 \times 229.6 \times 1000/(1.14 \times 14{,}400 \times 275))$$
$$= 0.98$$

$N_{sd}/N_{b,Rd,z} + k_{LT}M_{y,sd}/M_{b,rd,y} \leq 1.0$

$(229.6/4104) + (0.98 \times 222.5/783.9)$
$= 0.339 < 1.0$ (column satisfactory).

6.5 STABILITY OF THE RAFTER

Check is limited to the load combination shown in Figure 5.5. All the other load combinations must be checked too. Section properties 533 × 210 UB 82.

$h = 528.3$ mm, $b = 208.7$ mm, $t_w = 9.6$ mm, $t_f = 13.2$ mm,

$d = 476.5\,\text{mm}, \quad \lambda_1 = 93.9, \quad \varepsilon = 86.6, \quad i_{LT} = 50.1 = \left(I_z I_w / W_{pl,y}^2\right)^{1/2}$

$i_z = 43.8$ mm, $i_y = 213$ mm, $\alpha_{LT} = 1610 = (I_w/I_t)^{0.5}$,

$W_{pl,y} = 2058 \times 10^3$ mm^3, $A = 10{,}500$ mm^2,

$I_w = I_z h^2 s/4$, $h_s = h - t_f$,

$I_{zz} = 2004 \times 10^7$ mm^4, $I_t = 51.5 \times 10^4$ mm^4

6.5.1 Section classification

The section needs to be checked to make sure that it is class 1 (plastic) to accommodate plastic hinge formation (Figure 6.4).

As the web is under combined axial and bending moment, we need to calculate α as follows:

$d_c = N_{sd}/(f_y \times t_w/\gamma_{M0}) = 203.1 \times 1000/(275 \times 9.6/1.1) = 84.6$ mm

$\alpha d = d/2 + d_c/2$

$\alpha = 0.58$

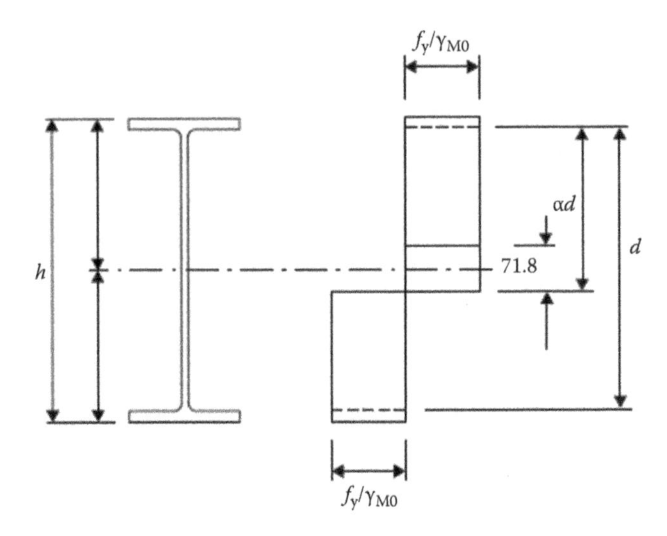

Figure 6.4 Plastic stress distribution, rafter.

$$d/t_w < 396\varepsilon/(13\alpha - 1.0)$$

$$\varepsilon = 0.92$$

$$d/t_w = 476.5/9.6 = 49.6 < 396\varepsilon/(13\alpha - 1.0) = 55.7$$

Therefore, the web is plastic.

$$c/t_f = 7.91 < 10\varepsilon = 9.2$$

Therefore, the flange is plastic.
Therefore, the whole section is plastic (ENV 1993-1-1, Table 5.3.1).

6.5.2 Moment of resistance

It is to be checked whether the plastic moment of resistance is reduced by

1. Coincident of shear force.
2. Coincident of axial force.

(a) *Check for reduction by coincident of shear force*

Check $V_{sd} < 0.5V_{pl,Rd}$

$$V_{pl,Rd} = A_v(f_y/\sqrt{3})/\gamma_{M1}$$
$$= 1.04 \times 528.3 \times 9.6(275/\sqrt{3})/1.1 = 921.2\text{ kN}$$

$$0.5 \times V_{pl,Rd} = 460.6\text{ kN} > \text{maximum shear force.}$$

Plastic moment resistance is not reduced by the coincident of the shear force.

(b) *Check for reduction by coincident of axial force*

(i) Check if:

N_{sd} < 0.5 × Plastic tensile resistance of the rafter web.

$N_{sd} = 203.1$ kN

Plastic tensile resistance of the rafter web

$= A_{web} \times f_y/\gamma_{M0}$

$= (A_{gross} - 2A_{flange}) \times 275/1$

$= (10,500 - 2 \times 208.7 \times 13.2) \times 275/1.1 = 1247.6$ kN

$N_{sd} = 203.1$ kN < 0.5 × 1247.6 (satisfactory)

(ii) Check if:

N_{sd} < 0.25 × Plastic tensile resistance of the rafter cross-section.

Plastic tensile resistance of the rafter cross-section

$= A \times f_y/\gamma_{M0}$

$= 10,500 \times 275/1.1 = 2625.0$

$N_{sd} = 203.1$ kN < 0.25 × 2625.0 < 656.25 kN (satisfactory)

∴ plastic moment of resistance is not reduced by coincident of axial force
∴ because both (a) and (b) above are satisfactory, we can neglect the effect of the shear and axial force on the plastic moment resistance.

6.5.3 Rafter check buckling between intermediate restraints

(a) Stable length check: Check if the maximum spacing 1500 mm between intermediate restraints at the rails (purlins) to ensure stability where the maximum bending moment occurs (Appendix D.4) (Figure 6.5).

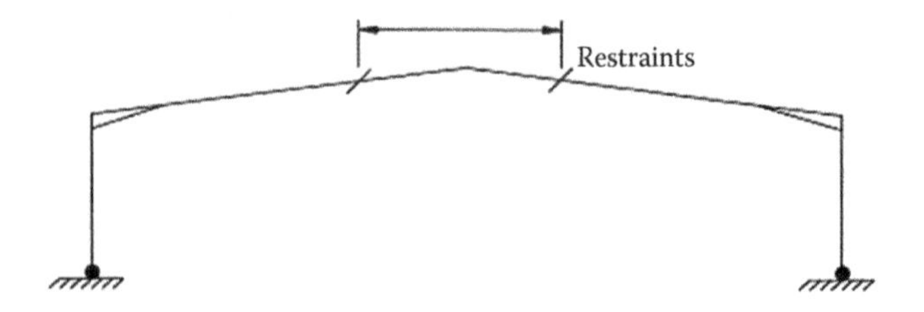

Figure 6.5 Highest bending moment – rafter.

$$L = 0.4[1/[M_{sd}/(C_1 W_{pl,y} f_y) + (4N_{sd}/Af_y)(i_{LT}/i_z)^2]]^{0.5} \lambda_1 i_{LT}$$

$$L = 0.4[1/[525 \times 10^6/(1.0 \times 2060 \times 10^3 \times 275)$$
$$+ 4 \times 168.4 \times 10^3/(105 \times 100 \times 275)(1/0.9)^2]]^{0.5} \times 86.850 \times 1$$
$$= 1580 \text{ mm} > 1500 \text{ mm (satisfactory)}$$

(b) Check for lateral torsional buckling between purlins for lower bending moments, see Figure 6.6.

Check if

$$N_{sd}/N_{b,Rd,z} + k_{LT} M_{y,sd}/M_{b,rd,y} \leq 1.0$$

$$M_{sd,max} = 303.5 \text{ kNm}$$

$$N_{sd,max} = 203.1 \text{ kN}$$

Calculate $N_{b,Rd,z}$

$$N_{b,Rd,z} = \chi_z A f_y/\gamma_{M1}$$

$$\chi_z = 1/[\phi + (\phi^2 - \bar{\lambda}^2)^{0.5}]$$

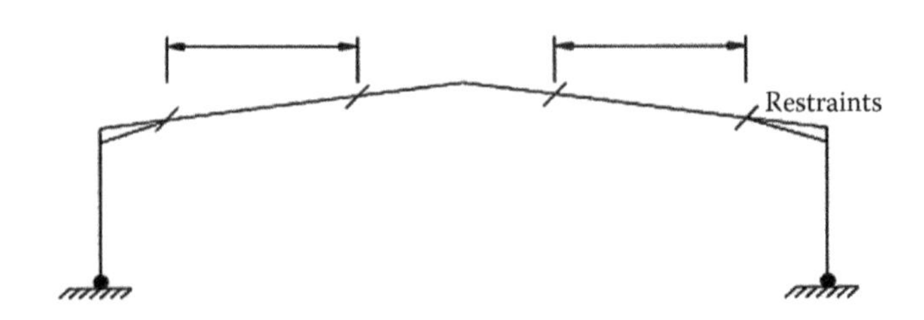

Figure 6.6 Lower bending moments – rafter.

Try rail spacing = 1500 mm

$\lambda_z = 1500/43.8 = 34.24$

$\bar{\lambda}_z = 34.24/86.8 = 0.4$

$$\phi = 0.5[1+\alpha(\bar{\lambda}_z - 0.2) + \bar{\lambda}^2]$$
$$= 0.5[1+0.34(0.4-0.2)+0.4^2] = 0.61$$

$$\chi_z = 1/[\phi + (\phi^2 - \bar{\lambda}^2)0.5]$$
$$= 1/[0.61 + (0.61^2 - 0.4^2)^{0.5}] = 0.93$$

$$N_{b,Rd,z} = \chi_z A f_y / \gamma_{M1} = 0.93 \times 10,500 \times 275/1.1$$
$$= 2441.25 \text{ kN} > 203.1 \text{ kN}$$

$M_{b,Rd} = \chi_{LT} \times W_{pl,y} \times f_y/\gamma_{M1}$

$$\chi_{LT} = 1/\left[\phi_{LT} + \left(\phi_{LT}^2 - \bar{\lambda}_{LT}^2\right)^{0.5}\right]$$

$\lambda_{LT} = L/i_{LT}/((C_1)^{0.5}[1 + (L/\alpha_{LT})^2/25.66]^{0.25})$

$C_1 = 1.0$ (conservative)

$\alpha_{LT} = (I_w/I_t)^{0.5}$

$\lambda_{LT} = 1500/50.1/((1.0)^{0.5}[1 + (1500/1610)^2/25.66]^{0.25}) = 26.6$

$\bar{\lambda}_{LT} = 26.6/86.8 = 0.31$

$$\phi_{LT} = 0.5\left[1+\alpha(\bar{\lambda}_{LT} - 0.2) + \bar{\lambda}_{LT}^2\right]$$
$$= 0.5[1+0.21(0.31-0.2)+0.31^2] = 0.56$$

$$\chi_{LT} = 1/\left[\phi_{LT} + \left(\phi_{LT}^2 - \bar{\lambda}_{LT}^2\right)^{0.5}\right]$$

$$\chi_{LT} = 1/[0.56 + (0.56^2 - 0.31^2)^{0.5}] = 0.98$$

$$M_{b,Rd} = 0.98 \times 2060 \times 10^3 \times 275/1.1 \times 10^6$$
$$= 555.17 \text{ kNm} > 303.5 \text{ kNm}$$

$N_{sd}/N_{b,Rd,z} + k_{LT}M_{y,sd}/M_{b,rd,y} \leq 1.0$

For k_{LT} = 1.0 (conservative)

$203.1/2441.25 + 1.0 \times 303/555.17 = 0.629 < 1.0$

Rafter buckling resistance between intermediate resistance is satisfactory.

6.5.4 Rafter check buckling between torsional restraints (stays)

For the loading shown in Figure 5.5, check the length between hunch tip and span rotational restraint, that is, 3500 and 15,500 mm (Figure 6.7).

M_{sd} (hunch) = 303.5 kNm

M_{sd} (3 m from hunch tip) = 131.05 kNm
(take hogging moment positive)

M_{sd} (12 m from hunch tip) = −525 kNm

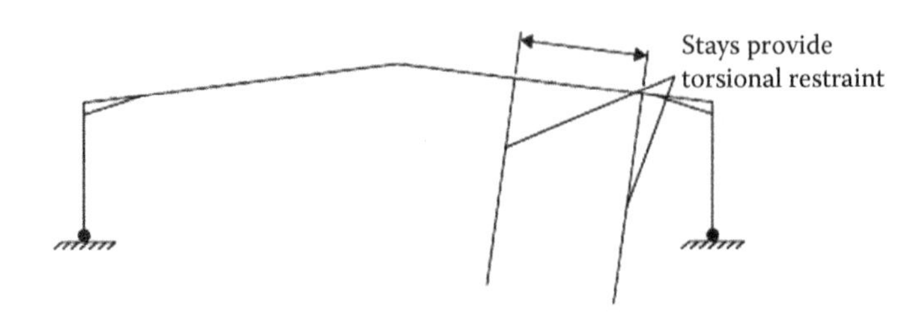

Figure 6.7 Worst buckling from gravity loads.

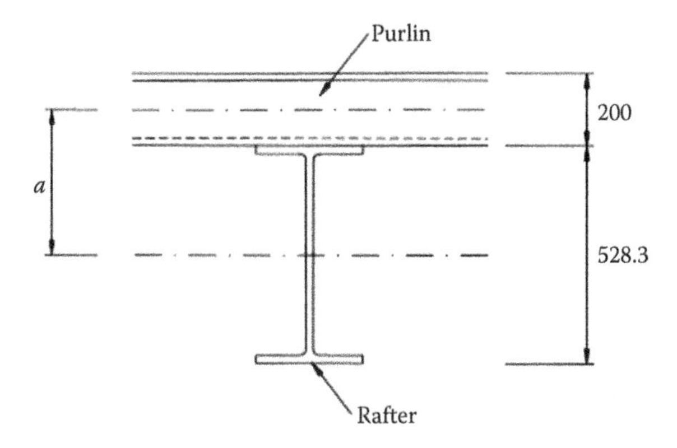

Figure 6.8 Rafter and stay.

Check if

$$N_{sd}/N_{b,Rd,z} + k_{LT}M_{y,sd}/M_{b,rd,y} \leq 1.0$$

For the stay, try 200 × 75 × 23 purlins with (Figure 6.8)

$$D = 200 \text{ mm}$$

$$a = 528.3/2 + 200/2 = 364.2 \text{ mm}$$

$$i_s^2 = i_y^2 + i_z^2 + a^2$$

$$i_y^2 = 213 \text{ mm}$$

$$i_z^2 = 43.8 \text{ mm}$$

$$i_s^2 = 18.0 \times 10^4 \text{ mm}^2$$

$$\alpha = [a^2 + (h_s/2)^2]/i_s^2$$

$$h_s = (D - t_f)/2$$

$$\alpha = 1.105$$

$$\lambda = (L_t/i_z)/\left[\alpha + I_t L_t^2/\left(2.6\pi^2 I_z i_s^2\right)\right]^{1/2}$$

$$= (12{,}000/43.8)/[1.105 + 515{,}000 \times 12{,}000^2/$$

$$(2.6\pi^2 \times 2010 \times 10^4 \times 18 \times 10^4)]$$

$$= 194.2$$

$\lambda_{LT} = \left(m_t^{0.5}c\right)\left[W_{pl,y} \times 2a/\left(A \times i_s^2\right)\right]^{0.5} \lambda$ (see Appendix F.3.4)

$\psi_t = -525/303.5 = -1.73$, take as -1.0

$y = 194.2/(12{,}000/43.8) = 0.7$

$m_t = 0.36$

$c = 1.0$ (see Appendix F.3.3, for load combinations including lateral loads, m_t should be obtained from Clause F.3.3.1.2)

$$\bar{\lambda}_{LT} = (0.36^{0.5} \times 1.0)[2060 \times 10^3 \times 2 \times 364.2/$$
$$(10{,}500 \times 18.0 \times 10^4)]^{0.5} \times 194.2$$
$$= 102.5$$

$$\bar{\lambda} = (\beta_A)^{0.5}\lambda/86.8(275/f_y)0.5,\ \beta_A = A_{eff}/A = 1.0$$
$$= 194.2/86.8 = 2.23$$

$\chi_z = 1/[\phi + (\phi^2 - \bar{\lambda}^2)]^{0.5}$

$\phi = 0.5[1 + \alpha(\bar{\lambda} - 0.2) + \bar{\lambda}^2]$

$\alpha = 0.21$

$\phi = 3.2$

$\chi_z = 0.34$

$N_{b,Rd} = 0.34 \times 10{,}500 \times 275/1.1 = 892.50$ kN

$\bar{\lambda}_{LT} = \bar{\lambda}_{LT}/86.6 = 102.5/86.8 = 1.18$

$$\chi_{LT} = 1/\left[\phi_{LT} + \left(\phi_{LT}^2 - \bar{\lambda}_{LT}^2\right)\right]^{0.5}$$

$$\phi_{LT} = 0.5\left[1 + \alpha(\bar{\lambda}_{LT} - 0.2) + \bar{\lambda}_{LT}^2\right] = 1.3,\ \alpha = 0.21$$

$\chi_{LT} = 79$

$M_{b,Rd,z} = 0.79 \times 2060 \times 1000 \times 275/(1.1 \times 10^6) = 406.85$

For $k_{LT} = 1$ (conservative approach)

$N_{sd}/N_{b,Rd,z} + k_{LT}M_{y,sd}/M_{b,rd,y} \leq 1.0$

$203.1/892.50 + 1 \times 303.5/406.85 = 0.97 < 1.0$

Use the same approach above and check the rafter for the other critical load combinations.

For comparison purpose see the following table:

	BS 5950-1-2000	*EC3*
Column	619 × 229 UB 101	610 × 229 × 113
Rafter	533 × 210 UB 82	533 × 210 UB 82

Chapter 7

Multistorey buildings

7.1 OUTLINE OF DESIGNS COVERED

7.1.1 Aims of study

Various methods are set out in BS 5950 for design of building structures. These methods were listed and discussed in general terms in Chapter 3.

To show the application of some of the methods, a representative example of a four-storey building is designed using different code procedures and compared.

7.1.2 Design to BS 5950

Designs are made to the following methods specified in BS 5950 for a braced frame:

1. Simple design
2. Rigid elastic design
3. Rigid plastic design
4. Semirigid plastic design

7.2 BUILDING AND LOADS

7.2.1 Specification

The framing plans for the multistorey office building selected for the design study are shown in Figure 7.1. The office space in the centre bays is of open plan with any offices formed with light partitions. The end bays contain lifts, stairs and toilets. Figure 7.2 shows relevant detail at walls and columns for estimation of loads.

The building specification is

- Steel frame with cast-*in-situ* floors;
- Dimensions – 35 m × 16 m × 18 m high with frames at 5 m centres;

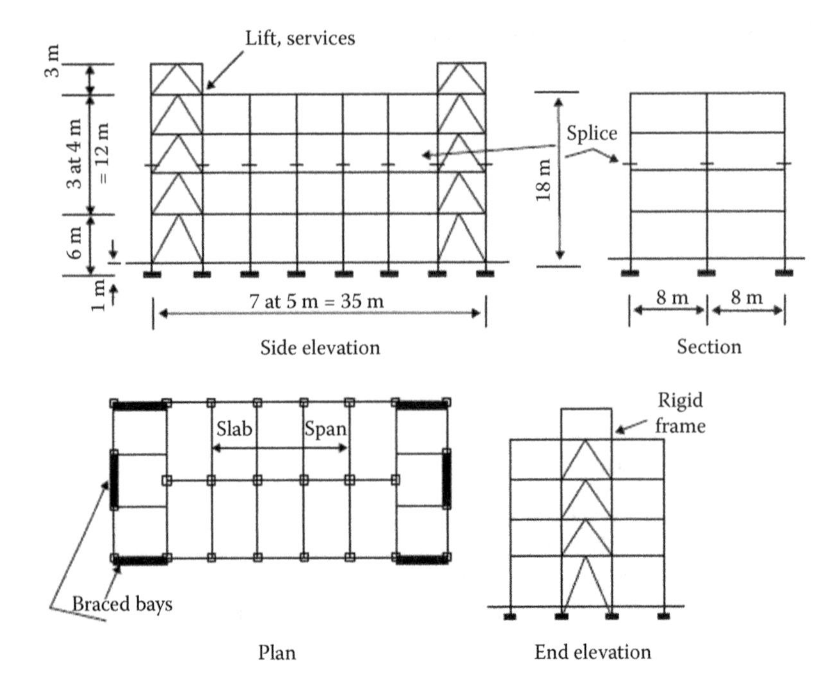

Figure 7.1 Fully braced building – framing plans.

- External cladding – brick/breeze block and double glazing;
- The building is fully braced in both directions.

7.2.2 Loads

(a) *Dead loads*

(i) ROOF

Topping (1.0 kN/m^2) + Slab (4.1 kN/m^2) + Steel (0.2 kN/m^2)
+ Ceiling (0.5 kN/m^2) + Services (0.2 kN/m^2) = 6 kN/m^2

(ii) FLOOR

Tiles, screed (0.7 kN/m^2) + Slab (4.3 kN/m^2) + Steel (0.3 kN/m^2)
+ Partitions (1.0 kN/m^2) + Ceiling (0.5 kN/m^2)
+ Services (0.2 kN/m^2) = 7 kN/m^2

(iii) COLUMNS AND CASING

Internal – 1.5 kN/m, external – 6.3 kN/m over 2.2 m height.

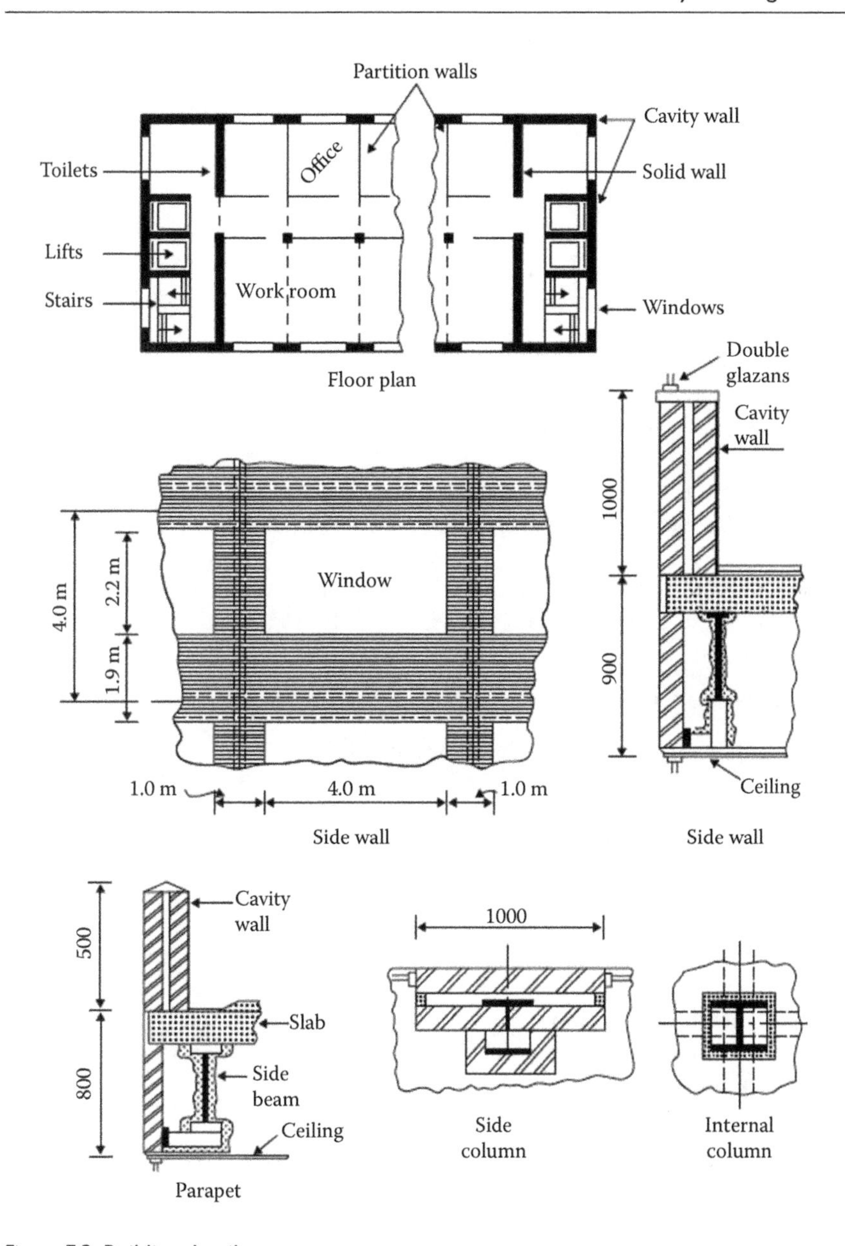

Figure 7.2 Building details.

(iv) WALL – PARAPET (FIGURE 7.2)

Cavity wall (4.8 kN/m^2) + Slab (4.1 kN/m^2) + Steel (0.5 kN/m) + Ceiling (0.5 kN/m^2) = 6.9 kN/m

(v) WALL – FLOOR LOADS (FIGURE 7.2)

Cavity wall (5.1 kN/m^2) + Slab (4.3 kN/m^2) + Steel (0.7 kN/m) + Double glazing (0.6 kN/m^2) + Ceiling (0.5 kN/m^2) = 10.6 kN/m

(b) *Imposed loads*

The imposed loads from BS 6399: Part 1 are

- Roof – 1.5 kN/m^2;
- Floors (offices with computers) – 3.5 kN/m^2.

The reduction in imposed loads with number of stories is 2–10%; 3–20%; 4–30% (Table 2 of the code).

7.2.3 Materials

Steel used is Grade S275, with bolts Grade 8.8.

7.3 SIMPLE DESIGN CENTRE FRAME

7.3.1 Slabs

Slabs can be designed to BS 8110. This conforms as satisfactory that:

- Roof slab – 170 mm;
- Floor slab – 180 mm with 10 mm diameter bars at 180 mm centres top and bottom.

7.3.2 Roof beam

Dead load = 6 kN/m^3 Imposed load = 1.5 kN/m^2

Design load = (1.4 × 6) + (1.6 × 1.5) = 10.8 kN/m^2

$M = 10.8 \times 5 \times 8^2/8 = 432$ kN/m

$S = 432 \times 10^3/275 = 1571$ cm^3

Select 457 × 191 UB 74, with $S = 1650$ cm³, $I_X = 33{,}300$ cm⁴, $T = 14.5$ mm, $p_y = 275$ N/mm².

Imposed load 1.5 × 5 × 8 = 60 kN

Live load serviceability deflection δ is given by

$$\delta = \frac{5 \times 60 \times 10^3 \times 8000^3}{384 \times 205 \times 10^3 \times 33{,}300 \times 10^4} = 5.85 \text{ mm}$$

$$< \text{span}/360 = 22.2 \text{ mm}$$

Thus section is satisfactory.

7.3.3 Floor beam

Dead load = 7.0 kN/m²

Imposed load = 3.5 kN/m²

Design load = 15.4 kN/m²

$M = 616$ kNm

$S = 2240$ cm³

Select 533 × 210 UB 92, with $S = 2360$ cm³.

δ = span/973 = 8.23 mm

< span/360 = 22.2 mm

7.3.4 Outer column – upper length 7–10–13 (Figure 7.3)

Check column above second floor level at joint 7:

Imposed load reduction = 10%

Load 7–10 = (6 + 7)20 × 1.4 + (6.9 × 5 × 1.4) + (2 × 2.2 × 6.3 × 1.4) + (10.6 × 5 × 1.4) + (1.5 + 3.5)20 × 1.6 × 0.9 = 669.3 kN

Load 7–8 = (7 × 20 × 1.4) + (3.5 × 20 × 1.6) = 308 kN

Assume 203 × 203 UC 46 eccentricity = 0.20 m (100 mm from column face). Initially assume moments divided equally.

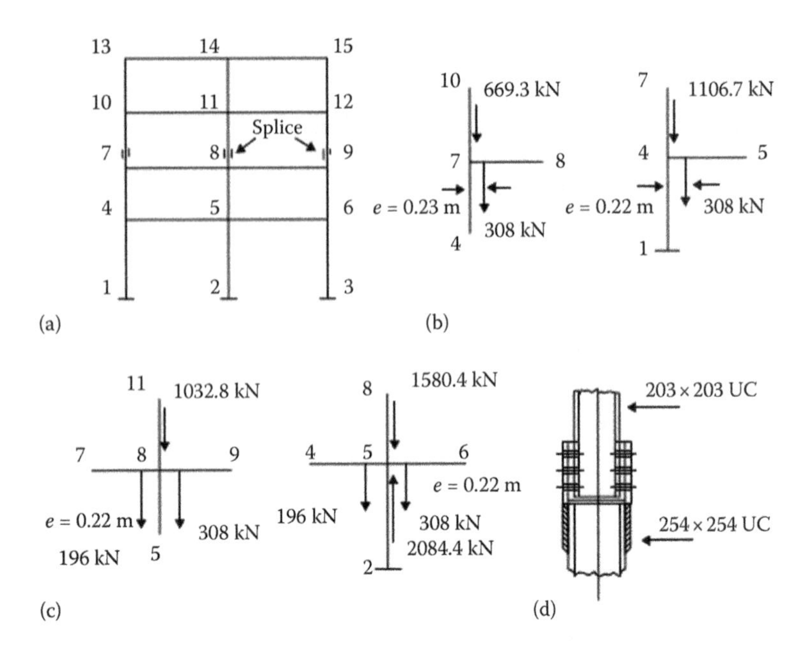

Figure 7.3 Columns – loads and details: (a) frame; (b) outer column; (c) centre column; (d) splice.

$M = 308 \times 0.20/2 = 31.08$ kNm

$F = 669.3$ kN

Try 203 × 203 UC 46 with $r_y = 5.13$ cm, $S = 497$ cm³, $A = 58.7$ cm², $I = 4570$ cm⁴, $T = 11.0$ mm, $p_y = 275$ N/mm².

$M = 308 \times 0.202/2 = 31.108$ kNm

$\lambda = 0.85 \times 400/5.13 = 66.3$

$p_c = 188$ N/mm² (Table 24(c))

$\lambda_{LT} = 0.5 \times 400/5.13 = 39.0$,

$p_b = 264$ N/mm² (Table 16)

$M_{bs} = 264 \times 497/10^3 = 131.2$ kNm

Combined:

$$(F_c + P_c) + (M_X / M_{bs}) \leq 1$$

$$\frac{669.3 \times 10}{188 \times 58.7} + \frac{31.108}{131.2} = 0.84$$

The section is satisfactory.

7.3.5 Outer column – lower length 1–4–7 (Figure 7.3)

Check column below first floor at joint 4:

Imposed load reduction = 20%

Load 4–7 = 1106.7 kN

Load 4–5 = 308 kN

$$(I/l)_{1-4} : (I/l)_{4-7} = \left(\frac{1}{6} : \frac{1}{4}\right) \Big/ \left(\frac{1}{6} + \frac{1}{4}\right) = 0.4 : 0.6$$

$\not>$ 1: 1.5–divide moments equally

M = 35.4 kNm

F = 1106.7 + 308 = 1414.7 kN

Try 254 × 254 UC 89, with r_Y = 6.55 cm, S = 1220 cm^3, A = 113 cm^2, I = 14,300 cm^4, T = 17.3 mm, p_y = 265 N/mm^2.

λ = 77.8, p_c = 166 N/mm^2

λ_{LT} = 45.8, p_b = 240 N/mm^2

M_b = 293 kNm

Combined:

$$\frac{1414.7 \times 10}{166 \times 113} + \frac{35.4}{293} = 0.87$$

The division of moments at joint 7 should be rechecked. The section is satisfactory.

7.3.6 Centre column – upper length 8–11–14 (Figure 7.3c)

Refer to BS 5950 Clause 5.1.2.1 which states that the most unfavourable load pattern can be obtained without varying the dead load factor γ_f. Check column above joint 8:

Load 8–11 = 1032.8 kN

Load 7–8 = 196 kN (imposed load zero)

Load 8–9 = 308 kN

For the column select 203 × 203 UC 52:

$M = 12.3$ kNm

$F = 1032.8$ kN

$(F/P_c) + (M_X/M_{bs}) \leq 1$

The section is satisfactory.

7.3.7 Centre column – lower length 2–5–8

Check section below first floor at joint 5 (Figure 7.3c).

Load above joint 5 = (6 + 2 × 7)4 × 1.4 + (12 × 1.5 × 1.4)
+ (1.5 + 2 × 3.5)40 × 1.6 × 0.8 = 1580.4 kN

Load 5–4 = 196 kN

Load 5–6 = 308 kN

Load below joint 5 = 2084.4 kN

$M = (308 - 196)0.23/2 = 12.9$ kNm

Try 254 × 254 UC 107, with r_Y = 6.59 cm, S = 1480 cm^3, A = 136 cm^2, T = 20.5 mm, p_y = 265 N/mm^2.

$\lambda = 0.85 \times 600/6.59 = 77.4$, $p_c = 161.8$ N/mm^2

$\lambda_{LT} = 0.5 \times 600/6.59 = 45.5$, $p_b = 240.5$ N/mm^2

$M_b = 1480 \times 240.5/10^3 = 355.9$ kNm

Combined:

$$\frac{2084.4 \times 10}{136 \times 161.8} + \frac{12.9}{355.9} = 0.98$$

The section is satisfactory.

7.3.8 Joint design (Figure 7.4)

Shear = 308 kN

Provide 4 No. 20 mm diameter Grade 8.8 bolts – shear capacity 91.9 kN/bolt.

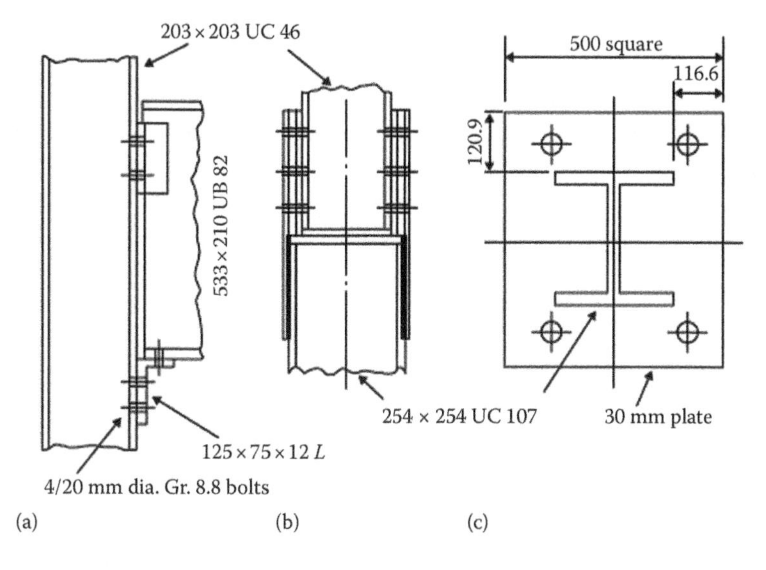

Figure 7.4 Frame joint details: (a) floor beam connection; (b) column splice; (c) centre column baseplate.

Bearing on 11 mm flange of 203 × 203 UC 46–101 kN/bolt. The beam/column joint and column spice is shown in the figure.

7.3.9 Baseplate – centre column (Figure 7.4)

Load = 2088 kN (30% imposed load reduction)

Using concrete grade 30, bearing pressure 0.6 f_{cu},

Area = $2088 \times 10^3/0.6 \times 30 = 11.6 \times 10^4$ mm^2

Provide 500 mm × 500 mm plate, with p = 8.35 N/mm^2. Thickness (BS 5950, Section 4.13.2.2) is given by

$$t = c[3w/p_{yp}]^{0.5},\ c = 86 \text{ mm}$$

$$\therefore t = 86\left[\frac{3\times 8.35}{265}\right]^{0.5}$$

$$= 26.4 \text{ mm}$$

Provide 30 mm thick baseplate (p_y = 265 N/mm^2).

7.4 BRACED RIGID ELASTIC DESIGN

7.4.1 Computer analysis

The building frame, supports, dimensions and design loads are shown in Figure 7.5a. The areas and moments of inertia of members taken from the simple design are listed in (b) in the figure. Loads on members 5–6, 8–9 and 11–12 are given for two cases:

1.4 × dead + 1.6 × imposed = 77 kN/m

1.4 × dead = 49 kN/m

Analyses are carried out for the following load cases:

1. All spans – 1.4 × dead + 1.6 × imposed;
2. Spans 8–9, 11–12 – 1.4 × dead, other spans – 1.4 × dead + 1.6 × imposed;
3. Spans 5–6, 8–9 – 1.4 × dead, other spans – 1.4 × dead + 1.6 × imposed;
4. Span 5–6 – 1.4 × dead, other spans – 1.4 × dead + 1.6 × imposed;
5. Spans 7–8, 5–6 – 1.4 × dead, other spans – 1.4 × dead + 1.6 × imposed.

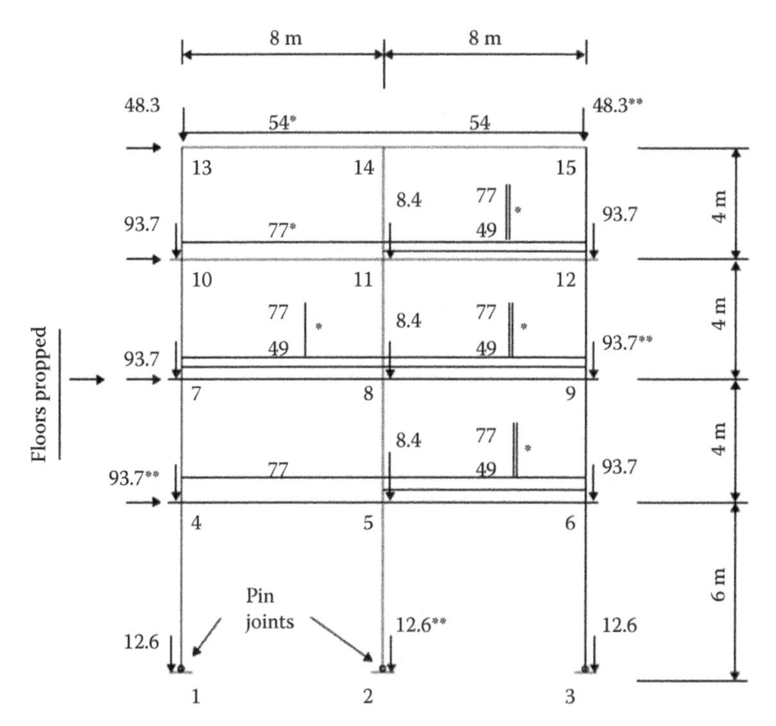

* Uniform loads: 54, 49, 77 kN/m
** Point loads: 48.3, 93.7, 12.6 kN
|| * Dead + imposed: 77 kN/m; dead only = 49 kN/m
External columns – a weight between floors = 19.4 kN

(a)

Members	A (cm^2)	I (cm^4)
4–5–6, 7–8–9, 10–11–12	118	55,400
13–14–15	96	33,400
1–4–7, 3–6–9	114	14,300
2–5–8	137	17,500
7–10–13, 8–11–14, 9–12–15	58.8	4560

(b)

Figure 7.5 Braced frame – computer analysis data: (a) dimensions and frame loads; (b) areas and moments of inertia.

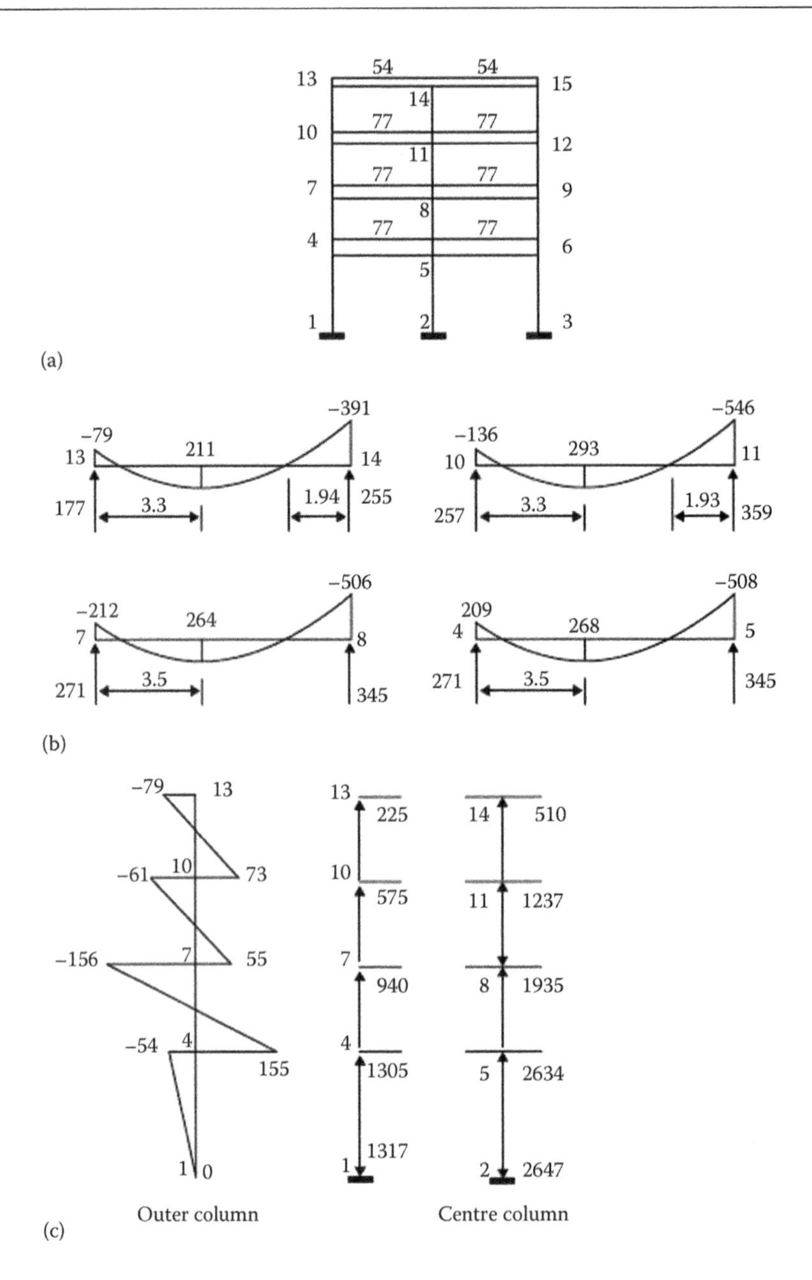

Figure 7.6 Computer analysis results for case 1: (a) beam loads (kNm); (b) beam moments (kNm) and shears (kN); (c) column moments (kNm) and thrusts (kN).

The load pattern that should be used to obtain the maximum design conditions is not obvious in all cases. Some comments are as follows:

1. All spans fully loaded should give near the maximum beam moments. Other cases could give higher moments.
2. The external columns are always bent in double curvature. Full loads will give near maximum moments though again asymmetrical loads could give higher values.
3. The patterns to give maximum moments in the centre column must be found by trials. The patterns used where the column is bent in double curvature give the maximum moments.

The results from the computer analysis for case 1 loads are shown in Figure 7.6. The critical values only of the actions from cases 1–5 are shown in Table 7.1.

7.4.2 Beam design

(a) *Roof beam 13–14–15*

Moment M_{14-13} = −391 kNm (Table 7.1)

Shear S_{14-13} = 255 kN

See floor beam design. Select 457 × 191 UB 67.

Table 7.1 Design action 5

Load case	*Member critical section*	*Moment, M (kNm)*	*Shear, F (kN)*	*Unrestrained length,[a] l (m)*
(a) Beam				
1	Roof beam 13–14	M_{14-13}–391	F_{14-13}255	1.94
5	Floor beam 10–11	M_{11-10}–546	F_{11-10} 359	1.94
Load case	*Member critical section*	*Moment, M (kNm)*		*Thrust, F (kN)*
(b) Column				
2	Outer column 10–7	M_{10-7} = −77	M_{1-10} = 68.3	F_{10-7} = 596.2
3	Outer column 7–4	M_{7-4} = −172.5	M_{4-7} = 170.9	F_{7-4} = 962.5
4	Outer column 4–1	M_{4-1} = 62.2	M_{1-4} = 0	F_{4-1} = 1322.9
2	Centre column 11–8	M_{11-8} = 23.1	M_{8-11} = −20.7	F_{11-8} = 1087.1
1	Centre column 5–8	M_{5-8} = 0	M_{8-5} = 0	F_{5-8} = 1223.4
1	Centre column 2–5	M_{2-5} = 0	M_{5-2} = 0	F_{2-5} = 2610.2
4	Centre column 2–5	M_{5-2} = 22.9	M_{2-5} = 0	F_{5-2} = 2473.2
5	Centre column 2–5	M_{5-2} = 25	M_{2-5} = 0	F_{5-2} = 2393

[a] The unrestrained length of compression flange occurs on the underside of the beam at the support (Figure 7.6).

(b) *Floor beam 10–11–12*

Moment $M_{11\text{–}10} = 546$ kNm (Figure 7.6b)

Shear $F_{11\text{–}10} = 359$ kN

$S = 546 \times 10^3/275 = 1985\ \text{cm}^3$

Try 533 × 210 UB 82, with $S = 2060\ \text{cm}^3$, $r_Y = 4.38$ cm, $u = 0.863$, $x = 41.6$, $I = 47{,}500\ \text{cm}^4$, $D = 528.3$ mm, $t = 9.6$ mm, $T = 13.2$ mm, $p_y = 275\ \text{N/mm}^2$.

Unrestrained length of compression flange = 1.94 m

$\lambda = 194/4.38 = 44.3$

$\lambda/x = 44.3/41.6 = 0.99$

$v = 0.99$ (Table 19)

$\lambda_{LT} = 0.863 \times 0.99 \times 44.3 = 37.8,\ \lambda_{LT} = uv\lambda\sqrt{\beta_w}$

$\beta_w = 1$ (Clause 4.3.6.9)

$p_b = 267\ \text{N/mm}^2$ (Table 16)

Shear capacity $P_v = 0.6 \times 275 \times 528.3 \times 9.6/10^3 = 836.8$ kN

$M_b = S \times P_b = 267 \times 2060/1000 = 550.02\ \text{kNm} > 546\ \text{kNm}$ (satisfactory)

The maximum shear, 359 kN, is therefore not high enough to affect the moment capacity (Clause 4.2.5.2).

Simple beam deflection due to the imposed load

$$= 3.5 \times 8 \times 5 = 140\ \text{kN}$$

$$\delta = \frac{5 \times 140 \times 10^3 \times 8000^3}{384 \times 205 \times 10^3 \times 47{,}500 \times 10^4} = 9.6\ \text{mm}$$

$$< \text{span}/360 = 22.2\ \text{mm}$$

The beam is satisfactory. All beams to be the same.

7.4.3 Column design

(a) *Outer column – upper length 7–10–3*

Critical actions from Table 7.1 are

Moment $M_{10-7} = 77$ kNm

Thrust $F_{10-7} = 596.2$ kN

Select 203 × 203 UC 47. See design for lower column length below.

(b) *Outer column – lower length 1–4–7*

Critical actions at sections 7–4 and 4–1 from Table 7.1 from cases 3 and 4 are

$M_{7-4} = -172.5$ kNm; $F_{7-4} = 962.9$ kN

$M_{4-1} = -62.2$ kNm; $F_{4-1} = 1322.9$ kN

Try 254 × 254 UC 89, with $r_Y = 6.55$ cm, $S_x = 1220$ cm^3, $A = 113$ cm^2, $x = 14.5$, $u = 0.851$, $T = 17.3$ mm, $p_y = 265$ N/mm^2.

For length 7–4 (= 4.0 m)

$\lambda = 51.8$

$p_c = 210$ N/mm^2

$\lambda/x = 3.57$

$v = 0.883$

$\lambda_{LT} = uv\lambda\sqrt{\beta_w} = \lambda_{LT} = 38.9, \beta_w = 1.0$

$p_b = 255.9$ N/mm^2

$M_b = 312.1$ kNm

$M_c = 323.3$ kNm

m_{LT} and m_X both = 0.44

Cross-section capacity check = 0.85

$m_X = m_{LT} = 0.6$

Member buckling resistance check = 0.73

For length 1–4 (= 6.0 m)

$\lambda = 77.8$

$p_c = 161.0\ \text{N/mm}^2$

$\lambda/x = 5.36\ v = 0.796$

$\lambda_{LT} = uv\lambda\sqrt{\beta_w} = 52.7$

$p_b = 224.3\ \text{N/mm}^2$

$M_b = 273.6\ \text{kNm}$

$\beta = 0$

m_X and $m_{LT} = 0.6$

Cross-section capacity check = 0.63

Overall buckling check = 0.86

The section is satisfactory.

(c) *Centre column – upper length 8–11–4*

The critical actions from Table 7.1 are

$M_{11\text{-}8} = 23.1\ \text{kNm};\ F_{11\text{-}8} = 1087.1\ \text{kN}$

$M_{5\text{-}8} = 0;\ F_{5\text{-}8} = 1223.4\ \text{kN}$

Select 203 × 203 UC 52.

(d) *Centre column – lower length 2–5–8*

The critical actions from Table 7.1 are

$M_{5\text{-}2} = 25\ \text{kNm};\ F = 2393\ \text{kN}$

$M_{5\text{-}2} = 22.9\ \text{kNm};\ F_{5\text{-}2} = 2473.2\ \text{kN}$

$M_{2\text{-}5} = 0;\ F_{2\text{-}5} = 2610.2\ \text{kN}$

Try 305 × 305 UC 118, with $r_Y = 7.77$ cm, $S_x = 1960$ cm^3, $A = 150$ cm^2, $x = 16.2$, $u = 0.851$, $T = 18.7$ mm, $p_y = 265$ N/mm^2. For length 5–2 (= 6.0 m):

$M = 22.9$ kNm

$F = 2473.2$ kN

$\lambda = 65.7$

$p_c = 184.6$ N/mm^2

$\lambda/x = 4.06$

$v = 0.858$

$\lambda_{LT} = uv\lambda\sqrt{\beta_w} = 47.9$

$p_b = 235.6$ N/mm^2 (Table 16)

$M_b = 459.4$ kNm

$\beta = -0.47$

$m_{LT} = 0.48$ (Table 18)

Cross-section capacity check = 0.64

Overall buckling check = 0.96

For length 2–5 (base):

$P_c = 150 \times 178.6/10 = 2679 \text{ kN} > 2610.2 \text{ kN}$

The section is satisfactory.

7.4.4 Joint design

One typical joint for the floor beams at the centre column only is designed.

(a) *Critical actions at joint II* (see Table 7.1)

Moment $M_{11-10} = 546$ kNm

Shear $F_{11-10} = 359$ kNm

(b) *Frame section and joint arrangement*

Column 8–11–14 uses 203 × 203 UC 52; beam 10–11–12 uses 533 × 210 UB 82. The proposed arrangement of the joint is shown in Figure 7.7.

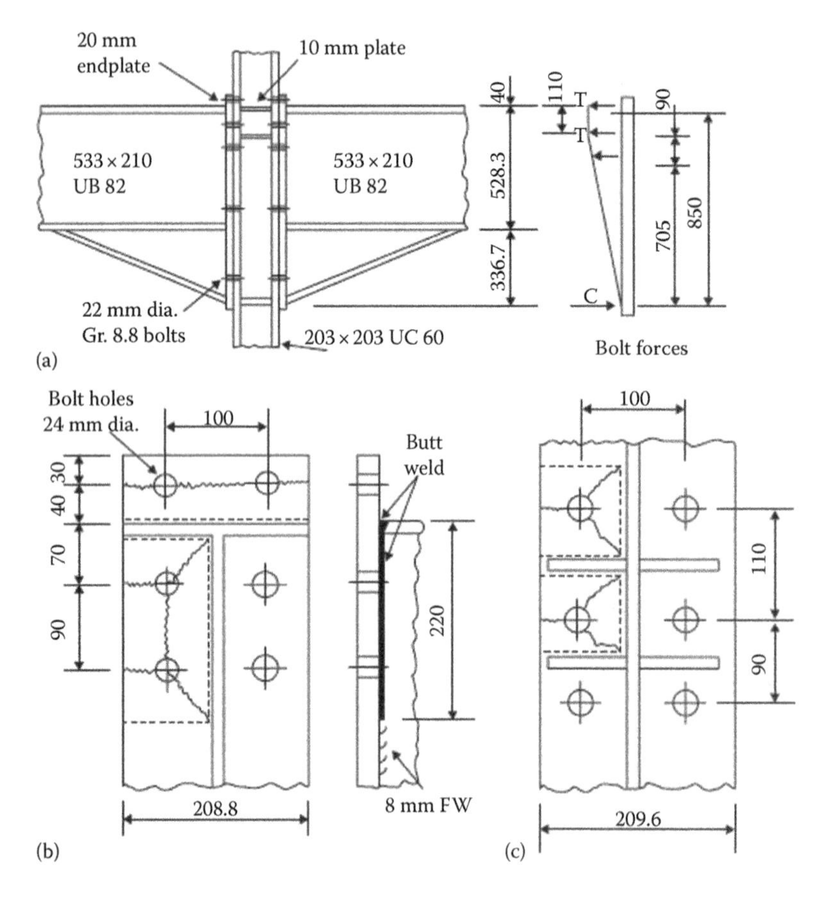

Figure 7.7 Floor beam – centre column joint: (a) joint arrangement; (b) beam endplate; (c) column flange.

(c) *Bolt size*

The top six bolts are assumed to resist tension with equal values T for the top two rows. Thus the moment

$$546 = (4T \times 0.85) + 2T \times 0.71^2/0.8$$

$$T = 117.2 \text{ kN}$$

Shear on bottom four bolts = 359/4

= 89.8 kN

Provide 22 mm diameter Grade 8.8 bolts, with

Tension capacity = $0.8P_tA_t$ = 136 kN

Single shear capacity = P_sA_s = 114 kN

(d) *Beam endplate thickness*

The yield line pattern is shown in Figure 7.8. For the top bolts:

$$[(208.8 \times 2) - 48]M'\phi = 2 \times 117.2 \times 40\phi \times 10^3$$

$$M' = 25\,381 = 265 \times t^2/4$$

$$t = 19.5 \text{ mm}$$

Provide 20 mm plate.

(e) *Column flange plate check*

Increase the upper column section to 203 × 203 UC 60. (Figures 7.7 and 7.8 show the dimension and the yield line pattern.) For 203 × 302 UC 60, A = 76.4 cm², B = 205.8 mm, t = 9.4 mm, T = 14.2 mm, r = 10.2 mm.

The check is made on the upper column length 11–14 above joint 11. The yield line analyses gives

$$[100 + (2 \times 40.2)]M'\phi + 2(87.75 + 28.4)0.7M'\phi + 40.6 \times 1.4M'\phi$$
$$= 117.2 \times 10^3 \times 35.15\phi$$

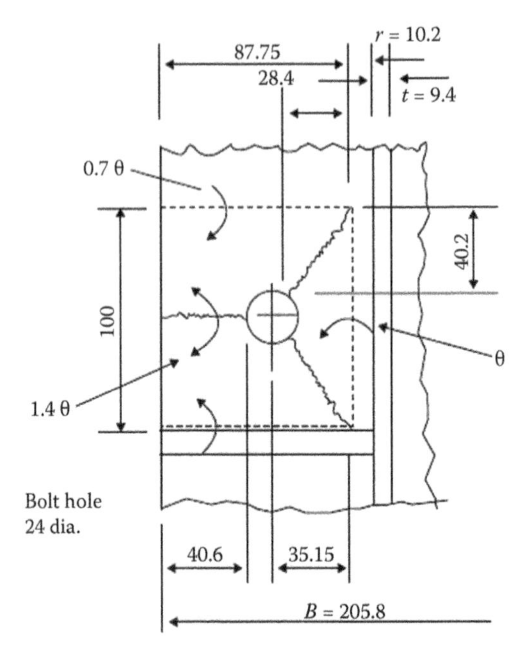

Figure 7.8 Column flange top bolt yield line pattern.

The yield line moment is

$$M' = 10{,}304 \text{ N mm/mm}$$

The flange resistance moment is

$$M_c' = 275 \times 14.2^2/4 = 13{,}863 \text{ N mm/mm}$$

The frame actions at joint II from computer analysis are

$$F_{11\text{–}14} = 518.4 \text{ kN}; \; M_{11\text{–}14} = 0$$

Combined:

$$\frac{518.4 \times 10}{76.4 \times 275} + \frac{10{,}304}{13{,}863} = 0.99$$

This is satisfactory. Note that the column flange could have been strengthened by a backing plate instead of increasing the column section.

7.5 BRACED RIGID PLASTIC DESIGN

7.5.1 Design procedure

The plastic design method for non-sway frames is set out in Clauses 7.7 and 7.7.2 of BS 5950. This clause specifies that the frame should have an effective bracing system independent of the bending stiffness of the frame members. The columns are to be checked for buckling resistance in accordance with Clause 4.8.3.3, Design of Compression Members with Moments.

The above code provisions indicate the following:

- Beams are designed for fixed end plastic moments

  $M_p = WL/16$

  where W is the design load.

- Columns are designed to provide sufficient resistance for plastic hinges to form in the beams. In pre-limit state terminology, they are designed to remain elastic.

Plastic hinges form at the beam ends. The beam-to-column connections can be made using either full-strength welds or high-strength bolts. Bolted connections designed to have a higher capacity than the beam plastic moment will be adopted in the design. An allowance is made for eccentricity of the joint and local joint actions are taken into account.

The braced frame specified in Section 7.2 above is designed to plastic theory. The frame dimensions and loads are shown in Figure 7.5.

7.5.2 Design loads and moments

(a) *Load patterns* (Figure 7.9)

Load patterns are arranged to give

- Total design load $W = (1.4 \times \text{dead}) + (1.6 \times \text{imposed})$ on all beams for beam design and beams in external bays for design of external columns;
- Total design load W on any beam and factored dead load on selected adjacent beams to give maximum moments in internal columns.

(b) *Analysis for beam and column moments*

(i) ROOF BEAM – EXTERNAL COLUMN (FIGURE 7.9A)

If the beam plastic moment M_{pb} exceeds the known column plastic moment M_{pc}, then the shear is zero at the centre hinge Y. Take moments about ends

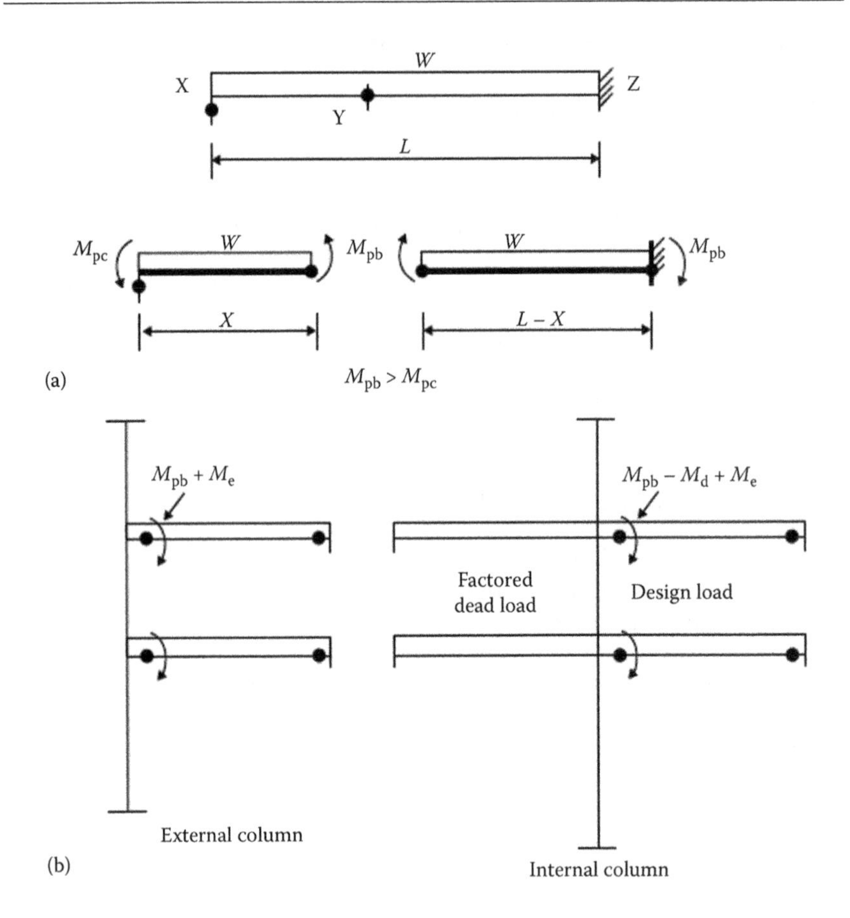

Figure 7.9 Load patterns: (a) external roof beam; (b) column subframes (M_e = moment due to eccentricity; M_d = moment due to factored dead load).

X and Z to give

$$M_{pb} = (wx^2/2) - M_{pc}$$

$$2M_{pb} = w(L - x)^2/2$$

where w is the design load on the roof beam (kN/m). Solve for x and calculate M_{pb}.

If $M_{pc} > M_{pb}$ then $M_{pb} = wL^2/16$.

(ii) ALL OTHER BEAMS

$$M_{pb} = wL^2/16$$

COLUMN SUBFRAMES

The subframes for determining the column moments are shown in Figure 7.9b for external and internal columns. Loads must be arranged in an appropriate pattern for the internal column. An allowance for eccentric end connections is made.

7.5.3 Frame design

(a) *Roof beam* (Figure 7.6a)

Assume that the moment capacity of the external column is greater than that of the roof beam:

$$M_{pb} = 54 \times 8^2/16 = 216 \text{ kNm}$$

$$S_x = 216 \times 10^3/275 = 785 \text{ cm}^3$$

Select 406 × 140 UB 46, with $S_x = 888$ cm^3.

(b) *Floor beam* (Figure 7.5a)

$$M_{pb} = 77 \times 8^2/16 = 308 \text{ kNm}$$

$$S_x = 308 \times 10^3/275 = 1120 \text{ cm}^3$$

Try 457 × 152 UB 60, with $S_x = 1290$ cm^3, $I_X = 25{,}500$ cm^4.

$$\text{Shear } F = 308 \text{ kN} = \frac{W}{2}$$

$$P_v = 0.6 \times 275 \times 454.6 \times 8.1/10^3 = 607.5 \text{ kN}$$

Deflection due to the imposed load is 17.5 kN/m on a fixed end beam.

$$\delta = \frac{17.5 \times 8000^4}{384 \times 205 \times 10^3 \times 25{,}500 \times 10^4} = 3.57 \text{ mm}$$

$$= \text{span}/2240$$

The beam would also be satisfactory if it were simply supported.

(c) *External column – upper length 7–10–13*

The loads and moments applied to the column are shown in Figure 7.10a.

Beam end plastic moment = 308 kNm

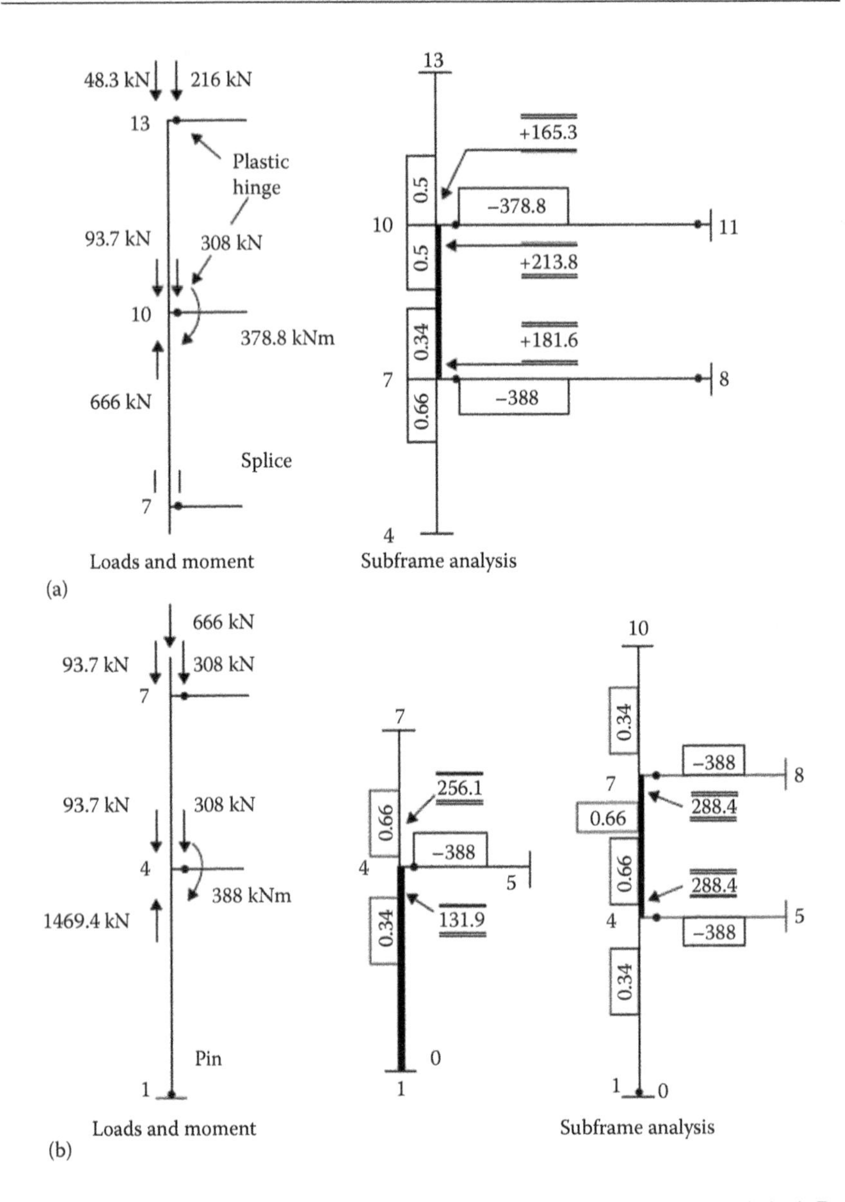

Figure 7.10 External column design: (a) upper length 7–10–13; (b) lower length 1–4–7.

This connection moment will be greater than 308 kNm because this moment is assumed to be developed at an eccentricity of 100 mm from the column face.

Assume column length 7–10–13 is 254 × 254 UC 73, with I = 11,400 cm^4 and length 1–4–7 is 305 × 305 UC 97, with I = 22,300 cm^4.

The beam moments are

$$M^{F}_{10-11} = 308 + (308 \times 0.23) = 378.8 \text{ kNm}$$

$$M^{F}_{7-8} = 308 + (308 \times 0.26) = 388 \text{ kNm}$$

The subframe for determining the column moments is shown in Figure 7.10a. The distribution factors for joints 10 and 7 are

$$K_{10-13} : K_{10-7} = 0.5 : 0.5$$

$$K_{7-4} : K_{7-10} = \frac{22,300 : 11,400}{33,700} = 0.66 : 0.34$$

The moment distribution is carried out. The design actions at joint 10 are

Thrust: F_{10-7} = 660 kN

Moments: M_{10-7} = 213.8 kNm

M_{7-10} = 181.6 kNm

Try 254 × 254 UC 89, with t = 10.3 mm, T = 17.3 mm, D = 260.3 mm, d/t = 19.4, A = 113 cm^2, S_x = 1220 cm^3, Z_x = 1100 cm^3, p_y = 265 N/mm^2.

Refer to column check in Section 7.4.3(b). For case here

p_c = 210 N/mm^2

M_c = 323.3 kNm (gross value, see below)

M_b = 312.1 kNm, m_{LT} and m_X = 0.44

Cross-section capacity check = 0.86

Moment buckling resistance check = 0.57

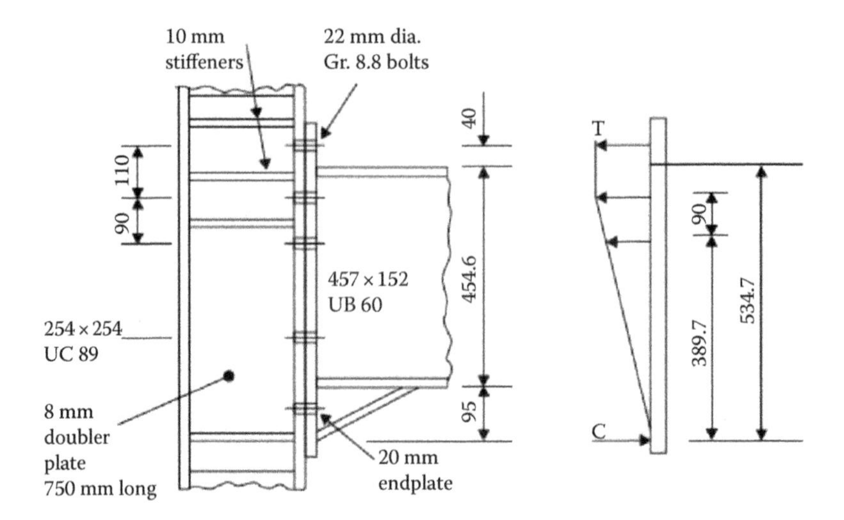

Figure 7.11 Joint to external column (joints to internal columns similar).

The proposed arrangement for joint 10 is shown in Figure 7.11. The bolt tension due to the plastic moment (308 kNm) and moment due to eccentricity of 100 mm (30.8 kNm) is

$$308 + 30.8 = 338.8 \text{ kNm}$$

$$338.8 = (4T \times 0.53) + (2T \times 0.39^2/0.48)$$

$$T = 123 \text{ kN}$$

Provide 22 mm diameter Grade 8.8 bolts. Tension capacity is 136 kN and shear 114 kN/bolt. Shear of lower four bolts = 4 × 114 = 456 kN. This ensures that the hinge forms in the beam. The haunch is small.

Refer to Section 7.4.4(e). The column flange can be checked in the same way:

$$\text{Yield line moment} = 123 \times 9779/111 = 10{,}836 \text{ N mm/mm}$$

$$\text{Flange capacity} = 265 \times 17.3^2/4 = 19{,}828 \text{ N mm/mm}$$

Above floor 10–11–12:

$$F_{10-13} = 283.7 \text{ kN}, M_{10-13} = 165.3 \text{ kNm}$$

Capacity check:

$$+\frac{283.7 \times 10}{265 \times 113} - \frac{165.3}{323.3} \pm \frac{10{,}836}{19{,}828} = 0.96$$

This is satisfactory.

Check column web for shear:

Web shear $F_v = 123 \times 4 + 0.81 \times 123 \times 2 = 691$ kN

Web capacity $P_v = 0.6 \times 265 \times 260.3 \times 10.3/10^3$

$= 434.7$ kN

Add 8 mm double plate to increase shear capacity of the column web.

(d) *External column – lower length 1–4–7*

Refer to Figure 7.10b. Assume column is 305 × 305 UC. Moment due to eccentricity is

$308 \times 0.26 = 80$ kNm

Total moment = 388 kNm

Check column length 1–4, length 6m. See subframe in Figure 7.10. The distribution factors allowing for the pin-end joint 1 are

$$K_{1-4} : K_{4-7} = \frac{\frac{3}{4} \times \frac{1}{6} : \frac{1}{4}}{0.375} = 0.33 : 0.66$$

The moment distribution may be carried out to give the design actions:

$M_{4-1} = 131.9$ kNm

$F_{4-1} = 1469$ kN

Try 305 × 305 UC 97, with $T = 15.4$ mm, $p_y = 275$ N/mm^2, $p_c = 188.2$ N/mm^2, $A = 123$ cm^2, $M_c = 437.2$ kNm, $M_b = 374.6$ kNm, $m_{LT} = 0.6$.

Cross-section capacity check = 0.73

Moment buckling resistance check = 0.83

Check column length 4–7, length 4 m. The design actions of joint 4 are

$F_{4-7} = 1087.1$ kN, $M_{4-7} = 288.4$ kN

Cross-section capacity check = 0.98

Moment buckling resistance check = 0.67

The column flange can be checked above joint 4. A check such as carried out in (c) above will show the section to be satisfactory.

The web will require strengthening with a doubler plate as set out in (c).

(e) *Centre column – upper length 8–11–14*

The loads and moments and subframe are shown in Figure 7.12a. Assume column 8–11–14 is 203 × 203 UC 60, $I = 6130\ \text{cm}^4$ and 2–5–8 is 305 × 305 UC 118, $I = 27\ 700\ \text{cm}^4$.

The beam fixed and the moments are found as follows. Beams 10–11 and 7–8 carry full design load of 77 kN/m:

$$M^{F}_{11\text{-}10} = (77 \times 64/16) + (77 \times 4 \times 0.21) = 372.7\ \text{kNm}$$

$$M^{F}_{8\text{-}7} = 308 + (308 \times 0.26) = 388\ \text{kNm}$$

Beams 11–12 and 8–9 carry the factored dead load of 49 kN/m:

$$M^{F}_{11\text{-}12} = (49 \times 64/12) + (49 \times 4 \times 0.21) = 302.5\ \text{kNm}$$

$$M^{F}_{8\text{-}9} = 261.3 + (196 \times 0.26) = 312.3\ \text{kNm}$$

The distribution factors are

$$K_{11\text{-}14} : K_{11\text{-}12} : K_{11\text{-}8} = \frac{6130/400 : 25{,}500/800 : 6130/400}{15.2 + 31.9 + 15.2}$$

$$= 0.24 : 0.52 : 0.24$$

$$K_{8\text{-}11} : K_{8\text{-}9} : K_{8\text{-}5} = \frac{6130/400 : 25{,}500/800 : 27{,}600/400}{15.2 + 31.9 + 69}$$

$$= 0.13 : 0.27 : 0.6$$

The results of the moment distribution are

$$M_{11\text{-}8} = -24.6\ \text{kNm}$$

$$M_{8\text{-}11} = -8.7\ \text{kNm}$$

$$F_{11\text{-}8} = 944.4\ \text{kN}$$

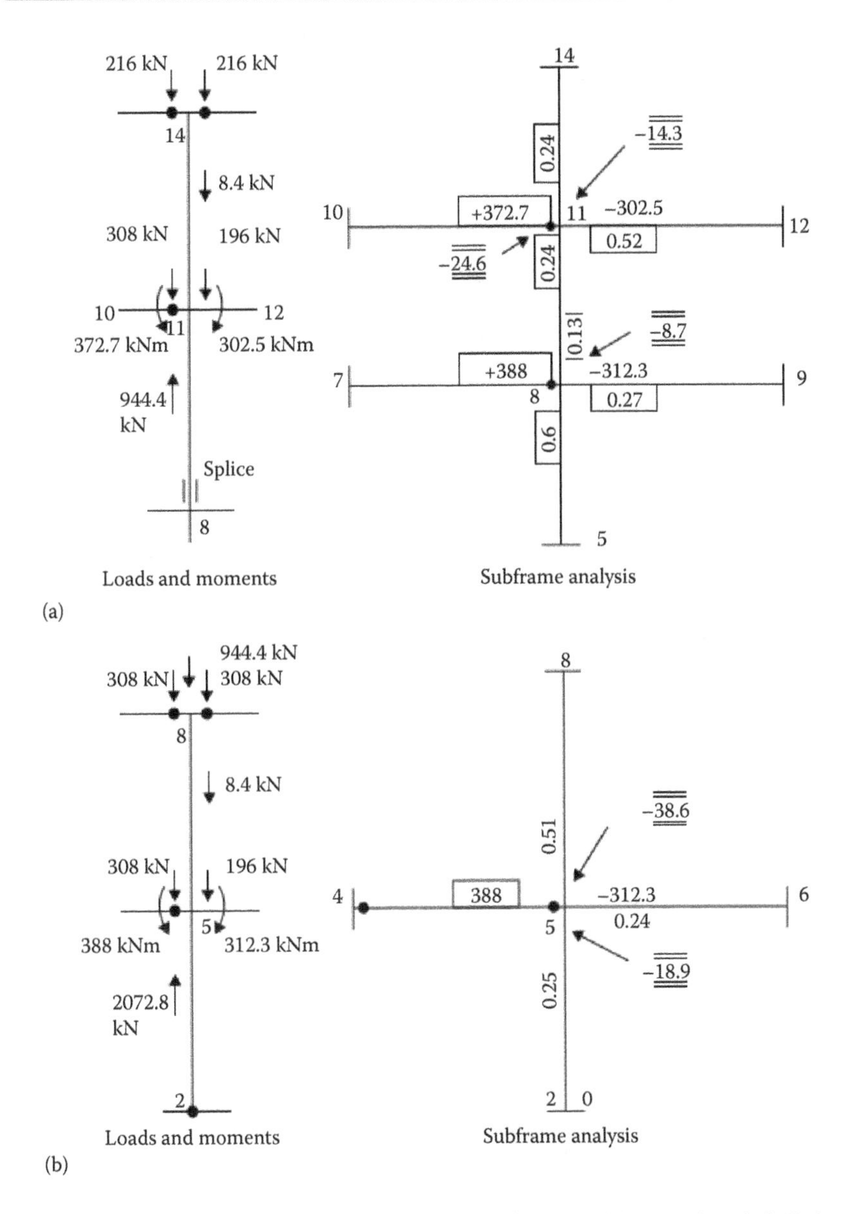

Figure 7.12 Internal column design: (a) upper length 8–11–14; (b) lower length 2–5–8.

Alternatively the hinge in the lower beams could be located in beam 8–9. This gives $M_{11\text{-}8}$ = +13.1 kNm, $M_{8\text{-}11}$ = –2.5 kNm.

Try 203 × 203 UC 60, with p_c = 190 N/mm², A = 76.4 cm², M_b = 160.9 kNm, M_c = 180 kNm, m_{LT} = 0.53, T = 14.2 mm, p_y = 275 N/mm².

Cross-section capacity check = 0.61

Moment buckling resistance check = 0.73

The beam end moment causing tension in the bolts is

$$M_{11\text{-}10} = 308 + 30.8 = 338.8$$

T = 123 kN (Section 7.6.3(c))

Refer to Sections 7.4.4(c) and (e), joint 11.

Yield line moment M' = 123 × 9779/111 = 10,836 N mm/mm

Flange capacity M'_p = 13,863 N mm/mm

For the column,

$$M_{11\text{-}14} = -14.3 \text{ kNm}$$

$$F_{11\text{-}14} = 440.4 \text{ kN}$$

Capacity check:

$$\frac{440.4 \times 10}{76.4 \times 275} - \frac{14.3}{180} \pm \frac{10,863}{13,862} = 0.96$$

This is satisfactory. The web shear capacity is also satisfactory.

(f) *Centre column – lower length 2–5–8*

Loads and moments and subframe are shown in Figure 7.12b. The beam fixed end moments are

$$M^F_{5\text{-}4} = 388 \text{ kNm}, M^F_{5\text{-}6} = 312.3 \text{ kNm}$$

The distribution factors allowing for the pin at 2 are

$$K_{5-8} : K_{5-6} : K_{5-2} = \frac{27,600/400 : 25,500/800 : 0.75 \times 27,600/600}{69 + 31.9 + 34.5}$$

$$= 0.51 : 0.24 : 0.25$$

The results of the moment distribution are

$$M_{5-2} = 18.9 \text{ kNm}$$

$$F_{5-2} = 2072.8 \text{ kN}$$

Try 305 × 305 UC 97, with p_c = 188.2 N/mm², A = 123 cm², M_c = 438 kNm, M_b = 374.6 kNm, m = 0.6.

Cross-section capacity check = 0.66

Moment buckling resistance check = 0.92

The column flange above first floor level is also satisfactory, as is the web shear capacity.

The sections are summarized in Table 7.2 (Section 7.7).

7.6 SEMIRIGID DESIGN

7.6.1 Code requirements

Semirigid design is permitted in both BS 5950 and Eurocode 3. The problem in application lies in obtaining accurate data on joint behaviour. A tentative semirigid design is made for the frame under consideration in order to bring out aspects of the problem. The design set out applies to braced structures only.

BS 5950 defines semirigid design in Clause 2.1.2.4 in terms that some degree of connection stiffners is assumed insufficient to develop full continuity. The clause specifies that:

> The moment and rotation capacity of the joints should be based on experimental evidence which may permit some limited plasticity providing the ultimate tensile capacity of the fastener is not the failure criterion.

The code also gives an alternative empirical method based on the rules of simple design where an end restraint moment of 10% of the beam free moment may be taken in design.

Eurocode 3 specifies in Clause 7.2.2.4 design assumptions for semi-continuous framing for structures. This states that:

> Elastic analysis should be based on reliably predicted design moment–rotation or force–displacement characteristics of the connections used.
> Rigid plastic analysis should be based on the design moment resistances of connections which have been determined to have sufficient rotation capacity.
> Elastic–plastic analysis can be used.

Eurocode 3 also permits in Clause 5.2.3.6 the use of suitable subframes for the global analysis of structures with semicontinuous framing.

Beam-to-column connection characteristics are discussed in Section 6.9 of Eurocode 3. The main provisions are summarized below:

- Moment–rotation behaviour shall be based on theory supported by experiment.
- The real behaviour may be represented by a rotational spring.
- The actual behaviour is generally nonlinear. However, an approximate design moment–rotation characteristic may be derived from a more precise model by adopting a curve including a linear approximation that lies wholly below the accurate curve.
- Three properties are defined by the moment–rotation characteristics (Figure 7.13):
    - Moment resistance M_R;
    - Rotational stiffness–use secant stiffness $J = M/\phi$;
    - Rotation capacity ϕ_C.

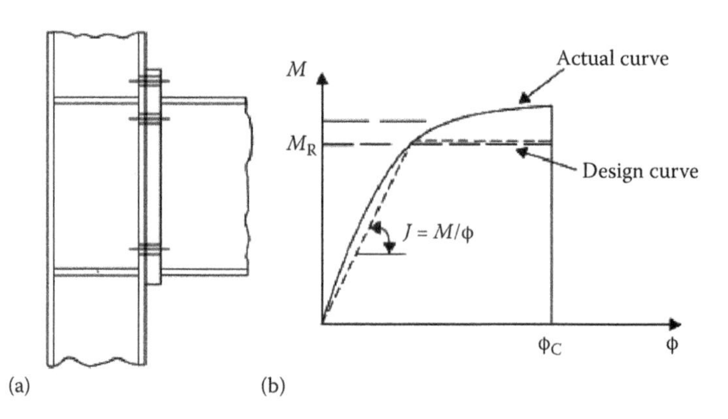

Figure 7.13 Beam-to-column connection: (a) details; (b) moment–rotation curves.

7.6.2 Joint types and performance

Two possible arrangements for semirigid joints – flush and extended endplate types – are shown in Figure 7.14. In both cases, the columns are strengthened with backing plates and stiffeners and the top bolts resisting moment are oversized. The endplates are sized as to thickness to fail at the design end moment.

The above provisions ensure that failure is not controlled by the bolts in tension and the major part of yielding occurs in the endplates and not in the column flange. Yield line failure patterns for the endplates in the two types of joint are shown in the figure, from which the failure loads can be readily calculated.

Analyses can be carried out to determine joint flexibility measured by the spring constant $J = M/\phi$. The endplate and column flange can be modelled

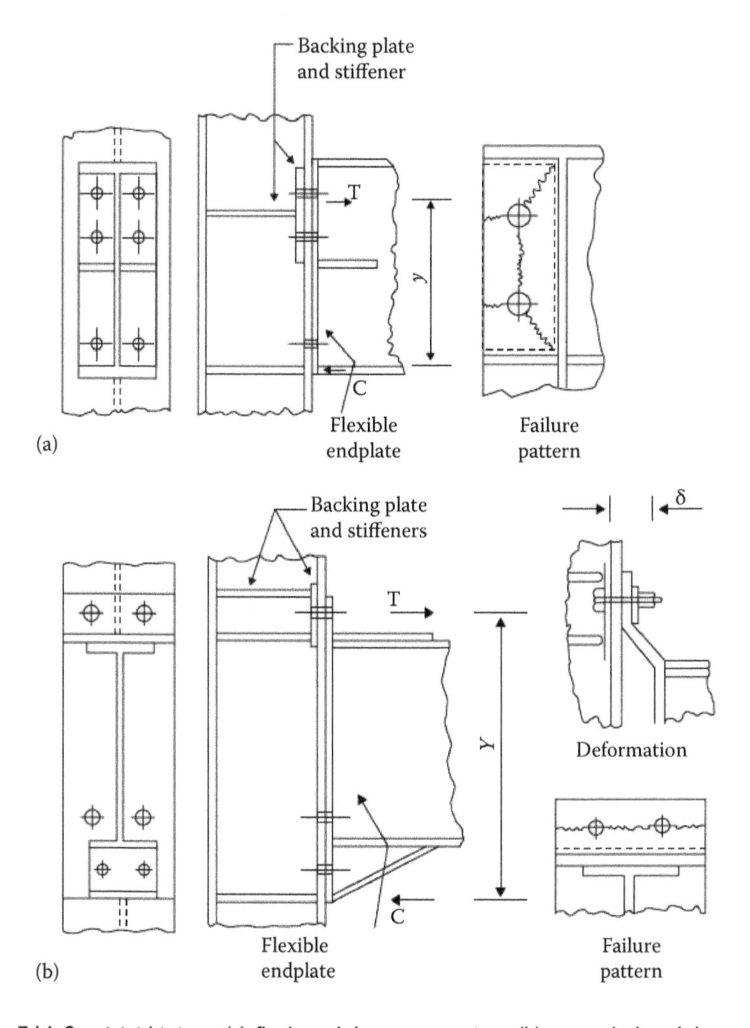

Figure 7.14 Semirigid joints: (a) flush endplate connection; (b) extended endplate.

using the finite element method and bolt extension included to give total deformation. Joint rotation is assumed to take place about the bottom flange or haunch (Lothers, 1960; Jenkins et al., 1986; technical papers in Dowling et al., 1987).

Only test results give reliable information on joint performance. The problem with using the data is that so many variables are involved that it is difficult to match the precise requirements of a given design problem to a test. BS 5950 is cautious in stressing that performance should be based on experimental evidence. Joint strengths can be higher than those predicted by yield line analysis due to plates resisting load in tension.

In the design example a value for joint flexibility, assessed from examining test results, will be used. Lothers (1960) and Jenkins (1986) quote values varying from 10,000 to 20,000 kNm/radian obtained by calculation and test for a variety of joints.

In the plastic design example given, the joint flexibility value is used only in the limited frame analysis for the centre column moments.

7.6.3 Frame analysis

(a) *Outline of methods*

Analysis of frames with semirigid joints may be carried out by the following methods:

- Elastic analysis where the joints are modelled by rotational springs, using
  - Moment distribution on the whole frame or on subframes
  - Computer stiffness analysis
- Plastic analysis using joint moment resistance which is much less than the beam plastic capacity. The columns are designed to have adequate buckling resistance.

Plastic analysis will be adopted for the design example. It is, however, necessary to distribute out-of-balance moments at internal column joints in proportion to elastic stiffnesses. Expressions are derived below for fixed end moments, stiffnesses and carryover factors for beams with semirigid joints.

(b) *Moment distribution factors*

The moment-area theorems are used to determine the moment distribution factors. The factors are derived for a uniform beam with the same joint type at each end and carrying only uniform load. Lothers (1960) gives a comprehensive treatment of the problem.

(i) FIXED END MOMENTS

There is no change in slope between the centre of the beam and the fixed ends (Figure 7.15a). That is, the area under the M/EI diagram plus spring rotation M_F/J is zero between stated points:

$$\frac{M_F l}{2EI} + \frac{M_F}{J} - \frac{wl^2}{8EI} \times \frac{2l}{3} \times \frac{1}{2} = 0$$

$$M_F = \frac{wl^3}{24EI} \Big/ \left(\frac{l}{2EI} + \frac{1}{J} \right)$$

where

M_F = fixed end moment
J = spring constant = M/ϕ
E = Young's modulus
I = moment of inertia of the beam
w = beam loading
l = span

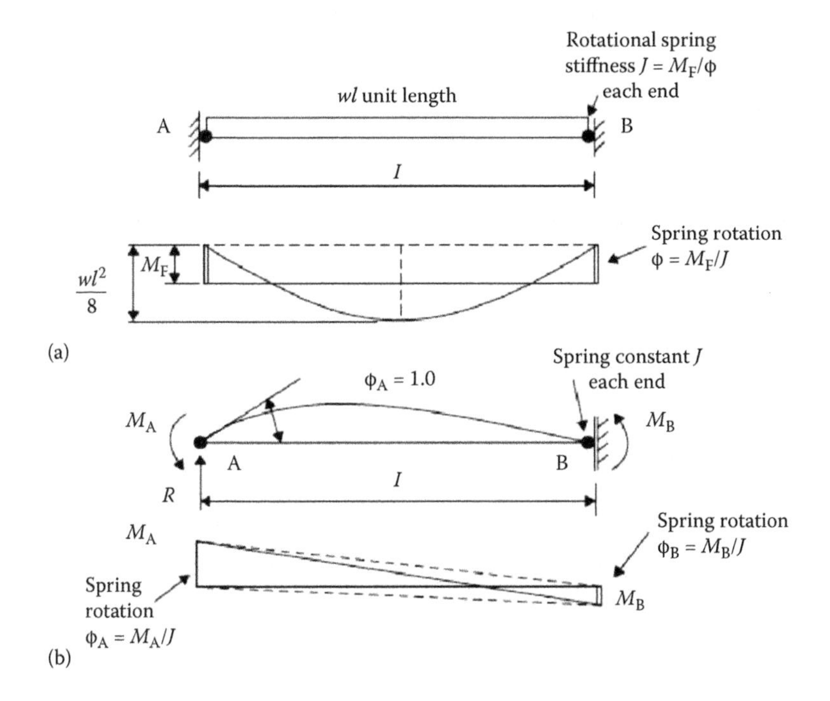

Figure 7.15 Beam moments: (a) fixed end moments; (b) stiffness and carryover factor.

(ii) STIFFNESS

Figure 7.15b shows a propped cantilever with springs at the ends subjected to a moment M_A causing a reaction R and slope ϕ_A at end A.

The stiffness at A is given by the value of M_A for rotation $\phi_A = 1.0$. Solve first for reaction R at A.

The deflection at A relative to the tangent at the fixed end B is zero. That is the moment of the areas of the M/EI diagram taken about A plus deflection due to the spring rotation at B caused by M_A and R is zero:

$$\frac{M_A l}{J} + \frac{M_A l^2}{2EI} - \frac{Rl^3}{3EI} - \frac{RI^2}{J} = 0$$

$$R = M_A\left(\frac{1}{J} + \frac{l}{2EI}\right) \bigg/ \left(\frac{l^2}{3EI} + \frac{I}{J}\right)$$

$$M_B = M_A - Rl$$

where M_B is the moment at end B.

The slope ϕ_A at A relative to the tangent at the fixed end B is equal to the area under the M/EI diagram between A and B plus the spring rotation at A due to moments M_A and M_B. This is given by

$$\phi_A = 1 = \frac{M_A}{J} + \frac{M_A l}{2EI} - \frac{M_B}{J} - \frac{M_B l}{2EI}$$

Solve for M_A to obtain the absolute stiffness value. Divide this by $4E$ to give the value in terms of (I/l) for a member without end springs.

(iii) CARRYOVER FACTOR A TO B

This is given by

$$-\frac{M_B}{M_A} = -\left(\frac{M_A - Rl}{M_A}\right)$$

Note that for a uniform member with no end springs:

Stiffness = $4EI/l \;\alpha\; I/l$

Carryover factor = $+1/2$

The above factors can then be used in a normal moment distribution process to analyse a complete braced frame or a subframe.

7.6.4 Frame design

(a) *Specification*

The internal frame of the multistorey building shown in Figure 7.1 is to be designed with semirigid joints of 50% fixity. The building is braced and the design loading is shown in Figure 7.5.

Full plastic moment $M_p = wl^2/16$

50% fixity $M_F = wl^2/32$

(b) *Roof beam*

Design load = 54 kN/m

At support, $M_F = 54 \times 8^2/32 = 108$ kNm

At centre, $M_c = wl^2/8 - 108 = 324$ kNm

$S = 324 \times 10^3/275 = 1178$ cm^3

Select 457 × 152 UB 60, with $S = 1290$ cm^3.

(c) *Floor beam*

Design load = 77 kN/m

At support, $M_F = 154$ kNm

At centre, $M_c = 462$ kNm

$S = 1743$ cm^3 for $p_y = 265$ N/mm^2

Try 457 × 152 UB 82, with $S = 1810$ cm^3, $I_x = 3660$ cm^4. Check the deflection due to the imposed load of 17.5 kN/m on a simply supported beam:

$\delta = 5w_il^4/384EI_x = 12.6$ mm

= Span/635 < Span/360

This is satisfactory.

(d) *External column – upper length 7–10–13*

The column loads and moments are shown in Figure 7.16a. Assume the following sizes for the columns: 7–10–13, 203 × 203 UC; 7–4, 254 × 254 × 89 UC. The fixed end moments allowing for eccentricity are

$$M^{F}_{10-11} = 154 + 308 \times 0.11 = 187.8 \text{ kNm}$$

$$M^{F}_{7-8} = 154 + 308 \times 0.13 = 194.0 \text{ kNm}$$

The subframe for determining the column moments for length 7–10 is shown in the figure. The distribution factors are

$K_{10-13} : K_{10-7} = 0.5 : 0.5$

For 10–7 (203 × 203 UC 60),

$I = 6130 \text{ cm}^4$, $I/L = 15.3$

For 7–4 (254 × 254 UC 89),

$I = 14{,}300 \text{ cm}^4$, $I/L = 35.8$

$K_{10-7} : K_{7-4} = 15.3 : 35.8/51 = 0.3 : 0.7$

The moment distribution is carried out. The design actions at joint 10 are

$F = 666$ kN

$M = 104.8$ kNm

Try 203 × 203 UC 60 (Section 7.4.3(a)), with $P_c = 190 \text{ N/mm}^2$, $A = 76.4 \text{ cm}^2$, $M_c = 180$ kNm, $M_b = 160.9$ kNm, $m = 0.53$.

Cross-section capacity check = 0.9

Moment buckling resistance check = 0.74

This is satisfactory.

(e) *External column – lower length 1–4–7*

Assume 254 × 254 × 89 UC where the moment due to eccentricity is 308 × 0.13 = 40 kNm.

Total moment = 194 kNm

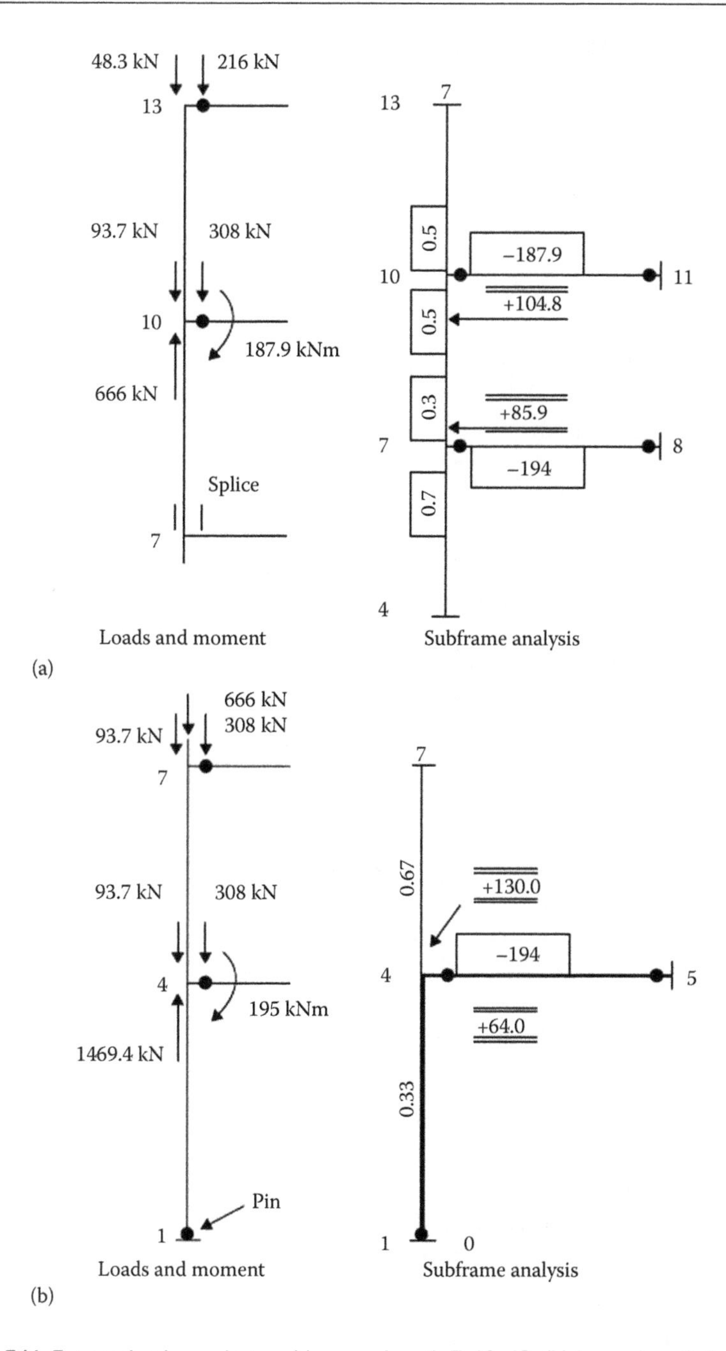

Figure 7.16 External column design: (a) upper length 7–10–13; (b) lower length 1–4–7.

The column loads and moment and subframe for length 1–4 are shown in Figure 7.16b. The distribution factors taking account of the pin end 1 are

$$K_{1-4} : K_{4-7} = \frac{\frac{3}{4} \times \frac{1}{6} : \frac{1}{4}}{0.375} = 0.34 : 0.66$$

The results of the moment distribution are

$$M_{4-1} = 64.0 \text{ kNm}$$

$$M_{4-7} = 130.0 \text{ kNm}$$

$$F_{4-1} = 1469.4 \text{ kN}$$

Try 254 × 254 UC 89, with $T = 17.3$ mm, $p_y = 265$ N/mm², $p_c = 160.2$ N/mm², $A = 113$ cm², $M_c = 324$ kNm, $M_b = 314.9$, $m = 0.6$ (Section 7.4.3(b)).

Cross-section capacity check = 0.69

Moment buckling resistance check = 0.93

Section is satisfactory. Length 4–7 is not as highly stressed.

(f) *Centre column – upper length 8–11–14*

Assume column 8–11–14 (Figure 7.17a) is 203 × 203 × 60 UC and 8–5 is 305 × 305 × 97 UC. The fixed end moments are

$$M^{F}_{11-10} = 154 + 308 \times 0.11 = 187.8 \text{ kNm}$$

$$M^{F}_{8-7} = 154 + 308 \times 0.16 = 203.3 \text{ kNm}$$

The fixed end moment for span 11–12 is calculated assuming a spring constant $J = M/\phi = 8000$ kNm/radian. The factored dead load on the span is 49 kN/m (Section 7.6.3(b)). The fixed end moment:

$$M_{11-12}\left(\frac{49 \times 8000^3}{24 \times 205 \times 10^3 \times 3660 \times 10^4}\right) \bigg/ \left(\frac{8000}{2 \times 205 \times 10^3 \times 3660 \times 10^4} + \frac{1}{8000 \times 10^6}\right)$$

$$= 78.8 \times 10^6 \text{ N mm} = 78.8 \text{ kNm}$$

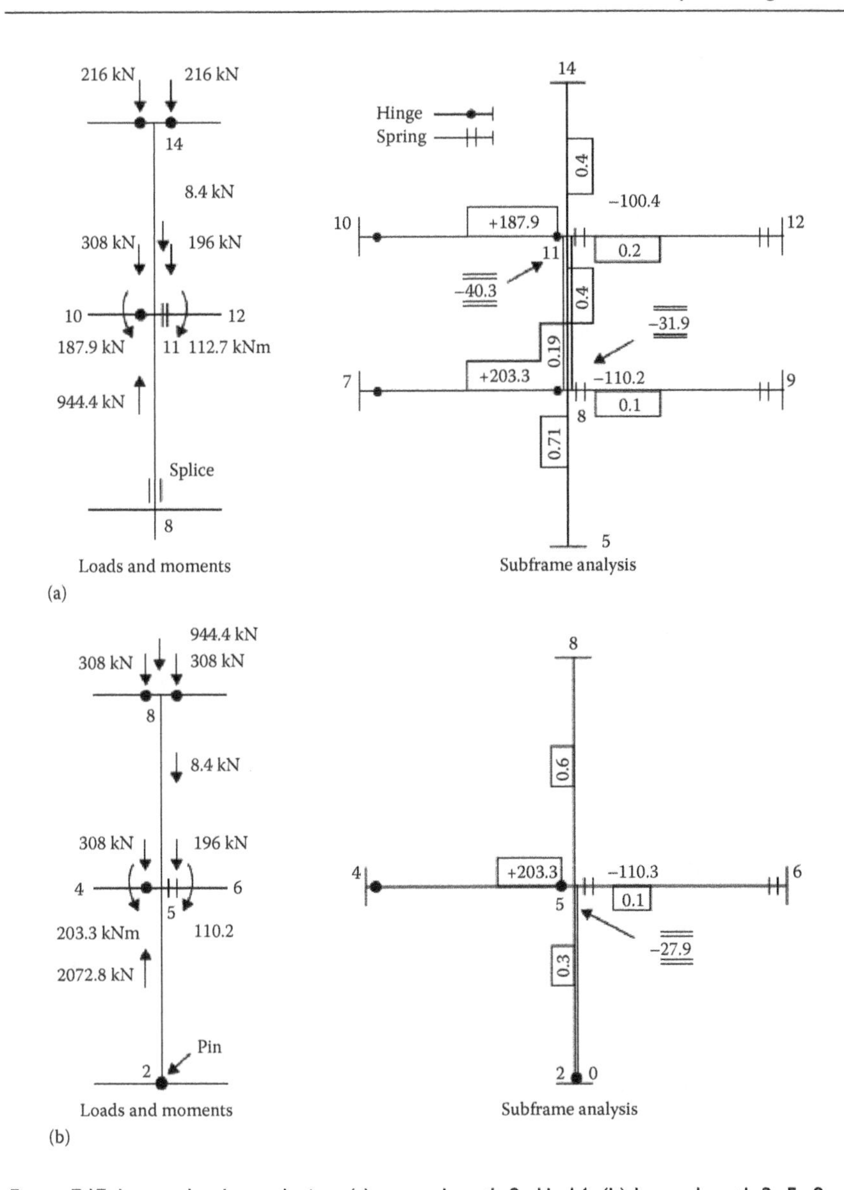

Figure 7.17 Internal column design: (a) upper length 8–11–14; (b) lower length 2–5–8.

The fixed end moments are

$$M^{F}_{11\text{–}12} = 78.8 + 49 \times 4 \times 0.11 = 100.4 \text{ kNm}$$

$$M^{F}_{8\text{–}9} = 78.8 + 49 \times 4 \times 0.16 = 110.2 \text{ kNm}$$

Solve for stiffnesses of beams 11–12 and 8–9 (Figure 7.17):

$$R = M_A\left(\frac{1}{8000 \times 10^6} + \frac{8000}{2 \times 205 \times 10^3 \times 3660 \times 10^4}\right) \Big/ \left(\frac{8000^2}{3 \times 205 \times 10^3 \times 3660 \times 10^4} + \frac{8000}{8000 \times 10^6}\right)$$

$$= 1.39 \times 10^{-4} M_A$$

$$M_B = M_A - 1.39 \times 10^{-4} M_A \times 8000 = -0.112 M_A$$

Solve equation for M_A:

$$1.0 = \frac{M_A}{J} + \frac{M_A l}{2EI} - \frac{0.112 M_A}{J} - \frac{0.112 M_A l}{2EI}$$

$$1.0 = \frac{0.888 M_A}{8000 \times 10^6} + \frac{0.888 \times 8000 M_A}{2 \times 205 \times 10^3 \times 3660 \times 10^4}$$

$$M_A = 6.29 \times 10^9 \text{ kNm}$$

In terms of I/l the stiffness for beam 11–12 is

$$\frac{6.29 \times 10^9}{4 \times 205 \times 10^3} = 7671$$

The subframe for column analysis is shown in Figure 7.17a. The assumed column sections and stiffnesses are

- 8–11,11–14 – 203 × 203 UC 60, $I/L = 6130 \times 10^4/4000 = 15{,}325$
- 5–8 – 305 × 305 UC97, $I/L = 22{,}300 \times 10^4/4000 = 5575$

Distribution factors – Joint 11:

$$K_{11\text{–}14} : K_{11\text{–}12} : K_{11\text{–}8} = \frac{15{,}325 : 7671 : 15{,}325}{38{,}321} = 0.4 : 0.2 : 0.4$$

Joint 8:

$$K_{8\text{–}11} : K_{8\text{–}9} : K_{8\text{–}5} = 0.19 : 0.1 : 0.71$$

The moment distribution analysis is carried out. The moments and axial load for column 8–11 are

$M_{11-8} = 40.3$ kNm

$M_{8-11} = 31.9$ kNm

$F_{11} = 944.4$ kN

Try 203 × 203 UC 46 (Section 7.4.3(a)), with $p_c = 188$ N/mm², $A = 58.8$ cm², $M_c = 137$ kNm, $M_b = 119.2$ kNm, $m_{LT} = 0.43$.

Cross-section capacity check = 0.88

Moment buckling resistance check = 1.0

This is satisfactory.

(g) *Centre column – lower length 2–5–8*

Loads and moments are shown in Figure 7.17b. Assume 305 × 305 UC 97 the beam end moments including moment due to eccentricity are

$M_{5-4} = 203.3$ kNm

$M_{5-6} = 110.2$ kNm

The subframe is shown in the figure. The stiffnesses allowing for the pin end 1 are

- For 5–2, $I/L = 0.75 \times 22{,}300 \times 10^4/6000 = 27{,}875$;
- For 5–8, $I/L = 22{,}300 \times 10^4/4000 = 55{,}750$;
- For 5–6, $I/L = 7671$ (Section 7.6.4(f)).

Distribution factors:

$$K_{5-2} : K_{5-8} : K_{5-6} = \frac{27{,}875 : 55{,}750 : 7671}{91{,}296} = 0.3 : 0.6 : 0.1$$

The moment distribution analyses is carried out. The moments and axial for columns 2–5 are

$M_{5-2} = 27.9$ kNm

$F_{5-2} = 2072.8$ kN

Try 305 × 305 UC 97 (Section 7.4.3(d)), with $p_c = 188.2$ N/mm², $A = 123$ cm², $M_c = 437.2$ kNm, $M_b = 374.6$ kNm, $m_{LT} = 0.6$, $T = 15.4$ and $p_y =$ 275 N/mm.

Cross-section capacity check = 0.68

Moment buckling resistance check = 0.94

A check shows that column length 5–8 is not as highly stressed as 5–2.

(h) *Joints*

The joint arrangements for the external and internal columns are shown in Figure 7.18. Only the plastic beam end moment causes tension in the bolts. The eccentric moment is due to bolt shear.

Tension $T = 154/2 \times 0.67 = 114.9$ kN

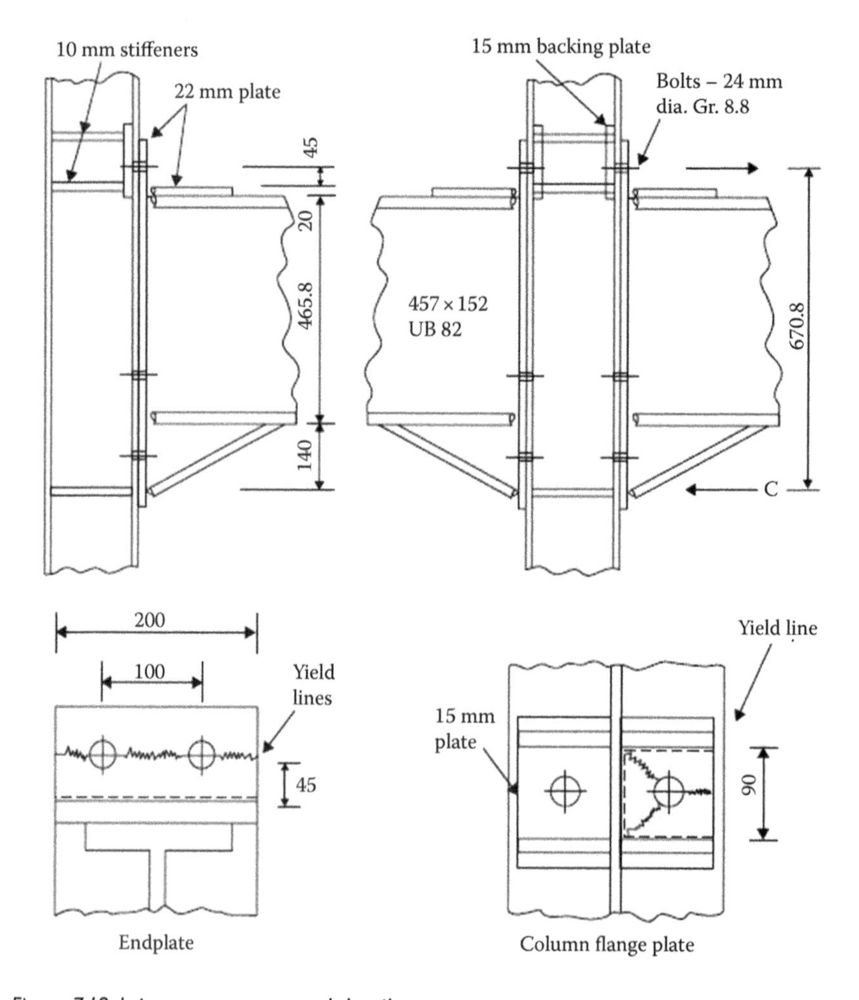

Figure 7.18 Joint arrangement and detail.

Table 7.2 Comparison of building designs

Design	Beams		External columns		Internal columns		Total steel weight[a] (KN)
	Roof	Floors	Upper	Lower	Upper	Lower	
Braced	457 ×191	533 × 210	203 × 203	254 × 254	203 × 203	254 × 254	10,031
simple	UB 74	UB 92	UC 46	UC 89	UC 46	UC 107	
Braced	457 ×191	533 × 210	203 × 203	254 × 254	203 × 203	305 × 305	10,348
rigid elastic	UB 67	UB 82	UC 60	UC 89	UC 60	UC 118	
Braced	406 × 140	457 × 152	254 × 254	305 × 305	203 × 203	305 × 305	9273
rigid plastic	UB 46	UB 60	UC 89	UC 97	UC 60	UC 97	
Braced	457 × 152	457 × 152	203 × 203	254 × 254	203 × 203	305 × 305	10,047
semirigid plastic	UB 60	UB 82	UC 60	UC 97	UC 46	UC 97	

[a] 5% is added for connections for simple design, 10% for rigid design.

Provide oversize bolts, 24 mm diameter, Grade 8.8, with tension capacity 158 kN, shear capacity 132 kN, holes 26 mm diameter.

For the endplate, yield line analysis gives

$$2 \times 114.9 \times 45 \times 10^3 = (200 + 148) \times 265t^2/4$$

Thus, thickness t = 21.2 mm. Provide 22 mm plate.

The flange backing plate shown is designed to resist bolt tension. The yield line pattern is shown in the figure. Analysis gives

Yield line moment = 12,100 N mm/mm

Thickness of plate required = 13.3 mm

Provide 15 mm plate.

The joint is such as to cause only low stresses in the column flange.

7.7 SUMMARY OF DESIGNS

The various designs carried out are summarized in Table 7.2. The designs are remarkably similar in weight of steel required. The rigid plastic design is about 10% lighter than the rigid elastic design.

Chapter 8

Multistorey buildings, simple design to EC3

8.1 OUTLINE OF DESIGN COVERED

8.1.1 Aims of study

A simple design of the four-storey building, see Section 7.2, has been repeated in this part of the book, so the results of application of BS 5950 and EC3 can be compared.

8.1.2 Design to EC3

(a) *Steel beams (Clause 5.1.5 EC3)*

Strength and serviceability checks are made to include the design and analysis of fully laterally restrained beams and unrestrained beams as given in Clause 5.1.5 EC3.

(i) Fully laterally restrained beams: Strength and serviceability checks are made to include:

1	Resistance of cross-sections-bending moment	(Clause 5.4.5 EC3)
2	Resistance of cross-section – shear	(Clause 5.4.6 EC3)
3	Resistance of cross-sections-bending and shear	(Clause 5.4.7 EC3)
4	Shear buckling resistance of web	(Clause 5.6 EC3)
5	Resistance to flanged-induced buckling	(Clause 5.7.7 EC3)
6	Resistance of web to transverse forces • Crushing resistance • Crippling resistance	(Clause 5.7 EC3)
7	Deflections	(Clause 4.2 EC3)

(ii) Lateral tensional buckling of beams LTB – unrestrained beams (Clause 5.5.2 EC3).

(b) *Column (Chapter 5 EC3)*

This chapter considers the design of

- Compression members,
- Members resisting combined axial force and moments,
- Column base plates.

(c) *Connections (Chapter 6 EC3)*

The results of bolting and welding appear slightly more conservative than BS 5950. This is mainly due to the use of large material factor of safety $\gamma_M = 1.25$.

To help comparison of the design procedures in BS 5950 and EC3 with regard to the design and analysis of beams with full lateral strength, beams without full lateral strength, columns, column baseplate and connections, the same roof beam, floor beam, columns, joint design and baseplate of Sections 7.3.2 through 7.3.5, 7.3.8 and 7.3.9 respectively, are redesigned following the EC3 procedures to give the reader a good opportunity to compare the results of application of these two relevant codes.

8.2 SIMPLE DESIGN CENTRE FRAME

The same building, loads, specifications and materials described in Section 7.2 are used in this chapter, so a comparison of the results of application of EC3 and BS 5950 can be made.

8.2.1 Roof beam with full lateral restraints

(a) *Section selection*

$$\text{Design action } F_{Ed} = \gamma_G g_k + \gamma_Q q_k$$

$$= 1.35 \times 6 + 1.5 \times 1.5$$

$$= 10.35 \text{ kN/m}^2 \quad \text{(Table 2.2 EC3)}$$

$$\text{Design bending moment } M_{Ed} = \frac{F_{Ed}L^2}{8} = \frac{10.35 \times 5 \times 8^2}{8} = 414 \text{ kNm}$$

$$W_{pl} = \frac{414 \times 10^3}{275} = 1505 \text{ cm}^3$$

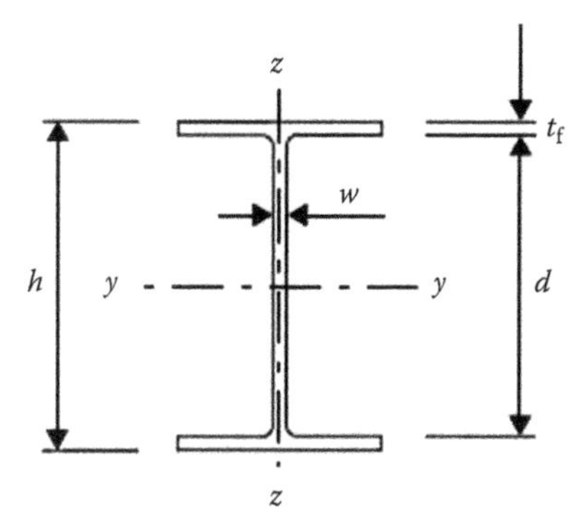

Figure 8.1 Steel beam cross-section.

Select 457 × 191UB74, with W_{pl} = 1650 cm^3, I_x = 3300 cm^4, t_f = 14.5 mm f_y = 275 N/mm^2 in grade F_e430 (Table 3.1 EC3)

(b) *Strength classification*

See Figure 8.1.

$$\varepsilon = (235/f_y)^{0.5} = 0.924$$

$$\frac{c_f}{t_f} = 6.57 < 10\varepsilon = 10 \times 0.924 = 9.24$$

$$\frac{c_w}{t_w} = 45.3 < 72\varepsilon = 72 \times 0.924 = 66.5$$

$\therefore$ 457 × 191UB74 section is classified as class 1 (Table 5.3.1 EC3)

For class 1, design moment of resistance

$$M_{pl,Rd} = \frac{W_{pl} f_y}{\gamma_{M0}}$$

$$= \frac{1650 \times 10^3 \times 275}{1.05} = 432.142 \times 10^6 \text{ N mm}$$

$$= 432.142 \text{ kNm} > 414 \text{ kNm} \quad \text{(satisfactory)}$$

(c) *Shear*

$$\text{Design shear force, } V_{Ed} = \frac{F_{Ed}}{2} = \frac{414}{2} = 207 \text{ kN}$$

For Class 1 section

Design plastic shear resistance = $V_{pl,Rd} = A_v(f_y\sqrt{3})/\gamma_{M0}$

$A_v = \eta h t_w = 1.0 \times 457.0 \times 9 = 4277.5 \text{ mm}^2$

$\therefore V_{pl,Rd} = [4277.5/(275/\sqrt{3})/1.05]/1000$

$= 646.804 \text{ kN} > 207 \text{ kN}$ (satisfactory)

(d) *Bending and shear*

If $V_{Ed} > 0.5V_{pl,Rd}$ (then reduce the theoretical plastic resistance moment of the section)

$V_{Ed} = 207 \text{ kN}$

$0.5\ V_{pl,Rd} = 0.5 \times 646.804 = 323.402 \text{ kN}$

$\therefore V_{Ed} < 0.5\ V_{pl,Rd}$

Therefore, no need to reduce the theoretical plastic resistance moment of the section.

(e) *Shear buckling resistance (Clause 5.6 EC3)*

Check if $h_w/t_w < 69\varepsilon$

$45.3 < 69 \times 0.924$

$45.3 < 64$

$\therefore$ no need to check shear buckling (Clause 5.6 EC3)

(f) *Flange induced buckling (Clause 5.7.7 EC3)*

$k(E/f_{yf})(A_w/A_{fc})^{1/2} > d/t_w$

$A_w = (h - 2t_{cf})t_w = (457.0 - 2 \times 14.5) \times 9 = 3852 \text{ mm}^2$

$A_{fc} = bt_{cf} = 190.4 \times 14.5 = 2760.8\ \text{mm}^2$

$k = 0.3$

$\therefore 0.3(210 \times 10^3/275)(3852/2760.8)^{0.5} = 270.327$

$\therefore 270.327 > h_w/t_w$

> 45.3

$\therefore$ satisfactory

(g) *Resistance of the web to transverse forces (Clause 5.7 EC3)*

(i) CHECK IF WEB CRUSHING RESISTANCE > V_{Ed}

$$R_{y,Rd} = (S_S + S_y)t_w f_{yw}/\gamma_{M1} \quad \text{(for I or H)}$$

$$S_y = 2.5(h-d)[1-(\sigma_{F,Ed}/f_{yf})^2]^{1/2}/[1.08S_S/(h-d)]$$

S_S = stiff bearing, in this case, it is assumed to be 100 mm

$$S_y = \frac{2.5(457.0-407.6)[1-(0)^2]^{1/2}}{[1+0.8\times 100/(457.0-407.6)]} = \frac{123.6}{2.619} = 47.185\ \text{mm}$$

At the end of the member S_y should be halved (Clause 5.7.3 EC3)

$\therefore R_{y,Rd} = (100 + 23.592) \times 9 \times 275/1.05$

$= 305.890\ \text{kN} > 207\ \text{kN}$ (satisfactory)

(ii) CHECK DESIGN CRIPPLING RESISTANCE OF WEB

$$R_{a,Rd} > V_{Ed}$$

For I or H sections

$$R_{a,Rd} = 0.5t_W^2(Ef_{yw})^{1/2}[(t_f/t_w)^{1/2} + 3(t_w/t_f)(S_S/d)]/\gamma_{M1}$$

$$= 0.5\times 9^2(210\times 10^3\times 275)^{1/2}[(14.5/9)^{1/2} + 3(9/14.5)(100/457)]/1.05$$

$$= [3{,}077{,}773.35][1.269+0.4]/1.05$$

$$= 489.051\ \text{kN} > 207\ \text{kN} \quad \text{(satisfactory)}$$

(h) *Check deflection*

$$M_{\max}=\frac{WL}{8}=\frac{8\times 5(g_k+q_k)L}{8}=\frac{40(6+1.5)\times 8}{8}=300\text{ kNm}$$

$$\text{Elastic moment resistance }(M_{c,Rd})\text{el}=\frac{W_{el}f_y}{\gamma_{M0}}=\frac{1460\times 10^3\times 275}{1.05}$$

$$=382.380\text{ kNm}>M_{\max}$$

Therefore, deflection can be calculated elastically.

(i) CHECK DEFLECTION DUE TO PERMANENT LOADING ($g_k + q_k$)

$$\delta_{\max}<\frac{\text{span}}{250}\quad\text{(Table 4.1 EC3)}$$

$$\frac{5}{384}\frac{WL^3}{EI}<\frac{8000}{250}$$

$$W=g_k+q_k=40(6+1.5)=3000\text{ kN}$$

$$\frac{5\times 300\times 10^3\times(8000)^3}{384\times 210\times 10^3\times 33{,}300\times 10^4}<32\text{ mm}$$

$$28.6\text{ mm}<32\text{ mm}\quad\text{(satisfactory)}$$

(ii) CHECK DEFLECTION DUE TO VARIABLE LOADING (q_k)

$$\frac{5}{384}\frac{WL^3}{EI}<\frac{\text{span}}{360}$$

$$q_k=5\times 8\times 1.5=60\text{ kN}=W$$

$$\therefore 5.72<\frac{8000}{360}$$

$$5.72\text{ mm}<22.22\text{ mm}\quad\text{(satisfactory)}$$

8.2.2 Floor beam – full lateral restraints

(a) *Selection of steel section*

$$q_k=7.0\text{ kN/m}^2$$

$$g_k=3.5\text{ kN/m}^2$$

Design action $(F_{Ed}) = \gamma_G g_k + \gamma_Q q_k = 1.35 + 1.5q_k$

$$= 1.35 \times 7\ (5 \times 8) + 1.5 \times 3.5 \times (5 \times 8)$$

$$= 588 \text{ kN}$$

$$M_{Ed} = F_{Ed}L/8 = \frac{588 \times 8}{8} = 588 \text{ kNm}$$

$$W_{pl} = 2138 \text{ cm}^3$$

Select 533×210 UB 92 with $W_{pl} = 2360 \text{ cm}^3$.

(b) *Check deflection*

$$M_{max} = \frac{(g_k + q_k) \times 8 \times 5 \times l}{8} = \frac{10.5 \times 40 \times 8}{8} = 420 \text{ kNm}$$

$$\text{Elastic moment resistance} = \frac{W_{el} f_y}{\gamma_{M0}}$$

$$= -\frac{207 \times 275}{1.05 \times 1000} = 542 \text{ kNm}$$

Greater than M_{max}, therefore, the deflection can be calculated elastically.

(i) CHECK DEFLECTION DUE TO PERMANENT LOADING $(g_k + q_k)$

$$\delta_{max} < \frac{\text{span}}{250} \quad \text{(Table 4.1 EC3)}$$

$$\frac{5}{384} \frac{WL^3}{EI} < \frac{8000}{250}$$

$$W = g_k = 5 \times 8(7 + 3.5) = 420 \text{ kN}$$

$$\frac{5}{384} \frac{420 \times 10^3 \times (8000)^3}{210 \times 10^3 \times 55{,}200 \times 10^4} = 24.154 \text{ mm} < 32 \text{ mm} \quad \text{(satisfactory)}$$

(ii) CHECK DEFLECTION DUE TO VARIABLE LOADING (q_k)

$$\frac{5}{384} \frac{WL^3}{EI} < \frac{8000}{360}, \quad W = 40 \times 3.5 = 140 \text{ kN}$$

$$\therefore 8.05 \text{ mm} < 22.2 \text{ mm} \quad \text{(satisfactory)}$$

8.3 BRACED RIGID ELASTIC DESIGN/FLOOR BEAM 10–11–12 (SEE SECTION 7.4.2 AND FIGURE 7.5)

EC3 approach to check lateral buckling restraint Clause 5.5.2.

8.3.1 Check buckling resistance of beam $M_{b,Rd} > M_{Ed}$

$$M_{b,Rd} = \chi_{LT}\beta_W W_{pl,y} f_y/\gamma_{M1}$$

$$\text{Buckling factor}, \chi_{LT} = 1/\left[\phi_{LT} + \left(\phi_{LT}^2 - \bar{\lambda}_{LT}^2\right)\right]^{1/2} \quad \text{(Clause 5.5.2 EC3)}$$

$$\phi_{LT} = 0.5\left[1 + \alpha_{LT}(\bar{\lambda}_{LT} - 0.2) + \bar{\lambda}_{LT}^2\right]$$

α_{LT} = Imperfection factor for lateral torsional buckling

= 0.21 for rolled sections

$\bar{\lambda}_{LT} = (\lambda_{LT}/\lambda_1)(\beta_W)^{1/2}$ = slenderness ratio

$\lambda_1 = \pi(E/f_y)^{1/2} = 93.3\varepsilon \quad \varepsilon = (235/f_y)^{1/2} = 0.93 \quad \lambda_1 = 87.327$

$\beta_W = 1$ for class 1 and 2 sections

$$\lambda_{LT} = \frac{L(W_{pl,y}^2/I_Z I_W)^{1/4}}{C_1^{1/2}\left[1 + \left(L^2 G I_t/\pi^2 E I_W\right)\right]^{1/4}}$$

L = length between the points, which have lateral restraints

= 1.94 m (see Figure 5.6)

$$G = \frac{E}{2(1+\upsilon)} = \frac{210\times10^3}{2(1+0.3)} = 80,769\ \text{N/mm}^2$$

$$W_{\text{pl,y}} = 2060\ \text{cm}^3,\quad I_t = 51.5\ \text{cm}^4,\quad I_W = 1.33\times10^{12}\ \text{mm}^6$$

$$I_Z = 2010\ \text{cm}^4$$

$$I_Y = 47,500\ \text{cm}^4$$

$$\therefore \lambda_{LT} = 1940\left[(2060\times10^3)/(2010\times10^4\times1.33\times10^{12})\right]^{1/4}/$$

$$1.879^{1/2}\left[1+(1920^2\times80,769\times51.5\times10^4)/\right.$$

$$\left.(\pi^2\times210\times10^3\times1.33\times10^{12})\right]^{1/4}$$

$$\therefore \lambda_{LT} = 38.219$$

$$\bar{\lambda}_{LT}(\lambda_{LT}/\lambda_1)\beta_W^{1/2}$$

$$= \frac{38}{87.327}\times1^{1/2} = 0.43$$

$$\alpha_{LT} = 0.21$$

$$\phi_{LT} = 0.5\left[1+\alpha_{LT}(\bar{\lambda}_{LT}-0.2)+\bar{\lambda}_{LT}^2\right]$$

$$= 0.5[1+0.21(0.43-0.2)+0.43^2] = 0.616$$

$$\chi_{LT} = 1/\left[\phi_{LT}+\left(\phi_{LT}^2-\bar{\lambda}_{LT}^2\right)^{1/2}\right] \le 1$$

$$= 1/[0.616+(0.616^2-0.43^2)^{1/2}] = 0.94$$

∴ Buckling resistance of the beam, $M_{b,Rd}$ is

$$M_{b,Rd} = \chi_{LT}\beta_W W_{pl,y} f_y/\gamma_{M1}$$
$$= (0.94 \times 1 \times 2060 \times 10^3 \times 275/1.05)/10^6$$
$$= 507.152 \text{ kNm} < 546 \text{ kNm} \quad \text{(not suitable)}$$

Try 533 × 210 UB 92, repeat the check above. This beam is satisfactory in terms of its lateral buckling resistance. Note that all the checks detailed in EC3 need to be done before the use of the beam, see Section 5.6.2.1.

8.4 COLUMN – UPPER LENGTH 7–10–13 (FIGURES 7.5, 7.10 AND 7.16), DESIGN AND CHECKING USING EC3

Imposed load reduction = 10%

$$N_{sd} = (6 + 7)20 \times 1.35 + (6.9 \times 5 \times 1.35) + (2 \times 2.2 \times 6.3 \times 1.35) + (10.6 \times 5 \times 1.35) + (1.5 + 3.5)20 \times 1.5 \times 0.9$$
$$= 641.541 \text{ kN}$$

Try 203 × 203 UC 46, eccentricity = 0.23 m (100 mm from column face). Initially assume moments divided equally

$$M = N_{sd}(\text{beam}) \times 0.20/2$$

$$N_{sd}(\text{beam reaction}) = \text{Load } 7\text{–}8 = (7 \times 20 \times 1.35) + (3.5 \times 20 \times 1.5)$$
$$= 294 \text{ kN}$$

$$\therefore M = 294 \times 0.20/2 = 29.4 \text{ kNm (at top) and}$$
$$M = -29.4 \text{ kNm (at bottom)}$$

For column 7–10–13, try 203 × 203 UC 46

Section properties	*Strength classification*
$i_z = 51.3$ mm	$t_f = 11.0$ mm, S275, steel grade Fe430
$W_{pl,y} = 497 \times 10^3$ mm³	∴ $f_y = 275$ N/mm² (Table 3.1 EC3)
$A = 58.7 \times 10^2$ mm²	$c_f/t_f = 9.24 < 10\varepsilon = 9.244$
$I_y = 4570 \times 10^4$ mm⁴	$h_w/t_w = 22.3 < 33\varepsilon = 30.5$
$t_f = 11.0$ mm	∴ column section is class 1
$f_y = 275$ N/mm²	

8.4.1 Check resistance of cross-sections, bending and axial force (Clause 5.4.8 EC3)

1. Check if $N_{pl,Rd} > N_{sd}$

$$N_{pl,Rd} = Af_y/\gamma_{M0} = 58.7 \times 10^2 \times 275/(1.05 \times 1000)$$

$$= 1537.38 \text{ kN} > 641.47 \text{ kN (satisfactory)}$$

2. Check if $M_{Ny} > M_{Ed}$

$$M_{Ny} = M_{pl,Rd,y}(1 - n)/(1 - 0.5a)$$

$$M_{pl,Rd,y} = W_{pl,fy}/\gamma_{M0} = 497 \times 10^3 \times 275/(1.05 \times 1000^2)$$

$$= 130.167 \text{ kNm}$$

$$n = N_{sd}/N_{pl,Rd} = 641.547/1537.38 = 0.417$$

$$a = (A - 2bt_f)/A = (5870 - 2 \times 203.6 \times 11)/5870 = 0.236$$

$$\therefore M_{Ny} = 130.167(1 - 0.417)/(1 - 0.5 \times 0.236)$$

$$= 86.04 \text{ kNm} > M_{Ed}(33.8 \text{ kNm}) \text{ (satisfactory)}$$

8.4.2 Resistance of member: Combined bending and axial compression (Clause 5.5.4 EC3)

For class 1 and 2 check that:

$$\frac{N_{sd}}{\chi_{min} A f_y / \gamma_{M1}} + \frac{k_y M_{y,Ed}}{W_{pl,y} f_y / \gamma_{M1}} + \frac{k_Z M_{z,Ed}}{W_{pl,z} f_y / \gamma_{M1}} \leq 1.0$$

$$L_{eff} = L_{effy} = L_{effz} = 1.0L = 1.0 \times 4000 \text{ mm} \quad \text{(Clause 5.5.1.5 EC3)}$$

$$\lambda_y = \frac{4000}{i_y} = \frac{4000}{88.2} = 45.352, \quad \lambda_z = \frac{4000}{i_z} = \frac{4000}{51.3} = 77.972$$

$$\lambda_z > \lambda_y$$

Therefore, the column will buckle about the Z–Z axis

$$N_{sd} = 641.547 \text{ kN}, \; M_{y,Ed} = 33.81 \text{ kNm}, \; M_{z,Ed} = 0.0$$

$A = 58.7 \times 10^2$ mm² $\gamma_{M1} = 1.05$

$f_y = 275$ N/mm² $W_{pl,y} = 497 \times 10^3$ mm³

χ_{min} = the smaller of χ_z and χ_y

$$\chi_z = 1/\left[\phi + \left(\phi^2 - \lambda_z^{-2}\right)^{1/2}\right]$$

$$\phi = 0.5\left[1 + \alpha\left(\lambda_z^{-} - 0.2\right) + \lambda_z^{-2}\right]$$

α = Imperfection factor from Table 5.5.1 EC3

For buckling curve, see Table 5.5.3 EC3.

$h/b = 203.2/203.6 = 1.0 < 1.2, \quad t_f = 11.0 \text{ mm} < 100 \text{ mm}$

$\therefore$ buckling curve C

$\therefore \alpha = 0.49$ (Table 5.5.1 EC3)

$$\lambda_z' = (\lambda_z/\lambda_1)(\beta_A)^{1/2}$$

$\lambda_1 = 93.9\varepsilon$ (Annex F, Clause F2.2 EC3) $\quad \varepsilon = (235/f_y)^{1/2}$

$= 93.9 \times 0.924 = 86.76$

$$\therefore \lambda_z' = (77.972/86.76) \times 1^{1/2} = 0.899$$

$$\therefore \phi = 0.5[1 + 0.49(0.899 - 0.2) + 0.899^2] = 1.075$$

$$\therefore \chi_z = 1/[1.075 + (1.075^2 - 0.899^2)^{1/2}] = 0.603$$

$$\chi_y = 1/\left[\phi + \left(\phi^2 - \lambda_y'^2\right)^{1/2}\right]$$

$$\phi = 0.5\left[1 + \alpha\left(\lambda_y' - 0.2\right) + \lambda_y'^2\right]$$

$$\lambda_y = 45.351$$

$$\lambda_y' = (\lambda_y/\lambda_1)(\beta_A)^{1/2}$$

$= (45.351/86.76)(1)^{1/2} = 0.523$

For $h/b = 1.0 < 1.2$ and $t_f = 11.0$ mm < 100 mm, from Table 5.5.1 EC3

$$\alpha = 0.34 \text{ (buckling about } y-y \text{ axis, curve } b)$$

$$\therefore \phi = 0.5[1+0.34(0.523-0.2)+0.523^2] = 0.69$$

$$\therefore X_y = 1/[0.69+(0.69^2-0.523^2)^{1/2}] = 0.875$$

$$\therefore \chi_{min} \text{ is the smaller of } \chi_z(0.603) \text{ and } \chi_y(0.875)$$

$$k_y = 1 - \frac{\mu_y}{\chi_y}\frac{N_{sd}}{Af_y}$$

$$\mu_y = \lambda'_y(2\beta_{My}-4) + \frac{W_{pl,y}-W_{el,y}}{W_{el,y}}$$

$$= 0.523(2\times 2.5-4) + \frac{497-450}{450}$$

$$= 0.523 + 0.104$$

$$= 0.627$$

$$k_y = 1 - \frac{0.627\times 641.547\times 1000}{0.875\times 58.7\times 100\times 275}$$

$$= 0.715$$

$$\therefore \frac{641.547\times 1000}{0.603\times 58.7\times 100\times 275/1.05} + \frac{0.715\times 33.81\times 10^6 + 0}{497\times 10^3\times 275/1.05} \leq 1$$

$$0.692 + 0.186 = 0.88 < 1 \quad \text{(satisfactory)}$$

This section is satisfactory.

8.5 OUTER COLUMN – LOWER LENGTH 1–4–7 (FIGURE 7.5)

8.5.1 Check column below 1st floor at joint 4

Imposed loads (variable actions) reductions = 20%

Factored axial loading 4–7 is

$$N_{sd} = (6 + 2 \times 7)20 \times 1.35 + (6.9 \times 5 \times 1.35) + (3 \times 2.2 \times 6.3 \times 1.35)$$

$$+ 3(10.6 \times 5 \times 1.35) + (1.5 + 2 \times 3.5) \times 20 \times 1.5 \times 0.8$$

$$= 1061.36 \text{ kN}$$

Factored loading 5–4 = 7 × 20 × 1.35 + 3.5 × 20 × 1.5 = 294 kN

$$(\mathrm{I}/l)_{1-4} : (\mathrm{I}/l)_{4-7} = \left(\frac{1}{6}\right)\left(\frac{1}{4}\right) / \left(\frac{1}{6}+\frac{1}{4}\right) = 0.4:0.6 < 1:15$$

Divide movement equally

$$M = N_{sd}(\text{beam}) \times 0.23/2 = 33.81 \text{ kNm} = M_{y,Ed}$$

$$N_{sd} = 1061.36 + 294 = 1355.36 \text{ kN}$$

Try 254 × 254 UC 89 in grade F_e 430, $N_{sd,Rd}$ = 1540 kN at L_E = 6.0 m. BS 5950-1: 2000, BS 4-1: 1993, Member buckling check:

Section properties	*Strength classification*
i_z = 113 cm, i_y = 11.2 cm	c_f/t_f = 6.31 < 10ε = 9.221
$W_{pl,y}$ = 1220 cm³	c_w/t_w = 15.6 < 33ε = 30.5
A = 113 cm³	∴ column section is class 1
I_y = 14,300 cm³	
t_f = 17.3 mm	
f_y = 275 N/mm² (Table 3.1 EC3)	

Check if

$$\frac{N_{sd}}{\chi_{min} A f_y / \gamma_{M1}} + \frac{k_y M_{y,Ed}}{W_{pl,y} f_y / \gamma_{M1}} + \frac{k_z M_{z,Ed}}{W_{pl,z} f_y / \gamma_{M1}} \leq 1$$

$$L_{effz} = L_{effy} = L_{effz} = 1.0L = 6.0\,\text{m} \quad \text{(Clause 5.5.1.5 EC3)}$$

$$\lambda_y = \frac{6000}{i_y} = \frac{6000}{112} = 53.57, \quad \lambda_z = \frac{6000}{65.5} = 91.69$$

$$\lambda_z > \lambda_y$$

Therefore, the column will buckle about the z–z axis

χ_{min}, the smaller of χ_z and χ_y

$$\chi_z = 1/\left[\phi + \left(\phi^2 - \lambda_z^{-2}\right)^{1/2}\right]$$

$$\phi = 0.5\left[1 + \alpha\left(\lambda_z^{-} - 0.2\right) + \lambda_z^{-2}\right]$$

α from Table 5.5.1 EC3

$$\frac{h}{b} = \frac{260.3}{256.3} = 1.015 < 1.2, \quad t_f < 100 \text{ mm}$$

∴ buckling curve C

∴ $\alpha = 0.49$ (Table 5.5.1 EC3)

λ'_z: $(\lambda_z/\lambda_1)(\beta_A)^{0.5}$

$$\lambda_1 = 93.9\varepsilon = 86.76, \quad \varepsilon = \left(\frac{235}{f_y}\right)^{1/2}, \quad \beta_A = 1$$

(see Annex F, Clause F.2.2 EC3)

$\therefore \lambda'_z = (91.6/86.76) \times 1$

$= 1.06$

$\therefore \phi = 0.5[1 + 0.49(1.06 - 0.2) + 1.06^2]$

$\phi = 1.27$

$\therefore \chi_z = 1/[1.27 + (1.27^2 - 1.06^2)^{1/2}]$

$= 0.508$ which is less than χ_y

For $k_y = 1.5$ (conservative)

$$\therefore \frac{1355.36 \times 10^3}{0.508 \times 11{,}300 \times 275/1.05} + \frac{0.55 \times 33.81 \times 10^6}{1220 \times 10^3 \times 275/1.05} + 0 \leq 1$$

$$0.058 = 0.96 < 1 \quad \text{(satisfactory)}$$

$$k = 1 - \frac{\mu_y}{\chi_y} \frac{N_{sd}}{Af_y}$$

$$\mu_y = \lambda_y'(2\beta_{My} - 4) + \frac{W_{pl,y} - W_{el,y}}{W_{el,y}}$$

$$\lambda_y' = (\lambda_y / \lambda_1)(\beta_A)^{0.5}$$

$$= (53.57/86.76) \times 1$$

$$= 0.617$$

$$\beta_{My} = 1.8 \text{ (Figure 5.5.3 EC3)}$$

$$\mu_y = 0.617(2 \times 1.8 - 4) + 0.109$$

$$= -0.138 < 0.9$$

$$\mu_y = 0.9$$

$$\therefore k = 1 - \frac{0.9 \times 1355.36 \times 10^3}{0.87 \times 11{,}300 \times 275} = 1 - 0.45 = 0.55$$

8.6 BASE PLATE (EC ANNEX L, CLAUSE L1 EC3)

8.6.1 Check bearing pressure and strength $N_{sd}/A_{ef} \leq f_j$

$$f_j = \beta_j k_j f_{cd} = \frac{2}{3} \times 1(30 \times 0.85) = 11.39 \text{ N/mm}^2$$

where

$f_{cd} = f_{ck}/\gamma_m$
f_{ck} = concrete cylinder comp. strength $0.85 f_{cu}$
$\gamma_m = 1.5$
N_{sd} = 2088 kN (N_{sd} will be <2088 kN if it is calculated according to EC3)
$A_{eff} = (2\chi + h)(2\chi + b) - 2(C - t_w/2)(h - 2t_f - 2\chi)$
h = 266.7 mm, b = 258.8 mm, t_w = 12.8 mm, t_f = 20.5 mm

Try $t = 30$ mm.

$$\chi = t(f_y/3f_j\gamma_{M0})^{1/2}$$

$$= 30(275/(3 \times 11.39 \times 1.05))^{1/2}$$

$$= 83 \text{ mm} < 0.5(500 - 266.7)$$

$$< 116 \text{ mm (satisfactory) (Figure 8.2)}$$

$$\therefore A_{eff} = (2 \times 83 + 266.7)(2 \times 83 + 258.8) - 2\,[129.4 - (12.8/2)] \times (266.7 - 2 \times 20.5 - 2 \times 83) - 438.38$$

$$= 183{,}372.58 \text{ mm}^2$$

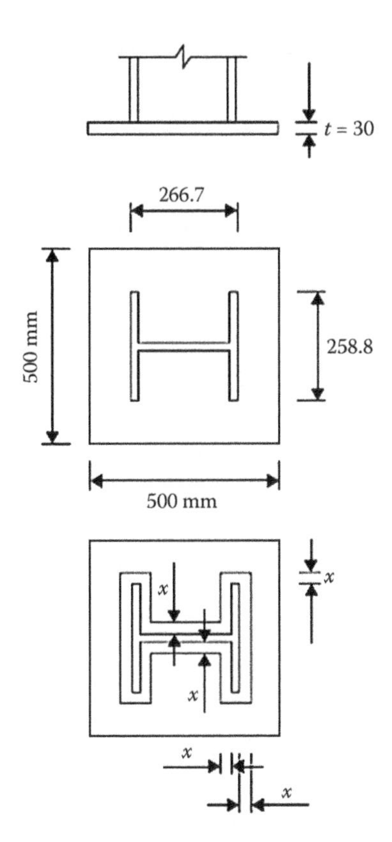

Figure 8.2 Base plate.

$$\therefore \frac{N_{sd}}{A_{eff}} = \frac{2088 \times 10^3}{183,372.58} = 11.38 \text{ N/mm}^2 < f_j (11.39 \text{ N/mm}^2)$$

∴ satisfactory

8.6.2 Check resisting moment $M_{Ed} < M_{Rd}$

$M_{Ed} = (X^2/2)N_{sd}/A_{eff} = (83^2/2) \times 2088 \times 10^3/183,372.58$

$= 39.221 \times 10^3$N mm/mm.run

$M_{Rd} = t^2 f_y/6\gamma_{M0} = 30^2 \times 275/(6 \times 1.05) = 39.285 \times 10^3$N mm/mm.run

$\therefore M_{Ed} < M_{Rd}$ (satisfactory)

8.7 JOINT DESIGN (FIGURE 5.4 EC3)

$V_{Ed} = (7 \times 20 \times 1.35) + (3.5 \times 20 \times 1.5) = 294$ kN

8.7.1 Check positioning for holes for bolts

Diameter of bolt, $d = 20$ mm

Diameter of bolt holes, $d_0 = 22$ mm (Clause 7.5.2 EC3)

End distance, $e_1 = 30$ mm $> 1.2d_0 = 1.2 \times 22 = 26.4$ mm (satisfactory)

Edge distance, $e_2 = 35$ mm $> 1.5d_0 = 1.5 \times 22 = 33$ mm (Clause 6.5.1 EC3)

See Figure 8.3.
e_1 and e_2 < the larger of

(a) $12t = 12 \times 11 = 132$ mm

(b) 150 mm

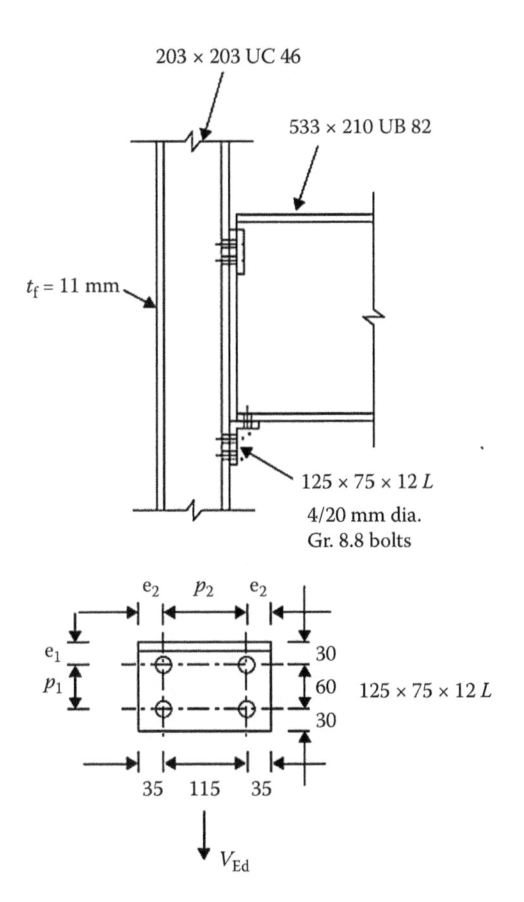

Figure 8.3 Joint design.

e_1 = 30 mm (satisfactory)

e_2 = 35 mm (satisfactory) (see Clause 6.5.1 EC3)

$P_1 = 60$ mm > ($2d_0 = 2 \times 22 = 44$ mm) (satisfactory)

$P_2 = 115$ mm > ($3d_0 = 3 \times 22 = 66$ mm) (satisfactory)

P_1 and $P_2 \leq$ the smallest of

$14t = 14 \times 11 = 154$ mm or 200 mm (satisfactory)

8.7.2 Check shear resistance of bolt group (Clause 6.5.5 and Table 3.3 EC3)

$V_{v,Rd}$ = no. of bolts × $F_{v,Rd}$

$$A = \frac{\pi d^2}{4} = \frac{\pi \times 20^2}{4}$$

$$= 4 \times \frac{0.6 f_{ub} A}{\gamma_{Mb}} = 4 \times \frac{0.6 \times 800 \times 314.16}{1.25}$$

$$= 482.59 \text{ kN} > 294 \text{ kN} \quad \text{(satisfactory)}$$

According to BS 5950-1-2000, shearing distance = 367.6 kN
Shear force = 308 kN, bearing capacity = 404 kN

8.7.3 Check bearing resistance

$t_f < t$ (angle cleat)

$$\text{bearing resistance of one bolt} = F_{b,Rd} = \frac{2.5\,\alpha F_u d t_f}{\gamma_{Mb}}$$

α is the smaller of

(a) $e_1/3d_0 = 30/(3 \times 22) = 0.45$

(b) $P_1/3d_0 - \frac{1}{4} = 60/(3 \times 22) - \frac{1}{4} = 0.45$

(c) $f_{ub}/f_u = 800/430 = 1.86$

(d) 1.0

$\therefore \alpha = 0.45$

$\therefore F_{b,Rd}$ (one bolt) = $2.5 \times 0.45 \times 430 \times 20 \times 11/(1.25 \times 1000) = 85.140$ kN/bolt

∴ bearing resistance of bolt group = 4 × 85.140
= 340.56 kN greater than 294 kN

8.7.4 Shear resistance of leg of cleat

Fastener holes need not be allowed for providing that

$$\frac{A_{v,net}}{A_v} \geq \frac{f_y}{f_u} = \frac{12(120-2\times22)}{12\times120} > \frac{275}{430}$$

$$0.63 \geq 0.63 \quad \text{(satisfactory)}$$

$$\therefore V_{pl,Rd} = A_v(f_y/\sqrt{3})/\gamma_{M0}$$

$$= 1440(275/\sqrt{3})/1.05 = 217.74 \text{ kN}$$

$$= 217.74 \times 2 = 435.48 \text{ (for two rows of bolts)}$$

$$> 294 \text{ kN}$$

BS 5950-1:2000	*EC3*
Shear force 308 kN	V_{Ed} = 294 kN
Shear resistance 367.6 kN	$V_{v,Rd}$ = 482.549 kN
Bearing resistance 404 kN	$F_{b,Rd}$ = 340.56 kN
	$V_{pl,Rd}$ = 435.48 kN

From the table it can be seen that BS is more conservative in calculating the design shear force and bolt joint shear resistance. When comparing the bolt joint bearing resistance, in the contrast, EC3 becomes more conservative in its approach.

The reader is highly recommended to understand both codes of practices (BS 5950 and EC3) and choose the most appropriate and safe design and analysis approach to the particular practical case concerned.

Chapter 9

Floor systems

9.1 FUNCTIONS OF FLOOR SYSTEMS

The floor system generally serves two purposes:

- Primarily the floor carries vertical dead and imposed loads and transmits these loads through beams to the columns/walls.
- The floor also has to act as a horizontal diaphragm that ties the building together, stabilizes the walls and columns and transmits horizontal wind load to rigid frames, braced bays or shear walls.

The aims in design of the floor system are

- To deliver the main vertical loads safely by the most direct and efficient route to the columns/walls without excessive deflection or vibration;
- To have the necessary horizontal strength/rigidity;
- To achieve a uniform arrangement and spacing of beams where possible to reduce costs – alternative layouts may need investigation;
- To keep construction depth to a minimum while accommodating necessary services – this reduces overall building costs;
- For all components to have adequate resistance to or protection against fire.

Types of floor systems are described below.

9.2 LAYOUTS AND FRAMING SYSTEMS

The layout of the floor framing depends on the shape and structural system used for the building. In steel-framed structures, the column arrangement defines the flooring divisions. Primary beams frame between the columns and may form part of the main vertical structural frames. Depending on spans, secondary beams may be provided to subdivide the intercolumn areas. Column spacings normally vary from 4 to 8 m in rectangular-shaped buildings but can be much greater. Secondary beams are normally spaced at 3 to 4 m centres.

Tall buildings generally have a central core and perimeter columns or tube-wall construction. The floor beams or girders frame between the core and outside wall. This arrangement allows maximum flexibility in the division of floor areas using lightweight partitions.

Some floor framing systems are shown in Figure 9.1. These include buildings with:

- One-way normally transverse framing where
  - Slab spans one way longitudinally;
  - Secondary beams span between frames and slabs span transversely;

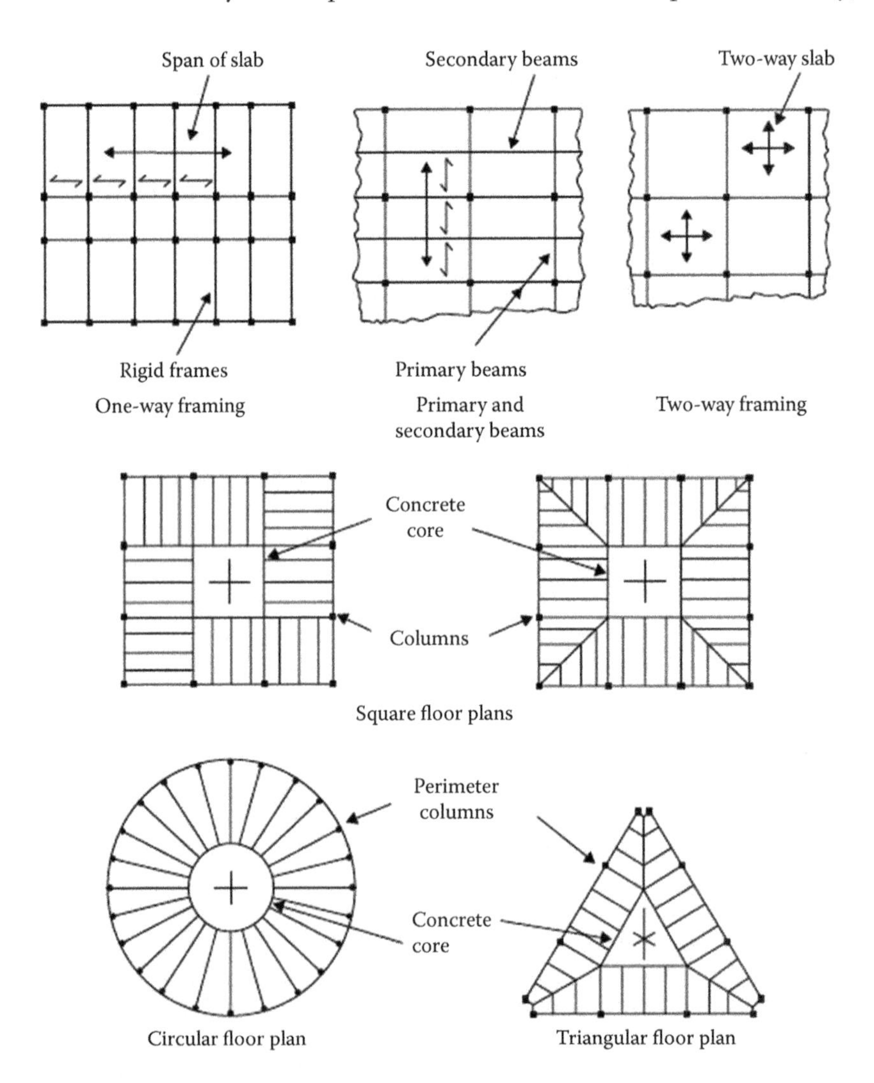

Figure 9.1 Floor framing systems.

- Two-way framing with two-way spanning slabs;
- Square, circular and triangular floor areas with beams spanning out from the core.

9.3 TYPES OF FLOOR CONSTRUCTION

Various types of floor construction in steel-framed buildings are shown in Figure 9.2. These can be classified as follows:

- Cast *in-situ* concrete slabs, one- or two-way spanning, supported on steel beams or lattice girders. Ribbed or waffle slabs can also be used for long spans.
- Precast, prestressed concrete slabs, one-way spanning, supported on steel beams. Slabs can be solid or hollow or double-T in form. Units can also be supported on shelf angles to reduce floor depth as shown in Figure 9.2.
- Composite deck, where the slab is poured on profiled steel sheeting which is embossed with ribs to ensure composite action. Design where the steel decking acts only as permanent formwork can also be made.
- Cast *in-situ* slab or composite concrete deck made to act compositely by stud shear connectors with the steel floor beams. This system gives considerable savings in weight of floor steel.

Lattice girders or castellated beams are more economical than universal beams for long spans. Lattice girder construction also permits services such as air conditioning ducts to be run through the open web spaces.

The stub girder floor is a special development aimed at giving long-span, column-free floor spaces. The system is only economical for long spans of 10–15 m. The construction gives up to 25% saving in weight of floor steel, a reduction in depth of floor and provides openings for services. Structurally, the stub girder acts like a modified form of composite Vierendeel girder. The validity of the system has been proved by extensive research and testing.

9.4 COMPOSITE FLOOR SLABS

9.4.1 General comments

The composite floor is cast on profiled steel sheets which act as permanent shuttering, supporting the wet concrete, reinforcement and construction loads. After hardening, the concrete and steel sheeting act compositely in carrying the loads.

Mesh reinforcement is provided over the whole slab. It is required to resist hogging moments. Alternatively, the concrete may be designed to

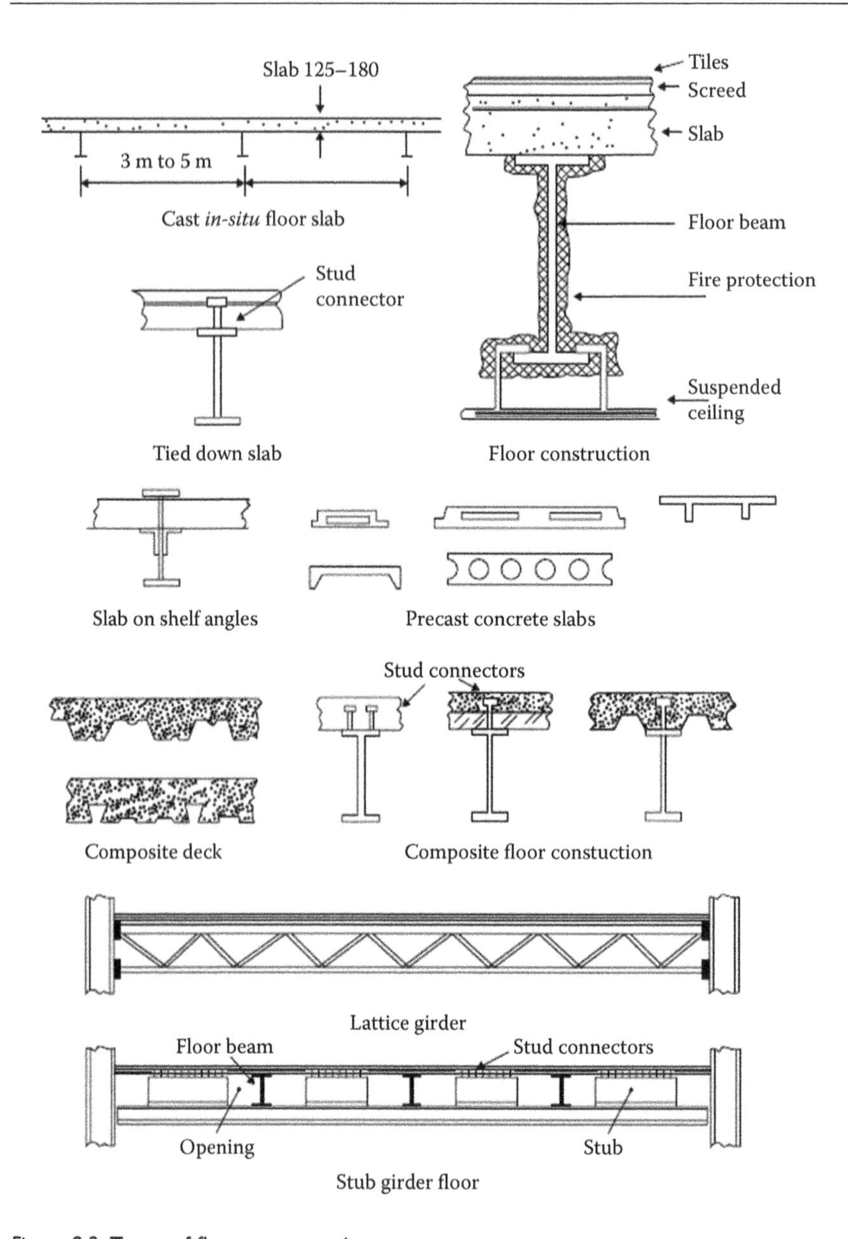

Figure 9.2 Types of floor construction.

carry the final loads without composite action when the sheeting acts as shuttering only.

Composite flooring is designed to BS 5950: Part 4 and EC4, see Chapter 10. Decking manufacturers load/span tables can be used to select the slab and sheeting for a given floor arrangement (John Lysaght, n.d.; Precision Metal Flooring, 1993). For example, Precision Metal Flooring (1993) gives the maximum span, slab thickness and metal decking gauge for single- and double-spanning slabs and propped slabs for various values of imposed load. The tables are based on:

- Construction load 1.5 kN/m^2;
- Deflection $\ngtr$ span/180 (construction), span/350 (composite slabs);
- Decking Grade Z 28 yield strength 280 N/mm^2;
- Concrete Grade 30;
- Mesh to BS 5950: Part 4;
- Shear connection – embossing and deck shape.

Fire load/span tables are also given to select slab thicknesses for various improved loads and fire rating times.

9.4.2 Design procedure

The design procedure from BS 5950: Part 4 is set out briefly as follows:

(a) Decking strength and serviceability. This depends on

- Effective section of compression flange and web – these are reduced due to buckling;
- Support capacity;
- Web strength;
- Deflection limit.

Design is to be in accordance with BS 5950: Part 6.

(b) Composite deck. Strength and serviceability checks are made to include:

- Moment capacity for sagging moments at mid-span and hogging moments over supports;
- Shear-bond capacity – some factors in the code expression must be obtained by tests on given decking;
- Vertical shear capacity;
- Deflection of the composite slab.

Expressions are given for simply supported and continuous slabs.

A composite deck section and sections for sagging and hogging moments are shown in Figure 9.3.

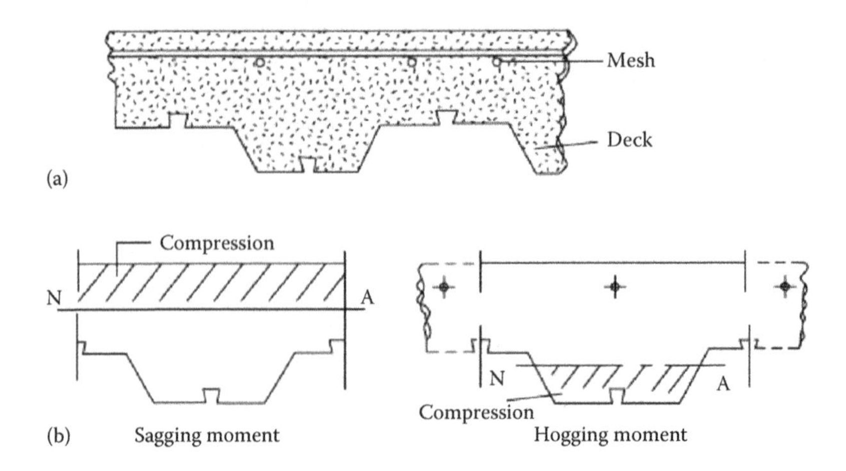

Figure 9.3 Floor sections: (a) composite deck; (b) moments in slabs.

9.5 COMPOSITE BEAM DESIGN

9.5.1 Design basis

Design of composite beams is to conform to BS 5950: Part 3. Or to EC4, see Chapter 10.

The composite beam is formed by connecting the concrete slab and the beam. The commonly used connector is the headed stud. The slab is to be a reinforced concrete floor slab or a slab supported on profiled steel sheeting.

The design process is outlined briefly. Detailed application of the code clauses is shown in the examples following.

9.5.2 Effective section

Referring to Clause 4.4.1 and Section 4.6 of the code and to Figure 9.4, the effective section for calculating moment capacity depends on the following.

(a) *Effective breadth of slab*

This depends on the direction of the slab span, whether perpendicular or parallel to the beam. For example, for a slab spanning perpendicular to the beam and according to EC4 and BS 5950: Part 3 the total effective breadth B_e is the sum of effective breadths b_e on each side: $b_e = L_z/8 \not> $ half the distance to the adjacent beam.

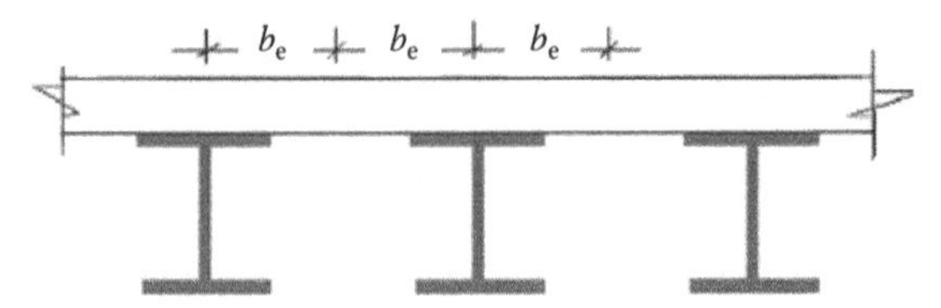

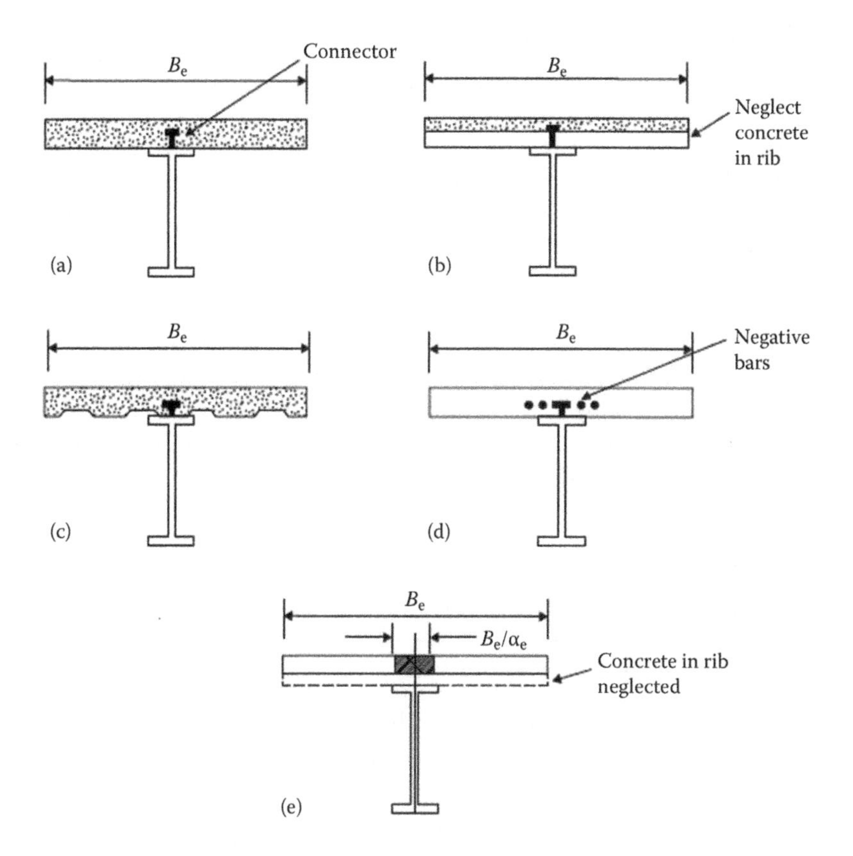

Figure 9.4 Effective section: (a) plain slab; (b) sheet perpendicular to beam; (c) sheet parallel to beam; (d) negative moment; (e) transformed section.

where L_z is the distance between points of zero moment, equal to the span of a simply supported beam.

Detailed provisions are given for other cases.

(b) *Composite slab*

For a slab spanning perpendicular to the beam, neglect ribs – use only concrete above ribs. For a slab spanning parallel to the beam – use full concrete section.

(c) *Portions neglected on the effective section*

Neglect concrete in tension, the profiled sheets in a composite slab and nominal mesh or bars less than 10 mm diameter.

9.5.3 Plastic moment capacity

The moment capacity of a composite section is based on (Clause 4.4.2):

- Concrete stress in compression = $0.45f_{cu}$, where f_{cu} is the concrete grade;
- Reinforcement in tension = $0.95f_y$, where f_y is the characteristic strength.

Only plastic and compact universal beam sections are considered.

9.5.4 Construction

The weight of wet concrete and construction loads is carried by the steel beam. For both propped and unpropped construction, beams may be designed assuming that at the ultimate limit state, the whole load acts on the composite member (Clause 5.1).

9.5.5 Continuous beam analysis

See Section 5.2 of the code, on which the analysis may be based.

(a) *Elastic analysis and redistribution*

The analysis is based on the value of the gross second moment of area of the uncracked section at mid-span (Figure 9.4e). Concrete in the ribs may be neglected (Clause 5.2.3 of the code).

Section 4.1 of the code gives an expression for calculating the effective modular ratio α_e for the concrete. This depends on the proportion of the total load that is long term.

The imposed load is to be arranged in the most unfavourable realistic pattern. The patterns to be investigated are

- Alternate spans loaded;
- Two adjacent spans loaded.

Dead load factors need not be varied. The resulting negative moment may be reduced by an amount not exceeding values given in Table 4 in the code. For plastic sections, 40% redistribution is permitted.

(b) *Simplified method*

The moment coefficients from Table 3 of the code can be used for uniform beams with uniformly distributed loads. Detailed requirements are given in Clause 5.2.2.

(c) *Plastic analysis (Clause 5.2.4)*

This may be used for uniform beams with uniform distributed load.

9.5.6 Design of members

(a) *Vertical shear (Clause 5.3.4)*

The vertical shear must be resisted by a steel beam web. The moment capacity is reduced by the high shear load.

(b) *Positive moment (Clause 5.3.1)*

The moment capacity is the plastic moment capacity of the composite section.

(c) *Negative moment (Clauses 5.3.2 and 5.3.3)*

The moment capacity is based on the steel section and effectively anchored tension reinforcement within the effective breadth of the concrete flange.

(d) *Stability of the bottom flange (Clause 5.2.5)*

In continuous beams, the stability of the bottom flange requires checking at supports for each span. Provisions for making the check are given in the clause. Lateral supports may be required. The unsupported length may be taken as the distance from the support to the point of contra-flexure.

9.5.7 Shear connectors (Section 5.4 of code)

The shear connector must transmit the longitudinal shear between the concrete slab and steel beam without crushing the concrete and without excessive slip or separation between the slab and beam. Headed studs welded to the beam are the main type of connector used.

(a) *Connector capacity*

In a solid slab the capacity of a connector is

- Design strength of shear connector (positive moments), $Q_p = 0.8Q_k$;
- Design strength of shear connector (negative moments), $Q_n = 0.6Q_k$.

Q_k is the characteristic resistance of a connector from Table 5 in the code, for example, for a 19 mm headed stud 100 mm high in normal weight Grade 30 concrete, $Q_k = 100$ kN.

(b) *Number of connectors required*

For positive moments the number $N_p = F_p/Q_p$, where F_p is the compressive force in the concrete at the point of maximum positive moment.

For negative moments, the number $N_n = F_n/Q_n$, where F_n is the force in the tension reinforcement.

The total number of connectors between the point of maximum moment and the support is $N_p + N_n$ connectors should be spaced uniformly. The minimum spacing is five times the stud diameter.

(c) *Characteristic resistance of headed studs*

Characteristic resistance values for solid slabs are given in Table 5 in the code (BS 5950–3.1). Formulas are given for modifying these values when profiled sheets are used.

9.5.8 Longitudinal shear (Section 5.6 of code)

Transverse reinforcement runs perpendicular to the beam span. Longitudinal shear from the connectors is resisted by the concrete flange, the transverse reinforcement and the steel sheeting if used.

(a) *Longitudinal shear*

The longitudinal shear force per unit length is

$$v = NQ/S$$

where N is number of connectors per unit length, S is the unit length and Q is Q_p or Q_n (Section 9.5.7).

(b) *Shear resistance*

An expression is given in the code for calculating the shear resistance per shear surface for normal and lightweight concrete. This includes contributions from the transverse reinforcement, the concrete slab and the profiled sheet if used. The formula and its application are described in the examples following Sections 9.6 and 9.7.

(c) *Shear surfaces*

Transverse shear surfaces for a solid slab and slab on profiled sheets are shown in Figures 9.5 (Section 9.6.2) and 9.15 (Section 9.7.6).

(d) *Profiled sheeting*

Clause 5.6.4 sets out the method of calculating the contribution of the profiled steel sheeting (Section 9.7.6).

9.5.9 Deflection (Section 6 of code)

(a) *Construction*

In unpropped construction, the total deflection

$$\delta_{total} = \delta_{dead} + \delta_{imposed}$$

where

δ_{dead} = deflection of steel beam carries concrete slab and beam;
$\delta_{imposed}$ = deflection composite section carries the imposed loads.

In propped construction, the composite section carries all loads. The behaviour is taken as linear elastic.

(b) *Simply supported beams*

Calculate deflection using the properties of the gross uncracked section (Section 9.5.5(a) above).

(c) *Continuous beams*

Allowances are to be made for

- Pattern loading – determine moments due to unfactored imposed load, on all spans then reduce support moments by 30%;
- Shakedown – described in the code.

Clause 6.1.3.5 of the code gives an expression for calculating the mid-span deflection δ for a continuous beam under uniform load or symmetrical point loads. In this expression, the simply supported beam deflection is modified according to the values of the span support moments as modified as noted above. (An example is given in Section 9.7.7.)

9.6 SIMPLY SUPPORTED COMPOSITE BEAM

9.6.1 Specification

Consider the simply supported steel floor beam in the structure shown in Figure 7.1. The characteristic loads on the beam are

- Dead load
  - o Slab and steel beam = 23 kN/m;
  - o Tiles, screed, partitions, ceiling, services = 12 kN/m.
- Imposed load = (3.5 kN/m^2) = 17.5 kN/m.

The design load on the composite section is 77 kN/m, and

$$M = 77 \times 8^2/8 = 616 \text{ kNm}$$

The materials are

- Concrete, with f_{cu} = 30 N/mm^2;
- Steel, with p_y = 275 N/mm^2.

The slab is 180 mm thick and the steel beam is 533 × 210 UB 92 with no composite action; the span is 8 m.

Redesign the beam as a composite section.

9.6.2 Moment capacity (Section 4.4 of code)

Concrete flange breadth:

$$B_e = \text{Span}/4 = 2 \text{ m}$$

Try 457 × 152 UB 52, with A = 66.6 mm^2, D = 449.8 mm, t = 7.6 mm, I = 21,400 mm^4. The composite section is shown in Figure 9.5.

Assuming the neutral axis lies in the slab, the depth is

$$x = (275 \times 66.6 \times 10^2)/(0.45 \times 30 \times 2000) = 67.7 \text{ mm}$$

$$\text{Lever arm } Z = 180 - (67.7/2) + (449.8/2) = 371.1 \text{ mm}$$

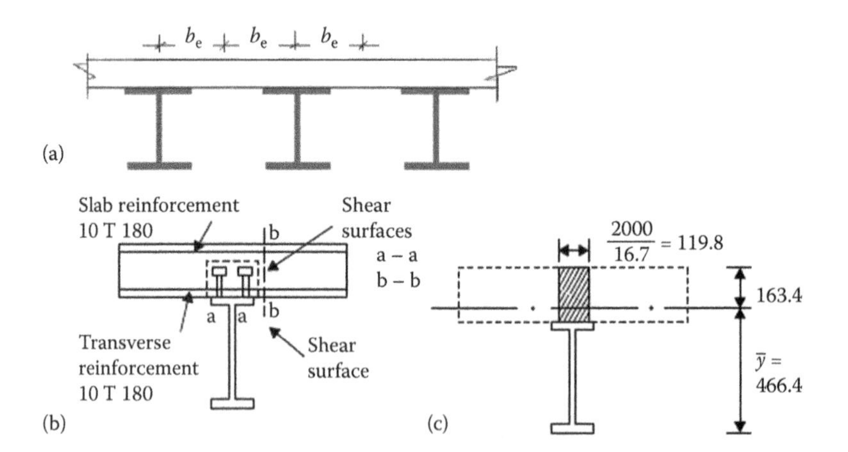

Figure 9.5 Composite beam: (a) moment capacity; (b) longitudinal shear; (c) transformed section.

Moment capacity $M_p = 275 \times 66.6 \times 10^2 \times 371.1/10^6 = 678.6$ kNm

> applied moment = 616 kNm

The section is satisfactory for the moment.

9.6.3 Shear (Section 5.3.4 of code)

The shear capacity is

$P_v = 0.5 \times 275 \times 449.8 \times 7.6/10^3 = 470$ kN

> applied shear = 308 kN

This is satisfactory.

9.6.4 Shear connectors (Section 5.4 of code)

Provide headed studs 19 mm diameter × 100 mm high, with characteristic resistance $Q_k = 100$ kN. The capacity in a solid slab under positive moment is

$Q_p = 0.8 \times 100 = 80$ kN

The number of connectors on *each side* of the centre of the beam is

$N_p = (275 \times 66.6)/(80 \times 10) = 23$

Spacing in pairs = 4000 × 2/22 = 364 (say 300 mm)

9.6.5 Longitudinal shear

The surfaces subjected to longitudinal shear from the connectors and the slab reinforcements are shown in Figures 9.4 and 9.5b. The top bars reinforce the slab for hogging moment. The bottom 10 mm diameter bars at 180 mm centres, $A_{sv} = 436$ mm²/m, resist shear due to composite action (Section 5.3.1 of code).

The longitudinal shear is

$v = 2 \times 1000 \times 80/300 = 533.3$ kN/m

The flange resistance (Clause 5.6.3), where the length of shear surface a–a [(2 × stud height) + 7ϕ = (2 × 100) + (7 × 19) = 333, say] 340 mm approximately and of shear surface b–b is 2 × 180 mm, is given by

V_r = [(0.7 × 436 × 460 × 2) + (0.03 × 340 × 1000 × 30)]/10^3
= 586 kN/m > v

This is satisfactory.

9.6.6 Deflection (Section 6.1 of code)

The beam is to be unpropped. The deflection of the *steel beam* due to self-weight and slab (23 kN/m) is

$$\delta_D = \frac{5 \times 23 \times 8000^4}{384 \times 205 \times 10^3 \times 21{,}400 \times 10^4} = 28.1 \text{ mm}$$

$$= \text{Span}/285$$

The composite section carries the imposed load plus finishes and partitions, total 29.5 kN/m. The deflection is calculated using the properties of the gross uncracked section.

The modular ratio α_e is determined using Clause 4.1. The imposed load is one-third long term. The proportion of the total loading which is long term is

P_e = (52.5 − 17.5/3)/52.5 = 0.89

α_e = 6 + (0.89 × 12) = 16.7

The transformed section is shown in Figure 9.5c. Locate the neutral axis

$$\bar{y} = \frac{(66.6 \times 10^2 \times 224.9) + (180 \times 119.8 \times 539.8)}{(66.6 \times 10^2) + (180 \times 119.8)} = 466.4 \text{ mm}$$

$$I_G = (21{,}400 \times 10^4) + (6660 \times 241.5^2) + (119.8 \times 180^3/12)$$
$$+ (180 \times 119.5 \times 73.4^2) = 7.75 \times 10^8 \text{ mm}^4$$

The deflection of the composite section is

$$\delta = \frac{5 \times 17.5 \times 8000^4}{384 \times 205 \times 10^3 \times 7.75 \times 10^8} = 5.84 \text{ mm}$$

$$= \text{Span}/1369$$

$$< \text{Span}/360$$

This is satisfactory.

9.7 CONTINUOUS COMPOSITE BEAM

9.7.1 Specification

(a) *Building*

A two-storey building and part floor plan for the first floor is shown in Figure 9.6. Design the end span of the continuous three-span floor beam ABCD. The beam is continuous over the ground floor columns. The loads are given below.

(b) *Flooring*

The floor construction is PMF Com Floor 70. This is double-spanning, unpropped, span 3 m, 1.2 mm gauge decking with normal weight concrete.

The decking is shown in Figure 9.7. From the manufacturer's load/span tables (Precision Metal Forming, 1993), for imposed loading of 6.7 kN/m^2, the maximum permitted span for a slab thickness of 150 mm is 3.39 m. The required span is 3 m. The fire load/span tables give a permitted span of 3.68 m for 90 minutes in fire rating, which is satisfactory. The imposed load on the slab including ceiling, finishes, etc. is given below. The slab dead load is 3.1 kN/m^2.

9.7.2 Floor loads

(a) *Separate characteristic loads*

(i) DURING CONSTRUCTION (FIGURE 9.9a, SECTION 9.7.3)

The dead load of the slab, deck and secondary beam is 3.2 kN/m, thus

Point load at $E = 3.2 \times 8 \times 3 = 76.8$ kN

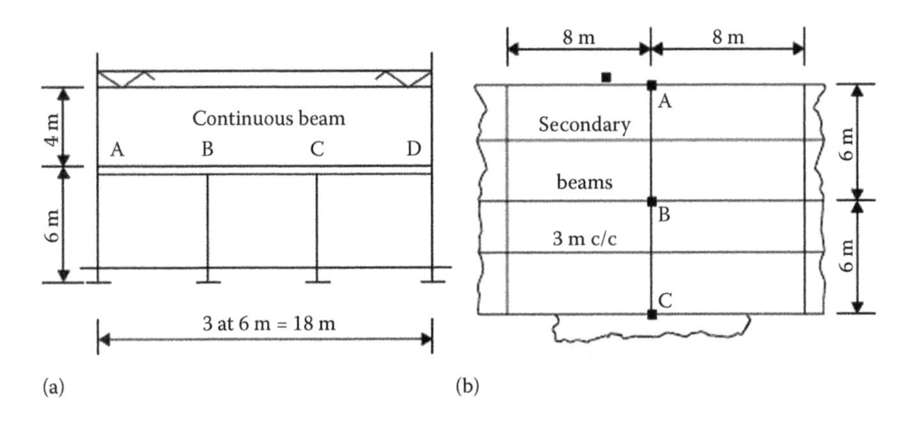

Figure 9.6 Two-storey building: (a) section; (b) part floor plan.

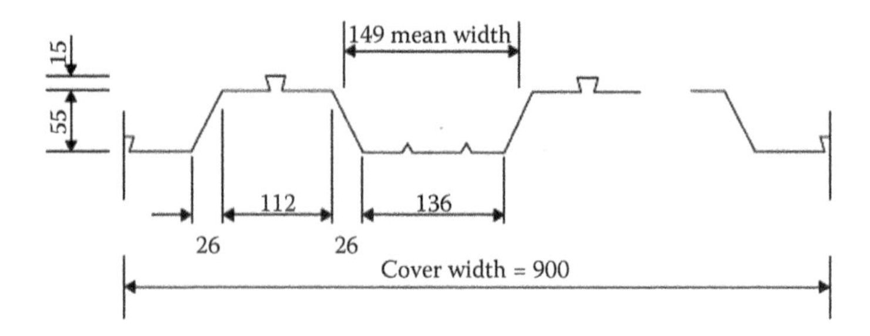

Figure 9.7 Com Floor 70 decking.

The dead load of the continuous beam and base is 0.5 kN/m^2.
The imposed load (Clause 2.2.3) is 0.5 kN/m^2, thus

Point load at $E = 0.5 \times 8 \times 3 = 12$ kN

The design loads are

Point load = $(76.8 \times 1.4) + (12 \times 1.6) = 126.7$ kN

Distributed load = $1.4 \times 0.5 = 0.7$ kN/m

(ii) DEAD LOAD ON COMPOSITE BEAM

The distributed load is 2.7 kN/m^2 (finish 1 kN/m^2, ceiling 0.5 kN/m^2, services 0.2 kN/m^2, partitions 1 kN/m^2), thus

Point load at $E = 2.7 \times 8 \times 3 = 64.8$ kN

The uniform load for beam and protection is 1.5 kN/m.

(iii) IMPOSED LOAD CARRIED BY COMPOSITE BEAM

The imposed load is 3.5 kN/m^2 (Figure 9.9), thus

Point load at $E = 3.5 \times 8 \times 3 = 84$ kN

(b) *Check imposed load on decking*

Imposed floor load (3.5 kN/m^2) + Finish (1 kN/m^2)
+ Ceiling and services (0.7 kN/m^2) + Partitions (1 kN/m^2)
= 6.2 kN/m^2 < 6.7 kN/m^2

(c) *Design loads on the composite beam*

(i) DEAD LOAD – PERMANENT LOADS

Point load = (76.8 + 64.8)1.4 = 198.2 kN

Distributed load = 1.5 × 1.6 × 6 = 14.4 kN/Span

The dead load factor is not varied.

(ii) IMPOSED LOAD

This is arranged to give maximum moments.

Point load = 84 × 1.6 = 134.4 kN

The design loads are shown in Figure 9.9b–d (Section 9.7.3).

9.7.3 Elastic analysis and redistribution

Elastic analyses are carried out for the end span AB (Figure 9.8) under pattern loading. A redistribution of 30% of the peak support moment is then carried out. (Table 4 in the code.)

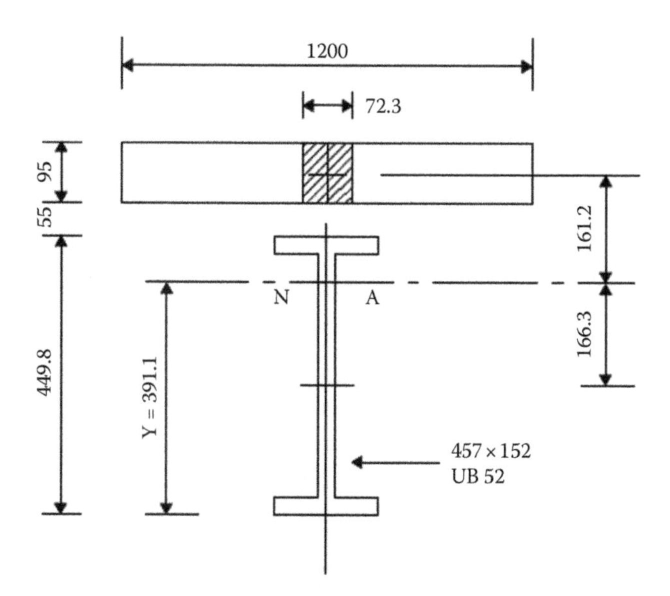

Figure 9.8 Span AB–section.

(a) *Approximate size of steel beam*

Design one 6 m span as a simply supported beam. Neglect the self-weight.

$$M = (198.2 + 134.4)6/4 = 498.9 \text{ kNm}$$

$$S = 498.9 \times 10^3/275 = 1814 \text{ cm}^3$$

Reducing by, say, 40% gives $S = 1088$ cm^3. Try 457 × 152 UB 52, with $S = 1100$ cm^3.

(b) *Distribution factors*

(i) EFFECTIVE MODULAR RATIO (CLAUSE 4.1)

$\alpha_e = \alpha_s + P_1(\alpha_1 - \alpha_s)$
$\alpha_1 =$ long-term modulus = 18
$\alpha_s =$ short-term modulus = 6
$P_1 =$ portions of total load which is long term (Section 9.7.2(b)) $= [9.4 - (3.5/3)]/9.4 = 0.88$
$\alpha_e = 6 + 0.88(18 - 6) = 16.6$

(ii) SECOND MOMENT OF AREA I_G

Use the gross uncracked for the elastic global analysis (Clause 5.2.3). For the gross value of I_G use the mid-span effective breadth uncracked, but neglect concrete in the ribs (Clause 4.2.2).

The effective breadth of concrete flange (Section 4.6 and Figure 2 of code) is

- For span AB,

$$B_e = 0.8 \times 6000/4 = 1200 \text{ mm}$$

- For span BC,

$$B_e = 0.7 \times 6000/4 = 1050 \text{ mm}$$

The gross and transformed sections are shown in Figure 9.8 for span AB.

For the steel beam of 457 × 152 UB 52, $A = 66.6$ cm^2, $I_X = 21{,}400$ cm^4 (dimensions on Figure 9.8).

For the transformed concrete, $A = 68.7$ cm^2, $I_X = 517$ cm^4. The neutral axis is

$$\bar{y} = \frac{(66.6 \times 22.5) + (68.7 \times 55.2)}{66.6 + 68.7} = 39.11 \text{ cm}$$

For span AB,

$$I_G = (66.6 \times 16.63^2) + 21{,}400 + (68.7 \times 16.12^2) + 517 = 58{,}188 \text{ cm}^4$$

For span BC,

$$I_G = 56{,}309 \text{ cm}^4$$

(iii) DISTRIBUTION FACTORS

$$K_{BA} : K_{BC} = \frac{(58{,}188/600):(56{,}309/600)}{96.9+93.8} = 051:0.49$$

The analysis should be based on the assumption of a uniform glass uncracked beam.

(c) *Elastic analysis*

The moment coefficients from the *Steel Designers Manual* (1986) are used in the analysis. Analyses are performed for

- Construction loads on the steel beam;
- Final loads on the composite beam (Clause 5.2.3.2, Pattern loads) for
  - o Dead load,
  - o Imposed load on spans AB, BC,
  - o Imposed loads on spans AB, CD.

The loads, moments and shears for the four load cases are shown in Figure 9.9.

The maximum design actions for the end span for elastic analysis are:

- For steel beam – construction loads, $M_E = 135.2$ kNm
- For composite beam,

$$M_B = 187 + 141.1 = 328.1 \text{ kNm}$$

$$M_E = 214.5 + 173.4 = 387.9 \text{ kNm}$$

$$V_{BA} = 137.5 + 90.7 = 228.2 \text{ kN}$$

(d) *Redistribution*

Reduce the peak support moment of –328.1 kNm by 15% to give

$$M_B = -278.9 \text{ kNm}$$

The redistributed moment and shears are shown in Figure 9.10.

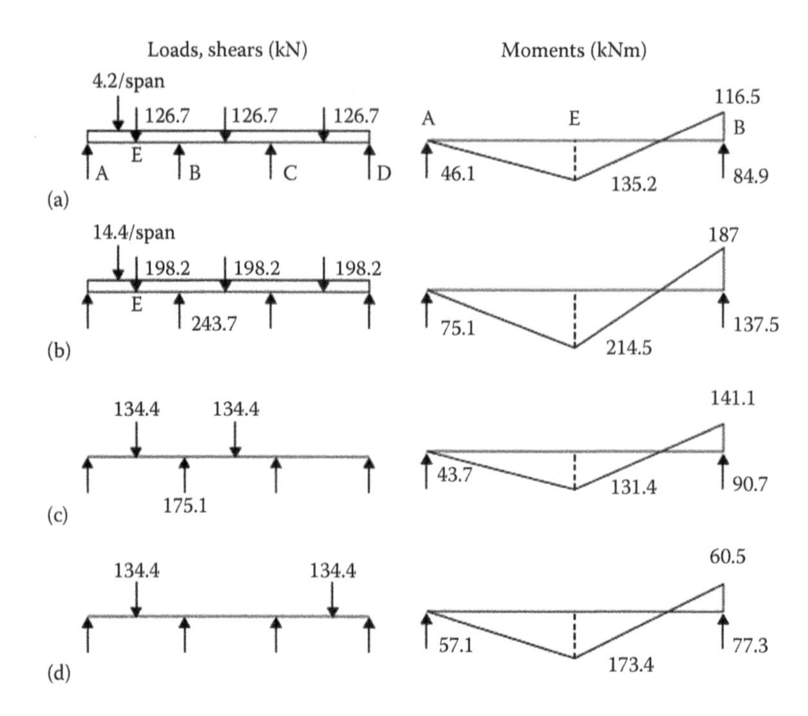

Figure 9.9 Design actions for end span AB: (a) construction loads on steel beam; (b) permanent dead load on composite beam; (c) imposed load on spans AB, BC; (d) imposed load on spans AB, CD.

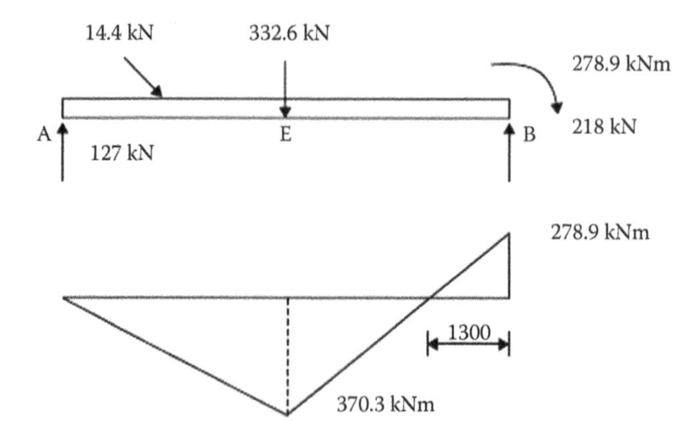

Figure 9.10 Redistributed shears and moments.

Sagging moment = 395.1 kNm

Shear = 211.7 kN

Unsupported length of compression flange = 1.1 m

9.7.4 Section design checks

(a) *Steel beam during construction*

Assume that the length of the bottom flange in compression from support B to the point of contraflexure is 2.5 m.

Try 457 × 152 UB 52, with $S = 1100\ \text{cm}^3$, $r_Y = 3.11$ cm, $x = 43.8$, $\lambda = 80$, $\lambda_{LT} = 66.6$, $\beta_w = 1$, $u = 0.859$, $v = 0.97$, $p_b = 190$ kNm (Table 16):

$$M_c = 190 \times 1100/10^3 = 207\ \text{kNm} > 135.2\ \text{kNm}$$

(b) *Composite beam – sagging moment (Clause 4.4.1)*

The sagging moment capacity of the composite beam at mid-span is based on

- Effective flange breadth B_e;
- The full concrete area including the ribs where the ribs are parallel to the beam;
- The sheeting, concrete in tension and reinforcement in compression is neglected.

The plastic moment capacity is found using (Clause 4.4.2):

- Concrete stress, $0.45f_{cu}$, where the concrete grade $f_{cu} = 30\ \text{N/mm}^2$;
- Design strength of steel $p_y = 275\ \text{N/mm}^2$.

The composite beam section is shown in Figure 9.11.

The trial steel beam is 457 × 152 UB 52, with $A = 66.6\ \text{cm}^2$.

Check location of neutral axis. The concrete flange capacity (Figure 9.7) is given by

$$0.45 \times 30[(1200 \times 95) + (136 + 26)4 \times 55]/10^3 = 1539 + 481 = 2020\ \text{kN}$$

The steel beam capacity is

$$275 \times 66.6/10 = 1829\ \text{kN}$$

Try locating a neutral axis in ribs. Assume that the neutral axis lays y_1 below the top of the rib (Figure 9.11). The depth of rib in compression is

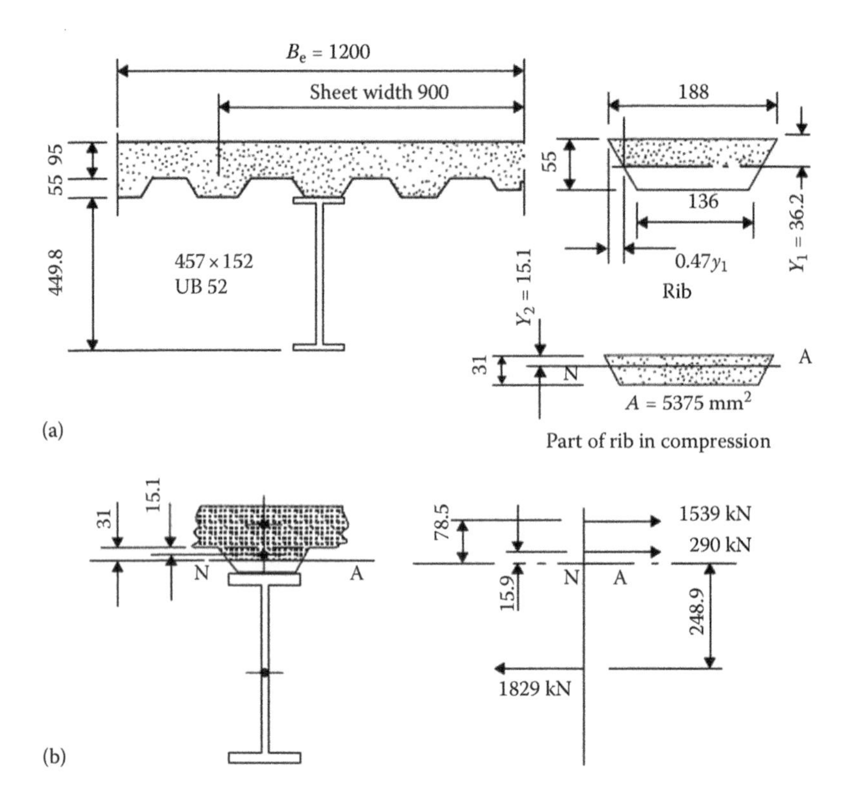

Figure 9.11 Composite beam: (a) sagging section; (b) forces.

$$1539 + 4[0.47y_1^2 + (188 - 0.94y_1)y_1]0.45 \times 30/10^3 = 1829$$

$$y_1 = 31 \text{ mm}$$

The portion of the rib in compression is shown in Figure 9.11a. The area of four ribs is 21,500 mm², with centroid 15.1 mm from top. The capacity is given by

$$0.45 \times 30 \times 21{,}500/10^3 = 290 \text{ kN}$$

The forces and their lever arms with respect to the neutral axis are shown in Figure 9.11b. The moment capacity is

$$M_c = (1539 \times 0.078) + (290 \times 0.016) + (1829 \times 0.249) = 580 \text{ kNm}$$

$$> \text{Sagging moment } 370.3 \text{ kNm}$$

In d and e below, the stability of the bottom flange is considered and the moment capacity is recalculated.

(c) *Composite beam – hogging moment (Clauses 4.4.1 and 4.4.2)*

The capacity is based on:

- Neglecting the concrete in tension and the profiled sheets;
- Including the stresses to design strength p_y and reinforcement in tension at design strength $0.95f_y$, where f_y is its characteristic strength.

The trial steel beam is 457 × 152 UB 52, where $S = 1100\ cm^3$, giving capacity for design strength p_y as

$$M_c = 275 \times 1100/10^3 = 302.5\ kNm$$

$$> \text{Support moment } 278.9\ kNm$$

The stability of the bottom flange requires investigation (Clause 5.2.5). The span is loaded with the factored dead load and the negative moment at the support is taken as M_c, the plastic design moment. This need not be taken as more than the elastic moment without redistribution, that is, 328.1 kNm from Section 9.7.3(c).

The loads, reactions and moments for span AB are shown in Figure 9.12. Solve for distance XB = 2.07 m, the unsupported length of the bottom flange in compression.

For 457 × 152 UB 52, $r_y = 3.11$ cm; $u = 0.859$; $x = 43.8$; $\lambda = 2070/31.1 = 66.6$; $\lambda/x = 1.51$; $v = 0.96$ (Table 19); $\lambda_{LT} = uv\lambda = 54.8$; $p_b = 226.5\ N/mm^2$ (Table 16); $S_x = 1100\ cm^3$.

Capacity of the steel beam = 249 kNm
< 278.9 kNm (redistribution moment)
shown in Figure 9.10

The trial section is 457 × 152 UB 52 with four 16 mm diameter bars, Grade 460 (Figure 9.13). Assume that the neutral axis lays in the web:

$$(0.95 \times 460 \times 804) + (152 \times 10.9 \times 226.5) + (y \times 7.6 \times 226.5)$$
$$= [(428 - y)7.6 \times 226.5] + (152.4 \times 10.9 \times 226.5)$$

$$y = 120.6\ mm$$

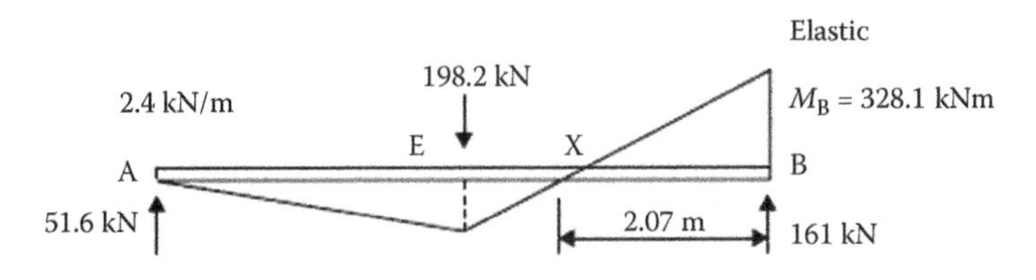

Figure 9.12 End span – dead load and elastic moment.

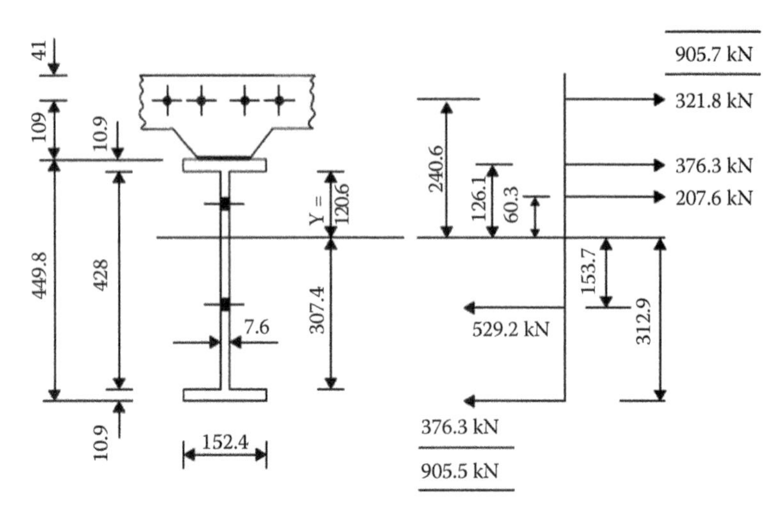

Figure 9.13 Forces in composite section.

The moment capacity is

$$M_c = (351.3 \times 0.241) + 376.3(0.126 + 0.313) + (207.6 \times 0.06) + (529.2 \times 0.154) = 336.8 \text{ kNm}$$

(d) *Joint – beam to column*

The joint arrangement is shown in Figure 9.14. This can be shown to be adequate.

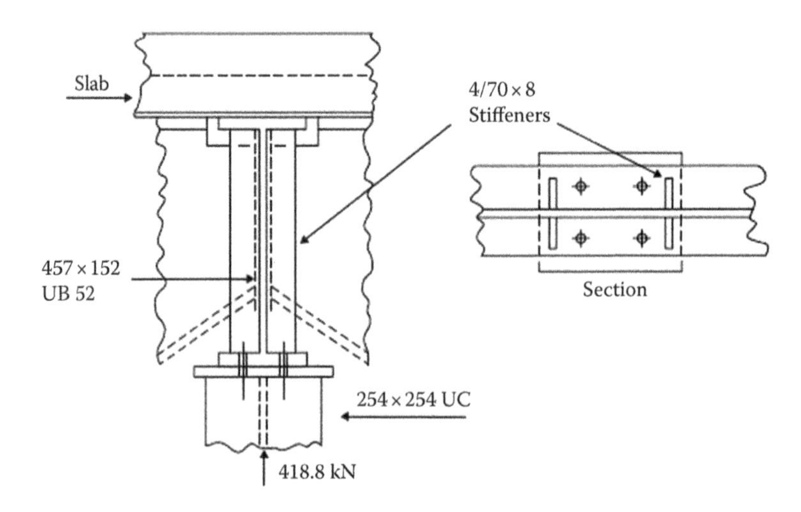

Figure 9.14 Joint–beam to column.

(e) *Beam shear*

From Figure 9.9, the shear is

$$137.5 + 90.7 = 228.2 \text{ kN}$$

The shear is resisted by the web of the steel beam. The shear capacity of 457 × 152 UB 52, with $d/t = 53.6$, is

$$449.8 \times 7.6 \times 0.6 \times 275/10^3 = 564 \text{ kN}$$

This is satisfactory.

9.7.5 Shear connectors

Stud connectors 19 mm diameter × 100 mm nominal height are to be provided. The characteristic load for Grade 30 concrete is $Q_k = 100$ kN per connector in a solid slab (Table 5 of code).

Referring to Clause 5.4.7.3, in slabs with ribs parallel to the beam where (Figure 5 in code)

Mean width of rib $b_r = 149$ mm

Overall depth of sheet $D_p = 55$ mm

$br/D_p = 2.7 > 1.5$ $(k = 1)$

There is no reduction in the value of Q_k.

Referring to Clause 5.4.3, the capacity of shear connectors is

Positive moment $Q_p = 0.8\ Q_k = 80$ kN

Negative moment $Q_n = 0.6\ Q_k = 60$ kN

Referring to Clause 5.4.4.1, from Figure 9.11b the longitudinal force for positive moments is 1539 + 290 = 1829 kN.

Total number of connectors = 1829/80 = 23

To develop the positive moment capacity, that is, the number required for each side of the point of the maximum moment.

Referring to Clause 5.4.4.2, from Figure 9.12 the longitudinal force for negative moment is 328.1 kN.

Number of connectors = 328.1/60 = 6 in span AB

Referring to Clause 5.4.5.1, the total number of connectors between a point of maximum positive moment and each support in span AB is 23 + 23 + 6 = 52. Spacing the studs equally,

Spacing $S = 6000/52 = 115$ mm

Minimum spacing $5d = 95$ mm along beam

9.7.6 Longitudinal shear

The arrangement of the decking for positive and negative moments is shown in Figure 9.15. The mesh reinforcement for the slab is discussed below.

(a) *Positive moment (Clauses 5.6.1 and 5.6.2)*

The sheeting and other transverse reinforcement can act as reinforcement to resist longitudinal shear from the shear connectors.

The longitudinal shear per unit length v at any point is determined by the connector spacing S:

$$v = NQ_p/S$$

where N is the number of connectors per unit length. Account is to be taken of the proportion of the effective breadth lying beyond the failure section in determining the shear at that section.

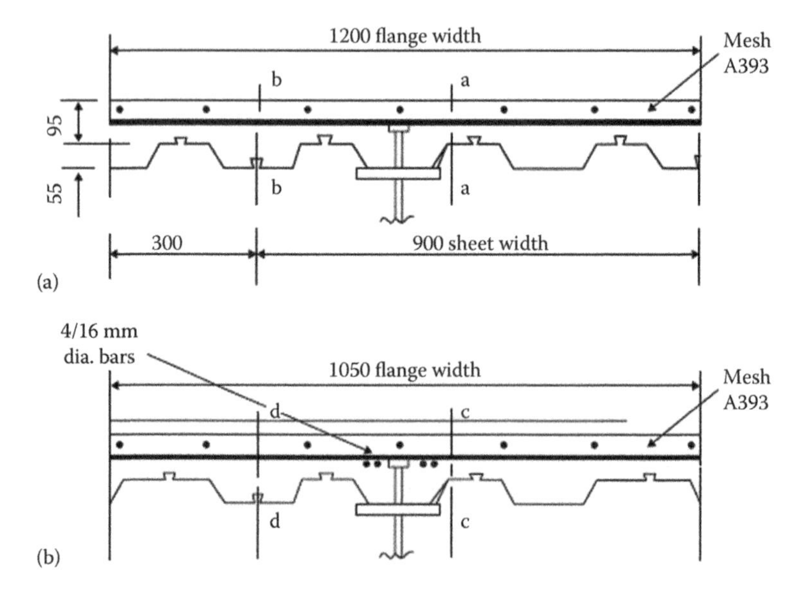

Figure 9.15 Moments: (a) positive; (b) negative.

From Figure 9.15a, the failure sections are aa and bb. For 1 m length:

$N = 1000/115 = 9$ say, $Q_p = 80$ kN

For section a–a:

$v = 9 \times 80 = 720$ kN/m

For section b–b:

$v = 720 \times 300/600 = 360$ kN/m

Referring to Clause 5.6.2, the resistance of the concrete flange for normal weight concrete is

$$u_r = 0.7A_{sv}f_y + 0.03A_{cv}f_{cu} + v_p$$
$$\subseteq 0.8A_{cv}\sqrt{f_{cu}} + v_p$$

where

$f_{cu} = 30$ N/mm^2
$f_y = 460$ N/mm^2
A_{cv} = concrete area per unit length
A_{sv} = area per unit length of reinforcement
v_p = contribution of sheeting = $t_p p_{yp}$
t_p = thickness of sheeting = 1.2 mm
$p_{yp} = 280$ N/mm^2 (PMF Com Floor 70)

Try steel mesh A252 – 8 mm wires at 200 mm centres, with A_{sv} = 252 mm^2/m. This complies with Clause 25 of BS 5950: Part 4. For section a–a, the resistance of two surfaces is

$v_r = 2[(0.7 \times 252 \times 460) + (0.03 \times 95 \times 1000 \times 30) + (1.2 \times 280 \times 1000)]/10^3 = 2(81.1 + 85.5 + 336) = 1005 \text{ kN/m} \not< 2 \times 0.8 \times 95 \times 10^3 \times 30^{0.5}/10^3 = 832.5 \text{ kN/m} > 720$ kN/m + (stud shear force)

For section b–b at a joint in the sheeting,

$v_r = 81.1 + (0.03 \times 150 \times 30) = 216.1 \text{ kN/m} > 180$ kN (half the stud shear force)

Both sections are satisfactory.

(b) *Negative moments* (Figure 9.15b)

For section c–c:

$$v = 9 \times 60 = 540 \text{ kN}$$

For section d–d:

$$v = 270 \text{ kN}$$

This is satisfactory.

9.7.7 Deflection (Clause 6.1.1)

The beam is unpropped. Deflections are based on properties of

- Dead load (self-weight of concrete and steel beam) on the steel beam;
- Imposed load on the composite section.

(a) *Dead load on steel beam*

Referring to Figure 9.16a and Section 6.7.1,

$$M_B = (0.15 \times 76.8 \times 6) + (0.1 \times 3 \times 6) = 70.9 \text{ kNm}$$

The deflection at the centre of the end span is

$$\delta_E = \frac{W_1 L^3}{48EI} + \frac{W_2 L^3}{384EI} - \frac{M_B L^2}{16EI} = 4.3 \text{ mm}$$

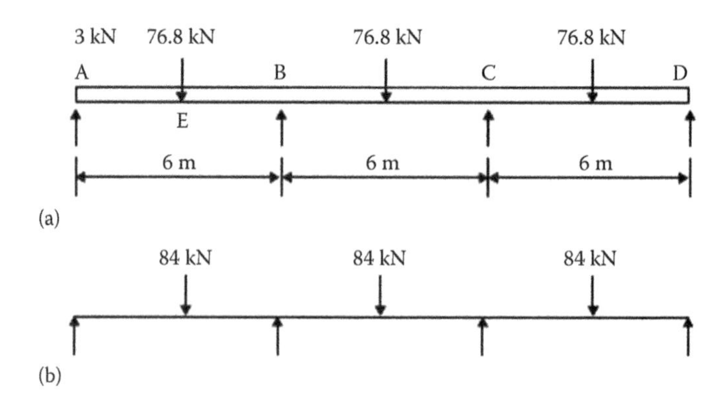

Figure 9.16 Loads: (a) dead–steel beam; (b) imposed–floor.

where, for 457 × 152 UB 52, W_1 = 76.8 kN, W_2 = 3 kN, M_B = 70.9 kNm, L = 6 m, E = 205 kN/mm², I = 21,400 cm⁴.

(b) *Imposed load on composite beam (Clause 6.13.2)*

Referring to Figure 9.16b and Section 9.7.1, to allow for pattern loading, the beam is loaded on all spans with the unfactored imposed load. The support bending moment is reduced by 30% to give

$$M_B = 0.7 \times 84 \times 0.15 \times 6 = 52.9 \text{ kNm}$$

No allowance is required for shakedown. The redistribution made was 15%.

(c) *Transformed section*

Referring to Section 9.7.3, I_G = 58188 cm⁴.

(d) *Deflection calculation (Clause 6.1.3.5)*

Simple beam deflection due to the imposed load

$$\delta_0 = \frac{84 \times 10^3 \times 6000^3}{48 \times 205 \times 10^3 \times 58{,}188 \times 10^4} = 3.17 \text{ mm}$$

The maximum moment in simply supported beam is

$$M_0 = 84 \times 6/4 = 126 \text{ kNm}$$

$$M_1 = M_B = 52.9 \text{ kNm}$$

for a continuous beam with symmetrical point loads,

$$\delta_c = 3.17(1 - 0.6 \times 52.9/126) = 2.37 \text{ mm}$$

This is very small. The beam is larger than required.

Chapter 10

Composite floor system to EC4

10.1 COMPOSITE FLOOR AND COMPOSITE BEAMS

In composite floor construction, concrete is cast on profiled steel sheets, which act as permanent shuttering, supporting the wet concrete, reinforcement and construction loads. After hardening, the concrete and steel sheeting act compositely by stud shear connectors with the steel floor beams. This system gives considerable savings in the weight of floor steel in carrying the loads; see Figure 10.1a–d.

The steel profiled sheets (decking) act as permanent shuttering for carrying the weight of the concrete in unpropped composite floor system and as tension reinforcement to carry the tension in the slab.

Steel profiled decks are fabricated with different forms of groves and slots to provide enough bonds with concrete. See manufacturing companies' publications and safe loads tables to decide the dimensions of the steel decking.

Steel beam is normally a universal I-beam. Lattice girders or castellated beams are more economical than universal beams for long spans. Lattice girder construction also permits services such as air conditioning ducts to be run through the open web spaces.

The stub girder floor is a special development aimed at giving long-span, column-free floor spaces. The system is only economical for long spans of 10–15 m. The construction gives up to 25% savings in the weight of floor steel and a reduction in depth of floor, and provides openings for services. Structurally, the stub girder acts like a modified form of composite Vierendeel girder. The validity of the system has been proved by extensive research and testing.

Steel reinforcement or mesh reinforcement is provided over the whole slab. It is required to resist cracks and moments (Figure 10.2). Alternatively, the concrete may be designed to carry the final loads without composite action when the sheeting acts as shuttering only.

Composite flooring is designed to EC4, BS EN 1994-1-1 2004(E). Decking manufacturer load/span tables can be used to select the slab and

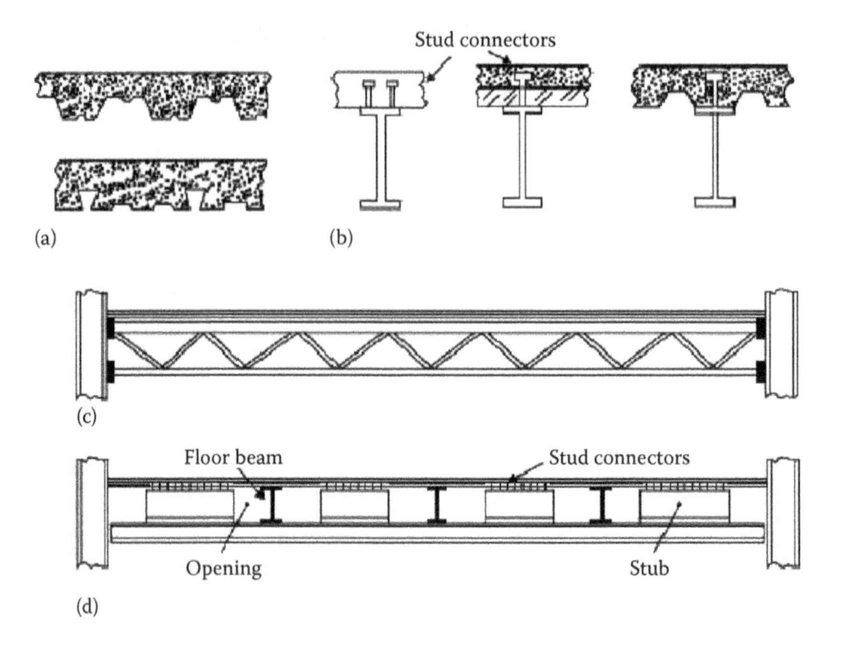

Figure 10.1 Composite floor and beam, ribs parallel to the beam. (a) Composite deck; (b) composite floor construction; (c) lattice girder; (d) stub girder floor.

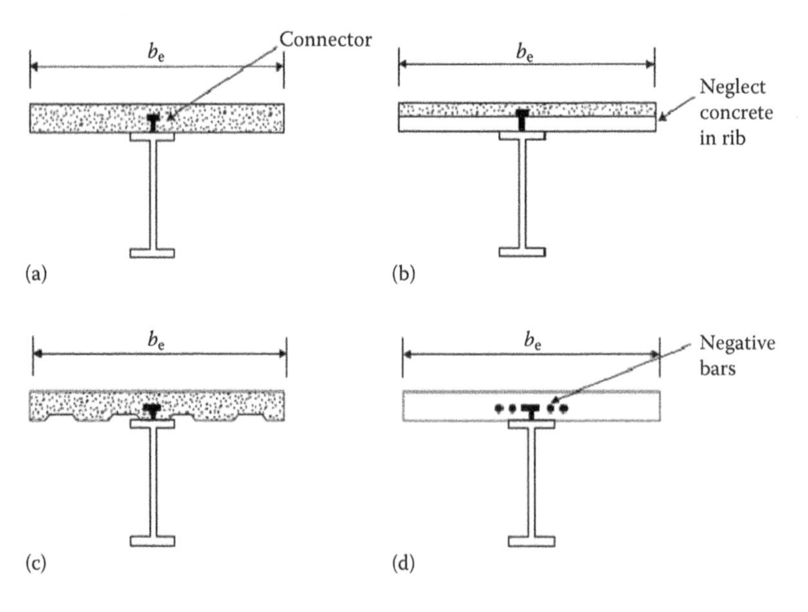

Figure 10.2 Composite beam-width of effective section. (a) Plain slab; (b) sheet perpendicular to beam; (c) sheet parallel to beam; (d) negative moment.

sheeting for a given floor arrangement (for example, the publications of TATA steel UK and SCI, UK).

10.1.1 The method of construction

The two methods of constructions include (1) unpropped construction or (2) propped construction where temporary props are placed under the steel beams to support the total loads from construction. Props construction has the disadvantage of causing extra cost due to the props cost and longer construction time together with causing difficulties in construction due to limited clear space under the flooring system. The props will be removed after the concrete is set to its proper compressive strength and then the composite beam carries the total loads. No such disadvantages exist when the unpropped construction method is used.

10.1.2 Effective width of the concrete flange (BS EN 1994-1-1 2004(E), Clause 5.4.1.2 (3–9))

Referring to Figure 5.1 and Clause 5.4.1.2 of BS EN 1994-1-1 2004(E) and to Figure 10.2, the effective section for calculating moment capacity depends on the *effective width of concrete*. This depends on the direction of the slab span, whether it is perpendicular or parallel to the beam. For example, at mid-span, for a slab spanning perpendicular to the beam and according to the code, the total effective breadth, b_e, is the sum of effective breadths b_e on each side and b_{eff} is given by

$$b_{eff} = b_0 + \Sigma b_{ci}$$

b_0 = distance between centres of shear connectors
b_{ci} = effective width of concrete flange on each side of the steel beam
$b_e = L_e/8 \not> $ half the distance to the adjacent steel beam

where L_e = the beam effective length = the distance between points of zero moment. L_e = the span of a simply supported beam. For a slab spanning perpendicular to the beam (Figure 10.2b), neglect ribs; use only concrete above ribs. For a slab spanning parallel to the beam (Figure 10.2c), use full concrete section.

For example, in a composite concrete floor made from continuous beams with span of steel beams of L_e = 12 m and distributed at 5 m centre to centre, the effective breadth, b_{eff}, of the concrete flange (ignoring b_0) is

$$b_{eff} = 2 \times L_e/8 = 2 \times 12/8 = 3.0 \text{ m} < 2 \times 5/2.$$

10.1.3 General comments on design

1. Refer to the following code of practices for the selection and capacity checks: use EC3 (EN1993-1-1) for steel structures, EC2 (EN1992-1-1) for concrete materials and structures and EC4 (EN1994-1-1) for composite steel and concrete structures.
2. The depth of a universal steel beam. This, at the early stage of design, may be assumed as

   - The (span of the beam)/20 (for simply supported steel beam)
   - The (span of the beam)/24 (for a continuous steel beam)

3. The yield strength, f_y, and the section classification of the cross section of the steel beam should be checked using nominal values of yield strength f_y and ultimate tensile strength f_u for hot rolled structural steel-EN 10025.2, Table 3.1 EC3.
4. During construction stage (for unpropped construction only)
   4.1 The loading on the composite beam is the total of

   - o Self-weight of the steel beam
   - o Steel decking or shuttering
   - o Construction imposed load ≥ 0.75 kN/m^2
   - o The load of the concrete slab (wet concrete)

   4.2 Check the following
   4.2.1 At ultimate limit state: check the capacity of the steel section in bending and shear.
   4.2.2 At the serviceability limit state: check that the deflection of the steel beam is within the acceptable limits.
5. Composite section at the ultimate limit state: bending and shear
   To decide the adequacy of the composite section, check the following:
   5.1 Ultimate moment of resistance of the composite section greater than the acting ultimate design moment.
   5.2 Shear resistance of the steel beam is greater than the acting maximum design shear forces.
   5.3 Adequate design of the shear connectors. At the ultimate limit state and for the composite section made from the steel beam and concrete slab to act as a composite unit, adequate numbers of shear connectors (studs) are required. Studs are normally connected to the surface of the steel beam to resist the horizontal shear force at the interface of the steel beam surface and concrete profiled steel deck. According to EC4, either full or partial shear connection can be used based on the detailing and design requirements for the particular case.
   5.4 Transverse reinforcement normally provided to (i) eliminate cracking of the concrete in close to the studs and to (ii) resist

the longitudinal shear force on the concrete in the flange of the composite section.

5.5 Check for serviceability limit state, that the deflection of the composite beam should be within the limit allowed by EC4 to prevent concrete and finishes from cracking.

10.2 INITIAL SELECTION OF THE STEEL BEAM SIZE

Design of composite beams is to conform to EC4 (BS EN1994-1-1 2004(E)). The composite beam is formed by connecting the concrete slab and the beam using a connector, normally headed studs (see Figure 10.1). The design process and detailed application of the code clauses are shown in the following design case study.

10.2.1 Design case study

Consider the simply supported steel beam in the structure shown in Figure 10.3. The designer decided to use steel decking span perpendicular to the steel beam and is directly resting on the steel beam. Cross section through generic profile is shown in Figure 10.3, unpropped construction throughout.

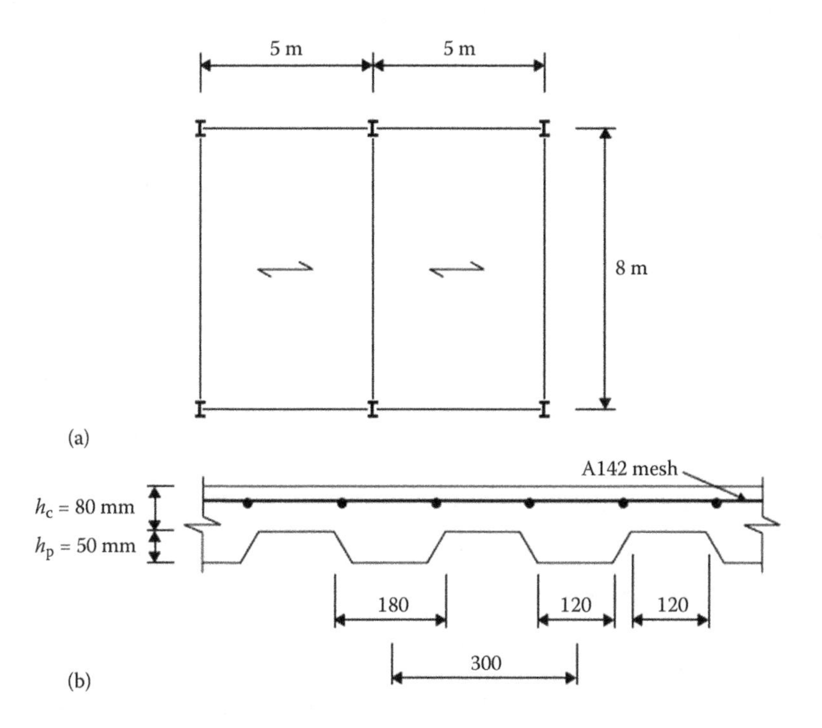

Figure 10.3 (a) Simply supported steel beams; (b) cross-section profile.

10.2.2 Design data

Concrete floor data

Spacing between beams centre to centre = b = 5.0 m
L = 8.0 m = beam span
Concrete floor depth, h = 130 mm

Steel deck data

Concrete depth h_c = 80 mm
Profile depth of deck, h_p = 50 mm; h_c = 80 mm
Diameter of shear connectors r =19 mm
Height of stud after welding, h = 95 mm

Concrete and steel beam data

Steel grade Fe 430, S275, f_y = 275 N/mm^2
Nominal thickness of element, $t \leq 40$ mm (Clause 3.3.2)
Partial factor of safety, γ_{M0} = 1.05
Design strength $f_d = f_y/\gamma_{M0}$ = 275/1.05 = 261.9 N/mm^2
Concrete, use normal-weight concrete with f_{cu} = 30 N/mm^2 ≈ C25
Cylinder strength class (Clause 3.1.2)
Density = 2400 kg/m^3 = 23.55 kN/m^3
Dead load of concrete slab = 130 × 10^3 × 23.55 – (50/0.3)(120 + 30) × 23.55 = 105 × 23.55/1000 = 2.47 kN/m^2

Construction stage loading data

Concrete slab (2.47 kN/m^2), steel profile deck (0.15 kN/m^2), mesh steel reinforcements (0.04 kN/m^2), steel beam (0.3 kN/m^2) = 2.96 kN/m^2
Construction load = 0.75 kN/m^2 (0.5 kN/m^2 in UK NAD)

Composite stage

Concrete slab (2.47 kN/m^2), steel profile deck (0.15 kN/m^2), mesh steel reinforcements (0.04 kN/m^2), steel beam (0.3 kN/m^2) = 2.96 kN/m^2, ceiling and services (0.5 + 0.2 = 0.7 kN/m^2), variable actions (imposed loads) = 3 kN/m^2, occupancy (2.0 kN/m^2-confirm with tables of the code of practices) + partitions 1.0 kN/m^2

10.2.3 Initial selection of beam size

From Table 8 (see reference composite beam design to EC4, SCI publication 121, 1994, tables for composite slabs and beams with steel decking), a 406 × 178 UB 54 grade S275 would be suitable for 4.5 kN/m^2 imposed load > 3 kN/m^2, the variable action used in this case study.

$h = 402.6$ mm	$b = 177.7$ mm	$t_w = 7.7$ mm
$t_f = 10.9$ mm	$d = 360.4$ mm	$c_f/t_f = 6.86$
$c_w/t_w = 46.8$	$A_a = 69.0$ cm^2	$I_{a,y} = 18{,}700$ cm^4
$W_{pl} = 1060$ cm^3		

$f_y = 275$ N/mm^2 ($t_f < 40$ mm), $\varepsilon = \sqrt{(235/f_y)} = 0.92$

$c_f/t_f = 86.86 < 9\varepsilon\ (= 9 \times 0.92 = 8.28)$

$c_w/t_w = 46.8 < 72\varepsilon = 66.4$

Therefore, the beam cross section is class 1, and plastic hinge can be developed.

10.2.4 Construction stage design

Check that the beam moment resistance at the construction stage is satisfactory, i.e.

Steel beam moment resistance > design moment (acting design moment)

$M_{a,pl,Rd} > M_{Ed}$

$M_{a,pl,Rd} = W_{pl,y} \times f_d = 1060 \times 10^3 \times 261.9/10^6 = 277.6$ kNm

$M_{Ed} = WL/8$

$W = (2.96 \times 8 \times 5 \times 1.35) + (0.75 \times 8 \times 5 \times 1.5) = 204.84$ kN

$M_{Ed} = 204.84 \times 8/8 = 204.84$ kNm < 277.6 kNm (satisfactory)

10.3 PLASTIC ANALYSIS OF COMPOSITE SECTION

Location of natural axis. The three possible locations for the neutral axis of the composite section are shown in Figure 10.4a–c.

(a) *Moment resistance of the composite section,* $M_{pl,Rd}$, Figure 10.4, case a.

Natural axis in the concrete flange.

In case a, where the neutral axis is located in the concrete flange, see Figure 10.4a.

This happens when

$R_c > R_s$

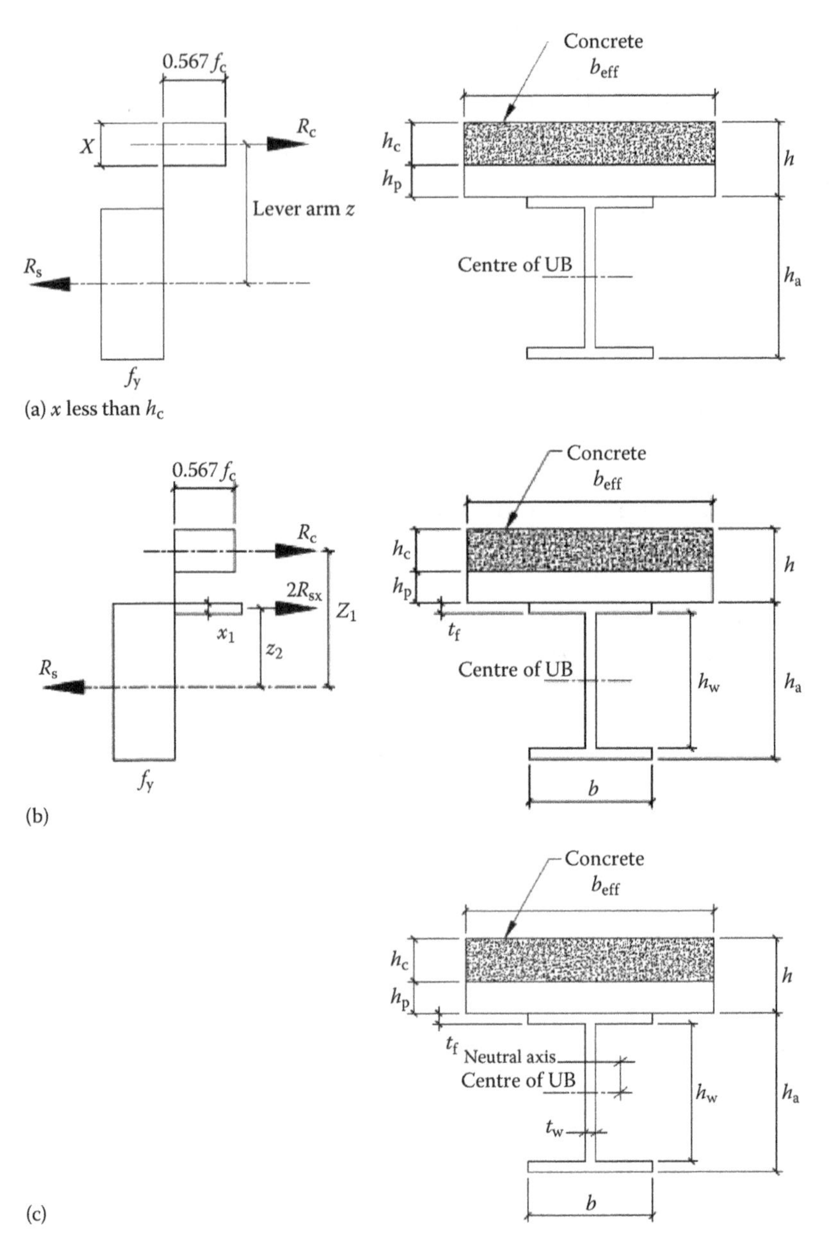

Figure 10.4 (a) Neutral axis in concrete flange; (b) neutral axis in the flange of the steel beam; (c) neutral axis in the web of the steel beam.

R_c = resistance of concrete flange, where

$R_c = 0.85 f_{ck} b_{eff} h_c / \gamma_c$, $\gamma_c = 1.5$, h_c = depth of concrete
R_s = resistance of steel cross section, where

$R_s = f_d A_a$

f_d = steel strength = f_y/γ_{M0}
A_a = cross-sectional area of the steel beam

and at equilibrium x (the depth of neutral axis) and $M_{pl,Rd}$ are given by

$$x = \frac{R_s}{0.85 f_{ck} b_{eff} / \gamma_c} = \text{depth to neutral axis}$$

$$M_{pl,Rd} = R_s \left[\frac{h_a}{2} + h_c + h_p - \frac{R_s h_c}{2R_c} \right]$$

Hint: z (lever arm) = $(h_a/2 + h_p + h_c - x/2)$

$M_{pl,Rd}$ = moment resistance of the composite section

(b) *Moment resistance of the composite section,* $M_{pl,Rd}$, case b. The neutral axis in the flange of the steel beam $x > (h_c + h_p) < (h_c + h_p + t_f)$

$$x_1 = \frac{(R_s - R_c)}{2f_y b} = \frac{(R_s - R_c)}{2f_y b} t_f$$

$$Z_c = (h_a + h + h_p)/2$$

$$Z_s = (h_a - x_1)/2$$

Resistance of

concrete flange, $R_c = 0.85 f_{ck/\gamma c}(b_{eff} h_c)$
steel section, $R_s = f_d A_a$
steel flange, $R_{sf} = f_d b t_f$

$$M_{pl,Rd} = \frac{R_s h_a}{2} + \frac{R_c(h + h_p)}{2} - \frac{(R_s - R_c)^2 t_f}{4R_{sf}}$$

$$M_{pl,Rd} = R_s h_a/2 + R_c(h_c + 2h_p)/2 - (R_s - R_c)^2 t_f/(4 f_y b t_f/\gamma_{M0})$$

(c) *Moment resistance of the composite section,* $M_{pl,Rd}$, case c. The neutral axis in the web of the steel beam.

$$M_{pl,Rd} = M_s + R_{cf}(h_a + h_c + 2h_p)/2 - R_{cf}^2 h_w/[4A_v f_y/(\gamma_s\sqrt{3})$$

10.3.1 Composite stage design, M_{Ed}

Permanent actions (dead loads) = 2.96 kN/m^2 + ceiling and services (0.5 + 0.2 = 0.7 kN/m^2) = 3.66 kN/m^2, variable actions (imposed loads) = 3 kN/m^2 (occupancy [2.0 kN/m^2-confirm with tables of the code of practices] + partitions 1.0 kN/m^2)

$$F_{Ed} = [1.35 \times 3.66 + 1.5 \times 3] \times 5 \times 8 = 377.64 \text{ kN}$$

$$M_{Ed} = 377.64 \times 8/8 = 377.64 \text{ kNm}$$

10.3.2 Compression resistance of concrete slab, R_c

$$R_c = 0.85 f_{ck/\gamma c}(b_{eff} h_c) \text{ or } 0.45 \times f_{cu} \times b_{eff} \times h_c$$
$$= 0.45 \times 30 \times 2000 \times 80/1000 = 2160 \text{ kN}$$

10.3.3 Compression resistance of steel section, R_s

$$R_s = (f_y/\gamma_s)\, A_a = 275/(1.05) \times 69.0 \times 100/1000 = 1807.14 \text{ kN}$$

$$R_s < R_c$$

10.3.4 Moment resistance of the composite beam, $M_{pl,Rd}$

$$M_{pl,Rd} = R_s\left[\frac{h_a}{2} + h_c + h_p - \frac{R_s h_c}{2R_c}\right] = 1807.14[402.6/2$$

$$+ 80 + 50 - 1807.14 \times 80/(2 \times 2160)] = 538.23 \text{ kNm}$$

$$M_{pl,Rd} > M_{Ed} \text{ (377.64 kNm)}$$

10.3.5 Location of neutral axis

$$b_{eff} = 2L_e/8 < 2\ (5/2)$$

$$b_{eff} = 2 \times 8/8 = 2\ \text{m}$$

$$x = \frac{R_s}{0.567 f_{ck} b_{eff}} = \frac{275 \times 6900/1.05}{0.567 \times 25 \times 2000} = 63.74\ \text{mm}$$

$x < 80$ mm, the neutral axis in the concrete flange, therefore case a.

10.4 THE SHEAR RESISTANCE OF THE COMPOSITE SECTION

For the composite section and as for the construction stage, the shear is resisted by the shear area A_v of the steel beam, and the shear resistance V_{Rd} is given by

$$V_{pl,Rd} = f_y A_v/(\gamma_a \sqrt{3})$$

$$A_v = d_w \times t_w$$

where high moments and shear forces occur at the same location of a steel beam, in which $V_{Ed} > 0.5V_{pl,Rd}$, the code recommended to use a reduced moment capacity for the composite section. This is normally achieved by reducing the bending stress in the web of the steel beam; see EC4, Section 6.2.2.4.

For the example above, pure shear (vertical shear) can be calculated as follows:

$$V_{pl,Rd} = f_y A_v/(\gamma_a \sqrt{3}) = 275 \times 3100.02/(\sqrt{3} \times 1.05) = 468.76\ \text{kN}$$

$$A_v = h_w \times t_w = 402.6 \times 7.7 = 3100.02\ \text{mm}^2$$

$$0.5V_{pl,Rd} = 234.38\ \text{kN} > V_{sd} = 233.82\ \text{kN}$$

Therefore, in this example, where the load is a UDL, the moment resistance of the section is not influenced by the shear force.

Shear connection—Shear connector resistance cl. 6.6 BS EN 1994-1-1-2004(E) (Clause 6.3.2.1 EC4)

The shear connectors are studs with heads welded to the surface of the steel beam. They are used to make sure that the concrete slab and steel beam act as a full composite unit and stop

1. Relative movement (slippage) between the surface of steel beam and the concrete. Slippage occurs due to the horizontal shear force at the interface between the steel beam surface and concrete slab.
2. Vertical lifting of the concrete slab away from the surface of the steel beam.

Note: Slippage commonly occurs where the rate of change of the shear and moment d_M/dx is maximum. For example, in a composite beam carrying a UDL, relative movement is maximum at the supports of the beam and zero at mid-span as the shear force is zero and the moment acting on the beam is maximum.

Resistance of a headed stud automatically welded. The lesser value of the following:

P_{Rd} = failure resistance (of concrete), where

$$P_{Rd} = 0.29\alpha d^2\sqrt{(f_{ck}E_{cm})}/\gamma_v$$

or

P_{Rd} = shear resistance at failure of stud at the location of weld collar, where

$$P_{Rd} = 0.8f_u\pi d^2/(4\gamma_v)$$

where

$\alpha = 0.2\alpha(h_{sc}/d + 1) < 1.0$, where

α is a factor taking into account the height of stud
d = stud diameter, $16 \leq d \leq 25$ mm
h_{sc} = stud height
f_u = ultimate tensile strength of steel beam ≤ 500 N/mm^2
E_{cm} = secant modulus of elasticity of the concrete in kN/mm^2

E_{cm} values for the following concrete strength classes are given below

	Concrete strength class			
	C25/30	*C30/37*	*C35/45*	*C40/50*
E_{cm} (N/mm^2)	31	33	34	35

$\gamma_v = 1.25 = 1/0.8$ (factor used to modify the basic resistance of shear connectors, BS 5950, Part 3).

P_{Rd} may be reduced as a result of the shape of the deck as follows:

For steel decking crosses the beams,

$$k_t = \frac{0.7}{\sqrt{n_r}} \frac{b_0}{h_p} \left(\frac{h_{sc}}{h_p} - 1 \right)$$

For ribs parallel to the supporting beam

$$k_1 = 0.6 \frac{b_0}{b_p} \left(\frac{h_{sc}}{h_p} - 1 \right) \le 1.0$$

n_r = the number of stud connectors in one rib ≤ 3
h_{sc} = height of the stud
h_p = the overall depth of the profiled steel sheeting
b_0 = the mean width of the concrete rib; see Figure 6.13 (BS EN 1994-1-1 2004(E))

Partial shear connection, Clause 6.6.1.2 BS EN 1994-1-1 2004(E)

The designer can use smaller numbers of shear studs, when the design moment is less than the moment resistance of the composite section. This provides simple details of the layout of the studs. In this case, the degree of shear connection is defined as

$$\eta = N/N_f = R_q/R_s$$

N = number of shear studs provided over the beam full length
N_f = required full number of shear studs over the full length of the beam
$R_q = NP_{Rd}$
$R_s = f_d A_a$

Degree of shear connection η: limits to BS EN 1994-1-1 2004(E)

1. If an equal flanges steel beam satisfies the following conditions
   - 16 mm ≤ d ≤ 25 mm, where d = the nominal diameter of the shank of the headed stud
   - The overall length of the stud after welding is ≥4d
   - $L_e \le 25$ m

   Then

$$\eta \ge 1 - (355/f_y)(0.75 - 0.03L_e) \qquad \eta \ge 0.4$$

for $L_e > 25$ m, $n \ge 1.0$

Composite section: Moment resistance for partial shear connection

$$M_{Rd} = M_{a,pl,Rd} + \eta(M_{pl,Rd} - M_{a,pl,Rd})$$

$M_{a,pl,Rd} = W_{pl} \times f_d$ = composite section moment resistance with full shear connection

$M_{a,pl,Rd}$ = steel section moment resistance

10.5 CASE STUDY – SHEAR CONNECTORS RESISTANCE

10.5.1 Partial shear connection

For the data given in the design case study of Section 10.2.1, design the number of the studs and accordingly check the adequacy of the composite section in terms of its moment resistance.

The resistance of headed stud shear connectors is the smaller of

$$P_{Rd} = 0.29\alpha d^2\sqrt{(f_{ck}E_c)}/\gamma_v \text{ (resistance of concrete at failure)}$$

or

$$R_d = 0.8 f_u \pi d^2/(4\gamma_v) \text{ (shear resistance of the stud at failure, at its weld collar)}$$

where

- α = a factor of safety takes into account the height of the stud = $0.2(h_{sc}/d + 1) < 1.0$
- d = stud diameter
- h_{sc} = the stud overall height
- f_u = 450 N/ultimate tensile strength ≤ 500 N/mm²
- E_{cm} = secant concrete modulus of electricity in kN/mm²
- $\alpha = 0.2(h_{sc}/d + 1) < 1.0$, $\alpha = 0.2 \times (95/19 + 1) = 1.2$
- $\therefore \alpha = 1.0$
- $\gamma_v = 1.25 = 1/0.8$ (factor used to modify the basic resistance of shear connectors, BS 5950, Part 3)

$$P_{Rd} = 0.29\alpha d^2 \sqrt{(f_{ck}E_{cm})}/\gamma_v$$

$$P_{Rd} = 0.29 \times 1 \times 19^2 \times (\sqrt{(25 \times 31/1000)})/1.25 = 73 \text{ kN}$$

or

$P_{Rd} = 0.8 f_u \pi d^2/(4\gamma_v)$

$P_{Rd} = 0.8 \times 450 \times (\pi \times 19^2)/(4 \times 1.25) = 81.7$ kN

$\therefore P_{Rd} = 73$ kN (Clause 6.6.3.2, BS EN 1994-1-1 2004(E))

P_{Rd} may be reduced as a result of the shape of the deck:

$h_{sc} = 95$ mm < (h_p + 7 mm)

For one stud per trough, $n_r = 1$

$k_t = [0.7/(\surd\eta_r)][b_o/h_p][(h_{sc}/h_p) - 1] \leq 1.0$ (rib transverse on supporting steel beam, Clause 6.6.6.2)

b_o = average trough width = (180 + 120)/2 = 150 mm

$k_t = [0.7/(\surd\eta_r)][b_o/h_p][(h_{sc}/h_p)]$,

$$k_t = \text{reduction factor} = \frac{0.7}{1} \times (150/50)[(95/50) - 1] = 1.89 > 1.0$$

$k_t = 1$, **no reduction and $P_{Rd} = 73$ kN**

η_r = number of studs per trough < 3 (Clause 6.6.4.2)

For two studs per trough, $\eta_r = 2$

$k_t = [0.7/(\surd\eta_r)][(b_o/h_p)][(s_{ch}/h_p) - 1] < 0.8$ (for $\eta_r = 2$)
$= 1.34 > 0.8$

$k_t = 0.8$ (Table 6.2, Clause 6.6.4.2)

$P_{Rd} = 0.8 \times 73 = 58.4$ kN

Therefore, the number of connectors on each side of the trough of the beam is given by

$= f_d \times 69.0 \times 100 \times 10^{-3}/P_{Rd}$
$= R_s/P_{Rd} = 261.9 \times 69.0 \times 100 \times 10^{-3}/73$
$= 24.75$, say 24.

Spacing in pairs = 4000 × 2/24 = 333.330 mm (say 300 mm). Spacing of studs is shown in Figure 10.5.

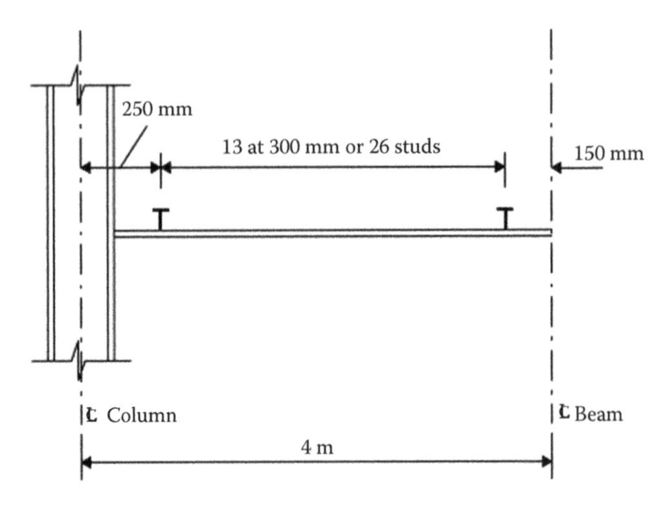

Figure 10.5 Spacing of shear studs.

10.5.2 Longitudinal shear force transfer, R_q

R_q, one stud = 13 × 73 = 949 kN

R_q, two studs = 13 × 2 × 58.4 = 1518.4 kN

10.5.3 Degree of shear connection, N/N_f (Clause 6.2.13): For $R_s < R_c$

N/N_f (one stud) > 0.4

$N/N_f = R_q/R_s = 949/1807.14 = 0.525 > 0.4$ (satisfactory)

10.5.4 Composite steel section with partial shear connection: Moment resistance

$M_{Rd} = M_{a,pl,Rd} + N(M_{pl,Rd} - M_{a,pl,Rd})/N_f$

$M_{pl,Rd} = W_{pl} \times f_d$ = moment resistance for composite section with full shear connection

$M_{a,pl,Rd}$ = moment resistance for steel section only

$M_{Rd} = 277.6 + 0.525(538.23 - 277.6)$
$= 414.430$ kNm < 377.64 kNm (satisfactory)

Hint: when the design moment > M_{Rd}, then before changing the beam size, try one of the following two options. Imagine $M_{Ed} = 450$ kNm, either increase the numbers of studs, say 18 studs using spacing, say 200 mm,

$N/N_f = R_q/R_s = 18 \times 73/1807.14 = 0.73 > 0.4$ (satisfactory)

$M_{Rd} = 277.6 + 0.73(538.23 - 277.6)$
$= 467.90 > 450$ kNm

or use 2 studs per trough

$n = N/N_f = R_q/R_s \geq (1 - (355/f_y)(0.25 + 0.03L)$ for $8 \text{ m} \leq L \leq 25 \text{ m}$
$= (1518.4/1807.14 = 0.84 > 1 - (355/f_y(275))\ (0.75 - 0.03L_e)$

$M_{Rd} = 277.6 + 0.84(538.23 - 277.6)$
$= 496.53$ kNm > 450 kNm (satisfactory)

Therefore, use 2 studs per trough.

10.6 CHECKS FOR SERVICEABILITY LIMIT STATE

At construction stage, the load is supported by the steel beam, while the concrete is still wet (not fully hardened).

10.6.1 Deflection for non-composite stage

$\delta_{const} = 5WL^3/(384E_aI_a)$,

where

W = total construction permanent actions
L = the beam length
E_a = steel elastic modulus
I_a = second moment of area of the section of the steel beam

For the above design case study, the total deflection of the composite steel beam

F = design load = $(2.96 + 0.75) \times 5 \times 8 = 148.4$ kN

δ (*dead + imp*) = $5FL^3/(384E_aI_{a,y})$
= $5 \times 148.4 \times 10^3 \times 8000^3/(384 \times 210 \times 10^3 \times 18{,}700 \times 10^4)$
= 25.2 mm < $L/200$

δ (dead) = 25.2 × (2.96 × 5 × 8/148.4) = 20.1 mm

10.6.2 Deflection for composite, at service

Calculate the modular ration to transfer the composite section into an equivalent steel section

$$n = E_a/(0.5E_{cm})$$

E_{cm} = modulus of elasticity of the concrete in kN/mm²

For the transformed section (Figure 10.6), the position of the centroid of the section is given by

$$\bar{x} = \frac{A_a n(h_a + h + h_p)}{2[A_a n + b_{eff}(h - h_p)]}$$

$$I_{com} = A_a(h + 2h_p + h_c)^2/[4(1 + nr)] + b_{eff} \times h^3 c/12n + I_{a,y}$$

$$r = A_a/(b_{eff} \times h_c)$$

For a class C25/30 concrete, E_{cm} = 31 kN/mm².

Variable actions (imposed loads) = 3 × 5 × 8 = 120 kN

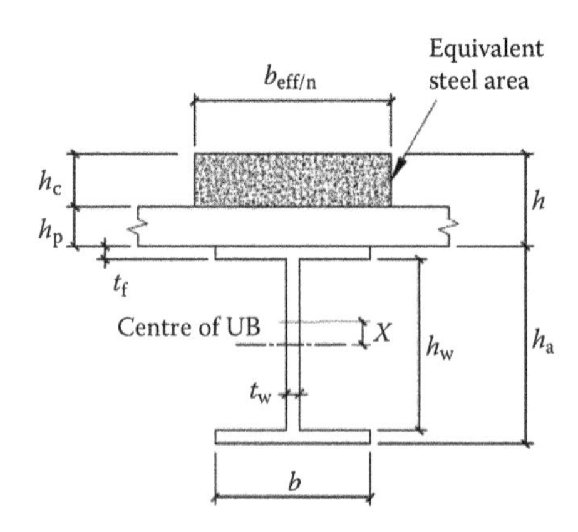

Figure 10.6 Transformed composite section.

I_c [composite section – elastic properties (uncraked section)]

$$I_{com} = A_a(h + 2h_p + h_c)^2/[4(1 + nr)] + b_{eff} \times h^3c/12n + I_{a,y}$$

$$r = Aa/(b_{eff} \times h_c) = 6900/(2000 \times 80) = 0.043$$
$$= \text{modular ratio} = n = E_a/(0.5E_{cm}) = 13.54 \text{ for normal concrete weight}$$

$$I_{com} = 6900(402.6 + 2 \times 50 + 80)^2/[(4(1 + 13.54 \times 0.043)] + 2000 \times 80^3/(12 \times 13.54) + 18{,}700 \times 10^4 = 5.63 \times 10^8 \text{ mm}^4$$

$$\delta_{com} \text{ (deflection with full shear)} = 5FL^3/(384 \times E_a \times I_c) = 5 \times (2.96 + 3) \times 5 \times 8 \times 10^3 \times 8000^3/(384 \times 210 \times 10^3 \times 5.63 \times 10^8) = 12.83 \text{ mm}$$

In this design case study, partial shear connection exists; therefore, the effect of slip has to be taken into account as below:

$$\delta/\delta_{comp} = 1 + 0.3[1 - N/N_f][(\delta_a/\delta_c) - 1]$$

$$\delta_a = 12.83 \times 5.63 \times 10^8/(18{,}700 \times 10^4) = 38.62 \text{ mm}$$

$$\delta = 12.83\{1 + 0.3(1 - 0.73)[(38.62/12.83) - 1]\}$$
$$= 14.9 \text{ mm} = L/536.6 < L/250 \text{ (satisfactory)}$$

10.6.3 Composite beam total deflection

= δ (deflection during construction stage) + δ (deflection due to variable actions) + δ (deflection due to ceiling and services)
= $20.1 + 12.83 + 0.7 \times 12.83/(2.96 + 3)$
= 34.43 mm = $L/232.356 < L/200$ (British practice)

10.7 CHECK TRANSVERSE REINFORCEMENT

Check resistance of the flange to splitting (Clause 6.6.2); use A142 mesh reinforcement in concrete slab.

$$v_{Rd} = 2.5A_{cv}\eta\tau_{Rd} + A_e f_{sk}/\gamma_s \le 0.2A_{cv}\eta f_{ck}/\gamma_c$$

Neglect contribution of decking

$$A_e = 142 \text{ mm}^2/\text{m}$$

$\eta = 1$ (normal weight concrete)

$\gamma_s = 1.15$

$\gamma_c = 1.5$

$f_{sk} = 460\ \text{N/mm}^2$

$\tau_{Rd} = 0.25 \times f_{ctk0,05}/\gamma_c = 0.25 \times 1.8/1.5 = 0.3$

$$\begin{aligned} v_{Rd} &= 2.5A_{cv}\eta\tau_{Rd} + A_e f_{sk}/\gamma s \leq 0.2A_{cv}\eta f_{ck}/\gamma_c \\ &= [2.5\times105\times10^3\times1\times0.3+(142\times460/1.15)]/1000 \\ &= 135.6\ \text{kN/m} < 0.2\times105\times10^3\times1\times25/(1.5\times1000) = 350\ \text{kN/m} \end{aligned}$$

therefore satisfactory.

10.8 CHECK SHEAR PER UNIT LENGTH, *v*

Using two shear connectors per trough,

$v = 58.4 \times 1 \times 0.5/0.3 = 97.33\ \text{kN/m} < 135.6\ \text{kN/m}$

Therefore, A142 mesh is satisfactory.

10.9 CHECK VIBRATION

Check if natural frequency $\cong 18/\sqrt{\delta_a} > 4$ Hz (building with no vibration machinery)

$\delta_a = 5FL^3/(384E_aI_{c1})$

$I_{c1} = I_{comp} + 0.1I_{comp} = 5.63 \times 10^8 \times 1.1 = 6.193 \times 10^8\ \text{mm}^4$

F = loading = 2.96 + 0.7 + 10%(imposed load × 3 = 3.96 kN/m²)

$F = 3.96 \times 5 \times 8 = 158.4\ \text{kN}$

$$\begin{aligned} \delta_a &= 5FL^3/(384E_aI_{c1}) \\ &= 5 \times 158.4 \times 10^3 \times 8000^3/(384 \times 210 \times 10^3 \times 6.193 \times 10^8) \\ &= 8.119\ \text{mm} \end{aligned}$$

Natural frequency $\cong 18/\sqrt{\delta_a} = 18/\sqrt{8.119} = 6.316 > 4$ Hz

Therefore, the composite beam is satisfactory against vibration.

Comparison between the design according to BS 5950-2000 and EC4 [BS EN 1994-1-1 2004(E)].

	BS 5950-2000	*EC4*
Beam	457 × 152 UB 52	406 × 178 UB 54
Concrete slab	180 mm	130 mm on re-entrant deck
Slab reinforcement	10T180	A142 mesh
No. of studs	2	2

Chapter 11

Tall buildings

11.1 GENERAL CONSIDERATIONS

The United States has always been the leader and chief innovator in the construction of tall buildings; a number of structures upward of 400 m with 100 stories have been built. Many tall buildings have arisen in recent times in Singapore, Hong Kong, Europe, UAE and Australia.

Three modern advances have contributed to making tall building architecture safe and such an outstanding success:

- The design of efficient lateral load-resisting systems which are an essential component of all such structures to resist wind and seismic loads, reduce sway and damp vibration – many ingenious systems have been devised;
- The power of modern methods of computer analysis including modelling the structure for static and dynamic analysis, coupled with model testing in wind tunnels and on shaking tables so that behaviour can be accurately predicted;
- The development of rapid construction methods in concreting, prefabrication techniques, drainage provision, etc.

Tall buildings are mainly constructed in city centres where land is in short supply; high population density coupled with high land prices and rents make their adoption economical. In provision of housing, one tall building can replace a large area of low-rise buildings, which can make way for other developments such as community centres, sports centres or open parkland. Tall buildings are used for offices, banks, hotels, flats, schools, hospitals, department stores, etc., and often for combined use, for example, offices/apartments.

Architects and engineers planning a tall building need to consider the following general constraints on design.

- Building regulations and planning laws for the city concerned – sometimes the maximum building height is limited.

- Intended occupancy – this governs the floor loading and influences the structural arrangement adopted. For example, in multistorey flats and hotels the division floor space can be the same on each floor and vertical loadbearing walls throughout the height of the building can be introduced. Masonry or concrete-framed structures are most often but not always adopted in these cases. In office buildings, banks, stores, etc. the core/perimeter wall building, where space division if required can be made by lightweight demountable partitions, is the most suitable solution.
- The transport of people is primarily vertical, requiring either a central core or shear wall/core areas at appropriate locations where lifts/stairs/services are provided. The design of tall buildings was only possible following the invention of the electric lift.
- Fire protection of the structural frame in steel buildings is mandatory, as is the provision of separate fireproof compartments for lifts and stairs. Limits are set on compartment sizes to prevent the spread of fire. All buildings must be fitted with sprinklers and be capable of easy and speedy evacuation. The design must comply with all relevant regulations.
- Heating or air conditioning is essential, depending on location. This requires space between floor slabs and suspended ceilings and in curtain walls and cores to accommodate ducts and pipes.
- Provision of services (lighting, power, telephone, television, computer networks, water, waste disposal) forms an important part of design and must be considered at the planning stage. Services can be incorporated in prefabricated wall and floor units during manufacture.

11.2 STRUCTURAL DESIGN CONSIDERATIONS

In the structural engineering sense, the multistorey building may be defined as tall when the horizontal loading due to wind or seismic effects becomes the most important consideration in design. This is particularly the case with modern buildings clad with lightweight curtain walling and using lightweight partitioning and fire protection.

The frame must be stiff enough to limit deflection to 1/300 of the height of each storey to prevent sway causing anxiety to the occupants. This limitation is of prime importance in very tall buildings and has led to the development of special structural forms such as the tube type of building described below. Multistorey rigid frame construction alone without shear walls/core is not suitable for very tall buildings because of excessive deflection.

Buildings can be entirely in steel framing with appropriate bracing or construction system to provide lateral load resistance. More often, buildings are a composite of steel framing and concrete shear walls or cores where one function of the concrete elements is to carry lateral loads. Concrete has the additional advantage of fire resistance. Steel requires protection with casing, intumescent coating or spray-on treatment.

The steel-framed building with composite steel deck/flooring, prefabricated cladding panels, lightweight demountable internal partitions, suspended ceilings and lightweight fire protection is ideally suited to the use of industrialized building techniques. The advantages of this type of construction are accuracy of shop fabrication of units and speed of erection with maximum site labour and few specialized skills.

The foundations can be expensive depending on site conditions because heavy loads are delivered onto small areas. Cellular rafts or multistorey basement foundations are commonly used where the space under the building provides car parking. The foundations may bear directly on the soil or be supported on piles or on caissons under thick cap slabs.

Very often the erection of the structure has to be carried out on a restricted site. This influences the design and limits the size of components to be fabricated. Very large transfer girders are needed in some designs where the plan and column arrangement change in the building height and special erection provisions are needed. Low- to medium-rise buildings can be erected with independent tower cranes located around the plan area. For tall buildings, erection must make use of the structure itself. In core buildings, the concrete core can be constructed first by slip forming and then used to erect the steelwork. Alternatively, climbing cranes can erect the steelwork on itself and the steel can progress ahead of the concrete shear walls or cores.

It is difficult to compare costs of different systems. Frame costs can be compared, but all factors, including foundations, flooring, cladding, partitions, fire protection, services, operating and maintenance costs should be included. The steel frame cost may not exceed 25% of the total building cost. All systems set out below continue to be used. Some preliminary comparative designs are given.

11.3 STRUCTURAL SYSTEMS

The main structural systems used for tall buildings are discussed below. The classification of the various types is based primarily on the method adopted to resist horizontal loading. A second classification is concerned with the method of construction used. Combinations of various types can be adopted. Both steel-framed only and composite structures are described (Hart et al., 1978; Orton, 1988; Taranath, 1988).

11.3.1 All-steel braced structure

In its simplest form, bracing forms a vertical cantilever which resists horizontal load. The simple method of design involving only manual analysis can be

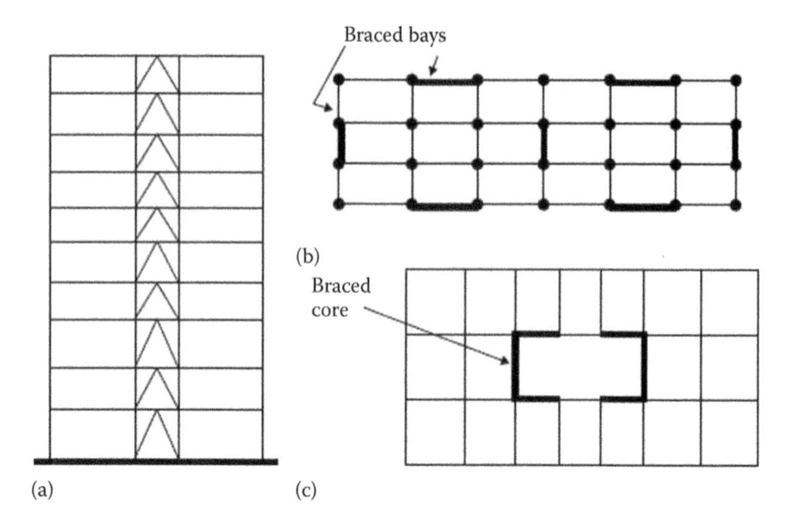

Figure 11.1 All-steel braced structures: (a) vertical bracing; (b) bracing on perimeter/interior walls; (c) bracing around core.

used for the whole structure for buildings braced with one or more cantilever trusses (Figure 11.1). The braced bays can be grouped around a central core, distributed around the perimeter of the building or staggered through various elevations. The floors act as horizontal diaphragms to transmit load to the braced bays. Bracing must be provided in two directions and all connections are taken as pinned. The bracing should be arranged to be symmetrical with respect to the building plan; otherwise, twisting will occur.

11.3.2 Rigid frame and mixed systems

(a) *Rigid frame structures* (Figure 11.2a)

In rigid frame structures, the horizontal load is resisted by bending in the beams and columns. The columns, particularly in the lower stories, must resist heavy moments so sections will be much larger than in braced buildings.

The frame normally has H-section columns. It is rigid in one direction only, across the short span, and is braced longitudinally. The connections are expensive, being welded or made with haunched beam ends and high-strength bolts. Frames rigid in both directions with box section columns have been constructed in areas subject to seismic loads.

The rigid frame structure deflects more than a braced structure. The deflection is made up of sway in each storey plus overall cantilever action. Due to excessive deflection, rigid frames are suitable only for low- or medium-rise buildings.

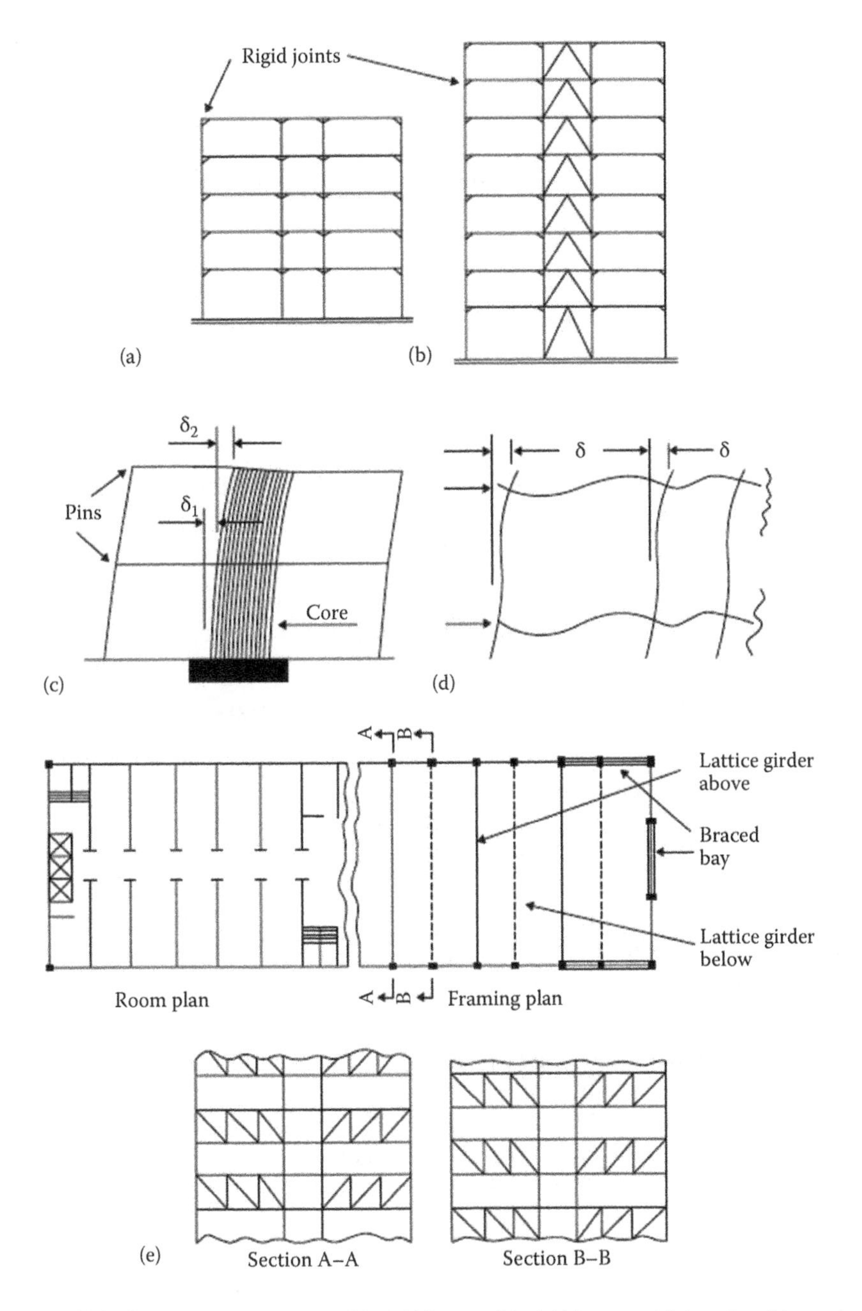

Figure 11.2 Construction systems: (a) rigid frame; (b) rigid frame with bracing; (c) cantilever deflection; (d) rigid frame deflection; (e) staggered lattice girder.

(b) *Mixed systems* (Figure 11.2b–d)

A mixed braced frame/rigid frame structure can also be adopted. This type occurs commonly in reinforced concrete constructions where the shear wall is combined with a concrete rigid frame.

The different modes of deflection for the cantilever/core/braced bay and rigid frame sway are shown in the figure.

(c) *Staggered lattice girder system* (Figure 11.2e)

This system, developed in the United States, is useful for long narrow buildings with a central corridor, for example, hotels or offices. Storey-deep lattice girders, staggered on adjacent floors, span between wall columns as shown in the figure. Lateral loads can be resisted in two ways:

- By end-braced bays with floors acting as rigid diaphragms;
- By rigid frame action in the transverse frames which can be analysed by a matrix computer program.

In the longitudinal direction, braced bays on the outside walls or shear walls at liftshafts/stairwells improve stability.

11.3.3 All-steel outrigger and belt truss system

In tall buildings, the lateral deflection can be excessive if the bracing is provided around the core only. This can be reduced by bringing the outside columns into action by the provision of outrigger and belt lattice girders, as shown in Figure 11.3. The tension and compression forces in the outer columns apply a couple to the core which acts against the cantilever bending under wind loads. The belt truss surrounding the building brings all external columns into action. A single outrigger and belt lattice girder system at the top or additional systems in the height of very tall buildings can be provided.

11.3.4 Composite structures

(a) *Concrete shear wall structures* (Figure 11.4)

The composite steel–shear wall structure consists of a steel-framed building braced with vertical reinforced concrete shear walls. The shear walls placed in two directions at right angles carry vertical and horizontal loads. The shear walls replace the braced bays in the all-steel building.

The shear walls can be located at the ends or sides or in appropriate locations within the building. They should be arranged to be symmetrical with respect to the plan; otherwise, twisting will occur. They provide fireproof walls at lifts and staircases. The walls may be reinforced concrete or concrete-cased steel sections and are designed in accordance with BS 8110 or EC2 and BS 5950 or EC3.

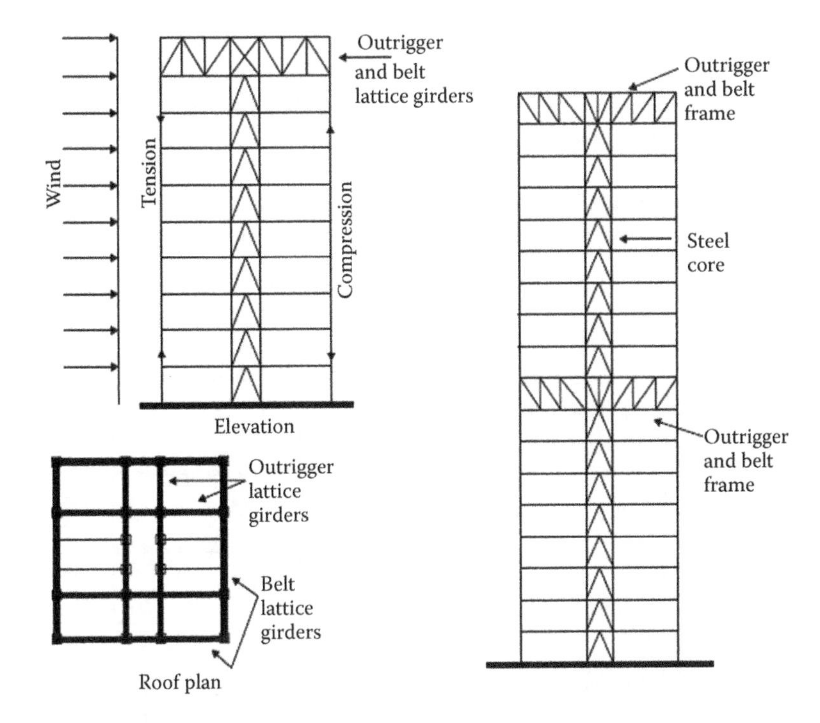

Figure 11.3 Outrigger and belt lattice girder system.

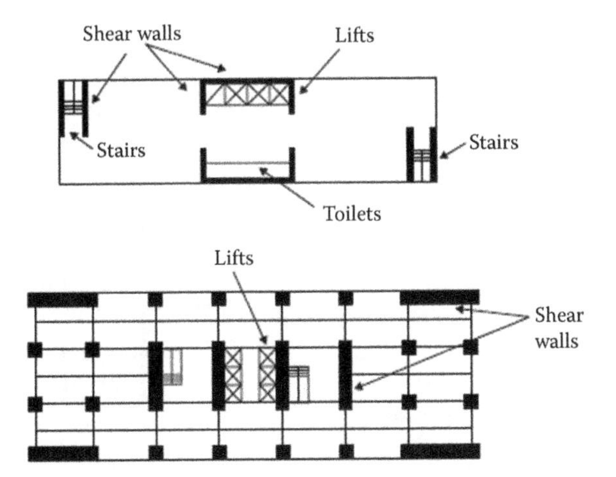

Figure 11.4 Steel building with concrete shear walls.

(b) *Concrete core structures* (Figure 11.5)

The steel frame with concrete core structure is a very common building type adopted for city centre offices. Many of the important features have already been mentioned. The main advantages are as follows.

- The space between core and perimeter is column-free, resulting in maximum flexibility for division using lightweight partitions.

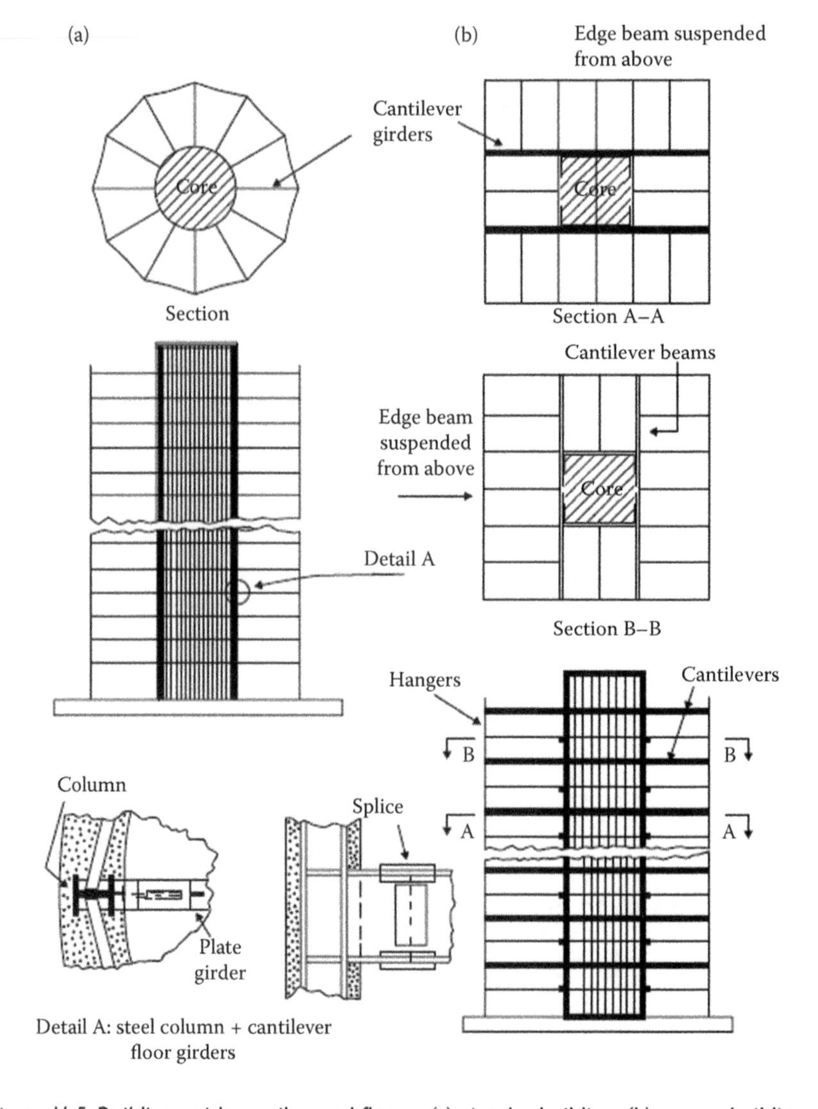

Figure 11.5 Buildings with cantilevered floors: (a) circular building; (b) square building.

- The core provides a rigidly constructed fire-resistant shaft for lifts and staircases.
- No bracing is required on the perimeter walls, so the facade treatment is uniform on all faces.

The structured action is clearly expressed in that the core is designed to resist all the wind loads on the building, the core loads from lifts, stairs, etc. and part of the floor loads. The floor plan may be square, rectangular, triangular, circular, etc. The floor steel may be:

- Supported on the core and perimeter columns;
- Cantilevered out from the core;
- Suspended from an umbrella girder at the top of the core (Section 11.3.5).

Construction is rapid, using slip forming for the core, which is then used to erect the building. Construction can be carried out within the area of the building.

The core may be open or closed in form. Closed box or tubular cores are designed as vertical cantilevers. Open cores, generally of channel or H-section, are designed as connected cantilever shear walls. Cores may be of reinforced concrete, composite steel and concrete with steel columns at extremities or cased steel sections.

Where floor girders are cantilevered out from the core, two possible arrangements are as follows:

- In a circular building, the girders can be supported from steel columns embedded in the core (Figure 11.5a). The concrete core resists wind load.
- In a square building, the cantilever girders on adjacent floors can be arranged to span at right angles to each other. This avoids the need for cantilevers to cross each other at right angles. The two edge beams on any floor parallel to the cantilevers on that floor are then suspended by hangers from the floor above (Figure 11.5b).

11.3.5 Suspended structures (Figure 11.6)

In suspended structures, an umbrella girder is provided at the top of the core from which hangers for the outer ends of the floor beams are suspended. In very tall buildings, additional umbrella girders can be introduced at intermediate locations in the height.

All loads, both vertical and horizontal, are carried on the core in suspended structures. Sections, bars in high-strength steel or cables are used for the suspension members. The time for erection can be shorter than for

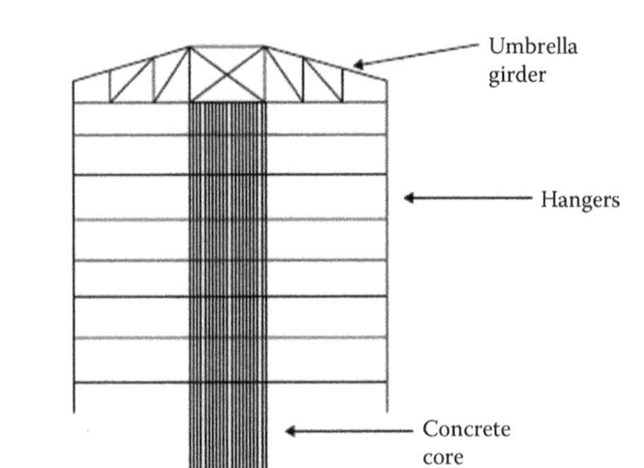

Figure 11.6 Suspended structure.

a conventional structure built upward. The core is constructed first by slip forming and used to erect the steelwork.

11.3.6 Tube structures

The tube type of structure was developed by Dr. Fazlur Khan of the United States for very tall buildings, say over 80 storeys in height. If the core type of structure is used, the deflection at the top would be excessive. The tube system is very efficient with respect to structured material used and results in a considerable saving in material when compared with conventional designs.

In this system, the perimeter walls are so constructed that they form one large rigid or braced tube which acts as a unit to resist horizontal load. The framed tube shown in Figure 11.7a consists of closely spaced exterior columns 1–3 m apart, tied at each floor level with deep floor beams. The tube is made up of prefabricated wall units of the type shown in Figure 11.7b. The small perforations form spaces for windows and the normal curtain walling is eliminated.

In the single-tube structure, the perimeter walls carry the entire horizontal load and their share of the vertical load. Internal columns and/or an internal core, if provided, carry vertical loads only. In the tube-within-a-tube system shown in Figure 11.7e, the core tube would carry part of the horizontal load. Very tall stiff structures have been designed on the bundled tube system shown in Figure 11.7f, which consists of a number of tubes constructed together. This reduces the shear lag problem which is more serious if a single tube is used.

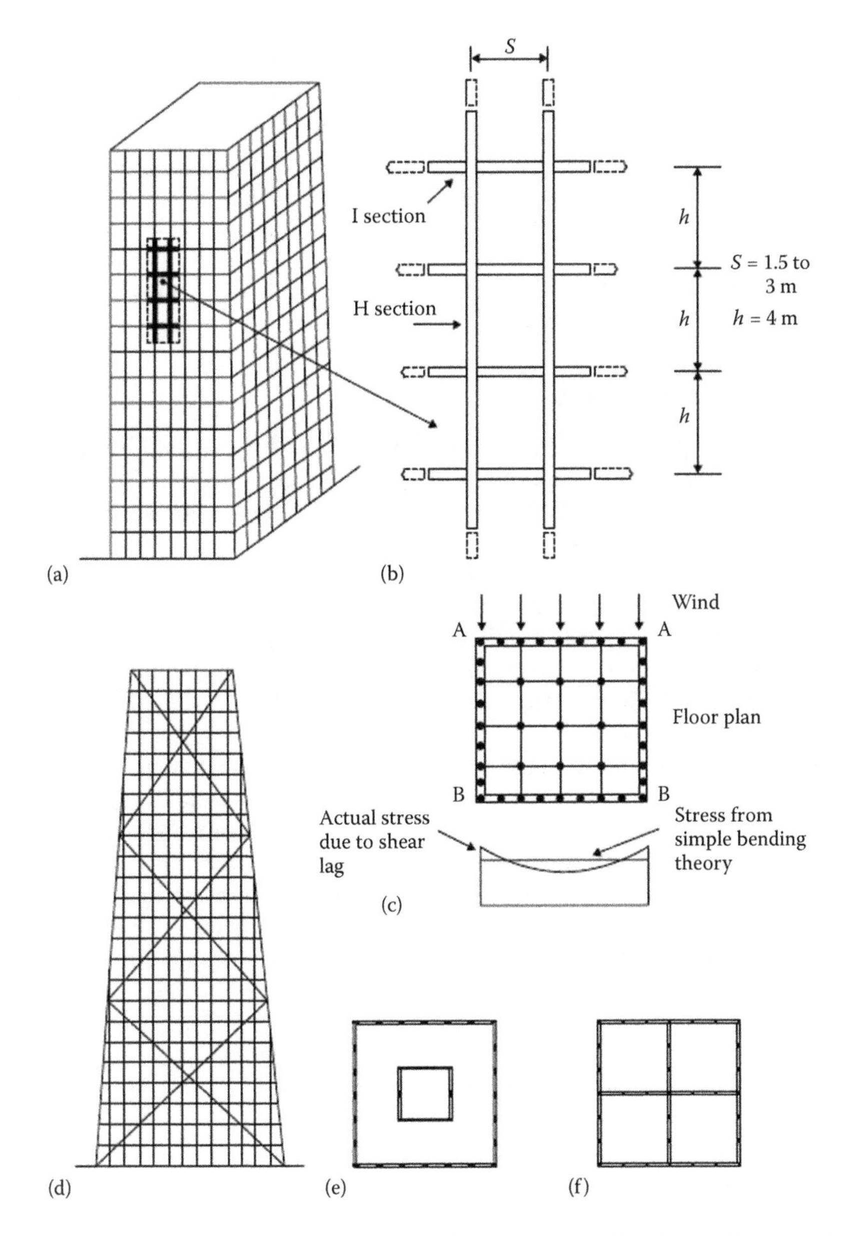

Figure 11.7 Tube structures: (a) framed tube; (b) prefabricated 'tree' unit; (c) stress distribution in walls AA, BB; (d) braced tube; (e) tube in tube; (f) bundled tube. (Buildings may have more storeys than shown.)

The analysis of a tube structure may be carried out on a space frame program. The main feature shown up in the analysis for horizontal load is the drop-off in load taken by the columns in the flange faces. This is caused by shear lag in the beam–column frame as shown in Figure 11.7c. Simple beam theory would give uniform load in these columns. Dr. Fazlur Khan developed preliminary methods of analyses which take shear lag into account (Khan and Amin, 1973).

The framed tube can be relatively flexible and additional stiffness can be provided by bracing the tube as shown in Figure 11.7c. This helps reduce shear lag in the flange tube faces.

11.3.7 SWMB structures

A brief description is given of a building system for tall buildings developed by Skilling Ward Magnusson Barkshire Inc. of Seattle, Washington (Skilling, 1988) – the SWMB system. The basis of the design is that it is far cheaper to use concrete rather than steel members to carry vertical loads. A dramatic saving in weight of steel used is possible.

Central to the system is the SWMB column which consists of a very large (up to 3 m diameter) concrete-filled tube. Three, four or more such columns connected by deep girders and moment frames extend through the height of the building and carry vertical and horizontal loads.

The concrete infill in the column is of very high strength (140 N/mm^2) from a mix with small aggregate, low (0.22) water/cement ratio and super-plasticizer giving a 300 mm slump. The concrete is pumped in under pressure from the bottom and rises up the column. Stud shear connections are welded to the inside to ensure composite action and efficient load transfer. Special stiffeners and connection plates are welded to the tubes for girder and moment connections. The tubes are shop fabricated in three-storey-high sections and welded together on site using splice rings.

11.4 CONSTRUCTION DETAILS

To assist in idealizing the structure for design, some of the common construction forms for the various building elements are set out briefly.

11.4.1 Roofs and floors (Figure 11.8a)

Floor construction was discussed in Chapter 9. The main types used for flat roofs and floors are

- Cast *in-situ* or precast concrete slabs or steel beams;
- Composite *in-situ* concrete on steel deck on steel beams.

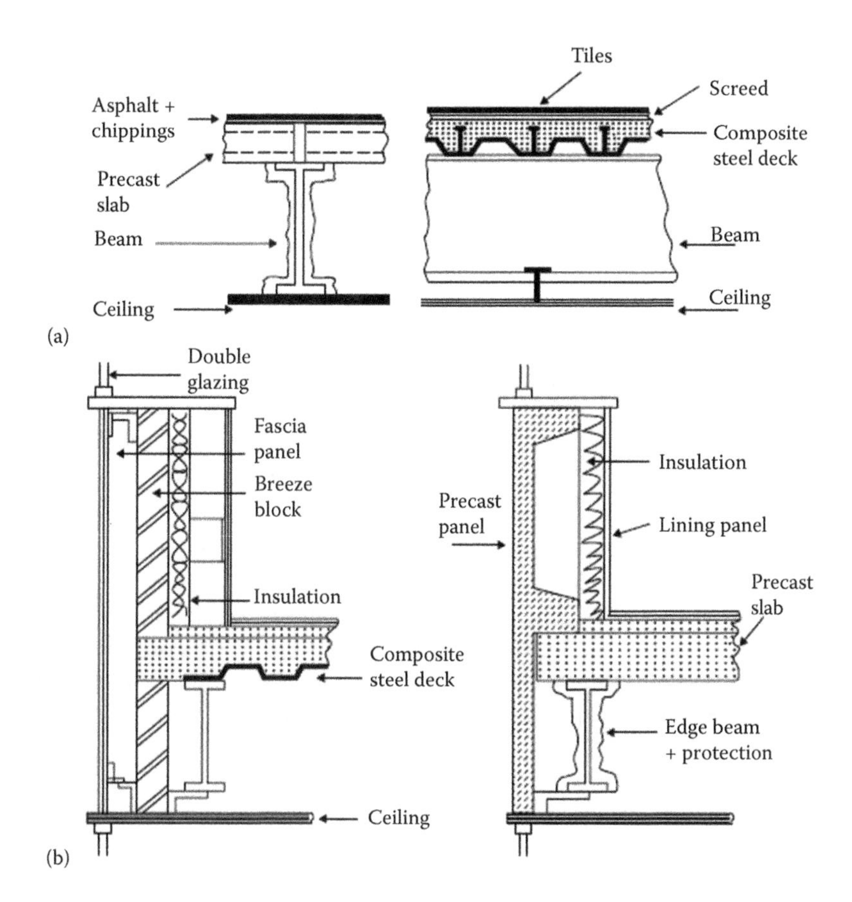

Figure 11.8 Typical building details: (a) roof and floor construction; (b) curtain walls.

The steel beams are usually designed to act compositely with *in-situ* slabs. The use of lattice or steel girders permits services to be run through floors. Floor beams must have fire protection.

11.4.2 Walls

Walls in steel-framed buildings may be classified as follows:

- Structural shear walls located in bays on the perimeter, around cores or in other suitable areas – these are of reinforced concrete or composite construction incorporating steel columns. All-steel braced bays with fireproof cladding serve the same purpose. These walls carry wind and vertical load.
- Non-load bearing permanent division and fire-resistant walls – these are constructed in brick and blockwork and are needed to protect lifts, stairs and to divide large areas into fireproof compartments.

- Movable partitions – these are for room division.
- Curtain walls – these include glazing, metal framing, metal or precast concrete cladding panels, insulation and interior panels. Typical details are shown in Figure 11.8b.
- Cavity walls with outer leaf brick, inner leaf breeze block – these are common for medium-rise steel-framed buildings.

11.4.3 Steel members

(a) *Floor beams*

Universal beams are generally designed as composite with concrete on steel deck floors. Compound beams, lattice girders, plate girders or stub girder construction are required for long spans. Heavy transfer girders are required where floor plans or column arrangements change.

(b) *Columns*

Universal columns, compound and built-up sections and circular and box sections are used. ARBED's heavy 'jumbo' sections are ideal for very tall buildings. (Trade ARBED Luxembourg, n.d. Also see TATA products.) Box columns external to the building can be protected from the effects of fire by circulating water to keep the temperature to a safe value.

(c) *Hangers*

Rounds, flats or sections in high-strength steel or steel cables are used.

(d) *Bracings*

All-steel, open or closed sections are used.

11.5 MULTISTOREY BUILDING – PRELIMINARY DESIGN

11.5.1 Specification

(a) *Design – steel core and perimeter columns*

The framing plans for a 20-storey office building are shown in Figures 11.9 and 11.10. The roofs and floors are cast-*in-situ* concrete slabs supported on steel beams. Curtain walls cover the external faces. The core walls are braced steel cantilevers enclosed in breezeblock fire protection. The internal steel-framed walls in the core are of similar construction. Lightweight partitions provide division walls in the office areas. Services are located in the core.

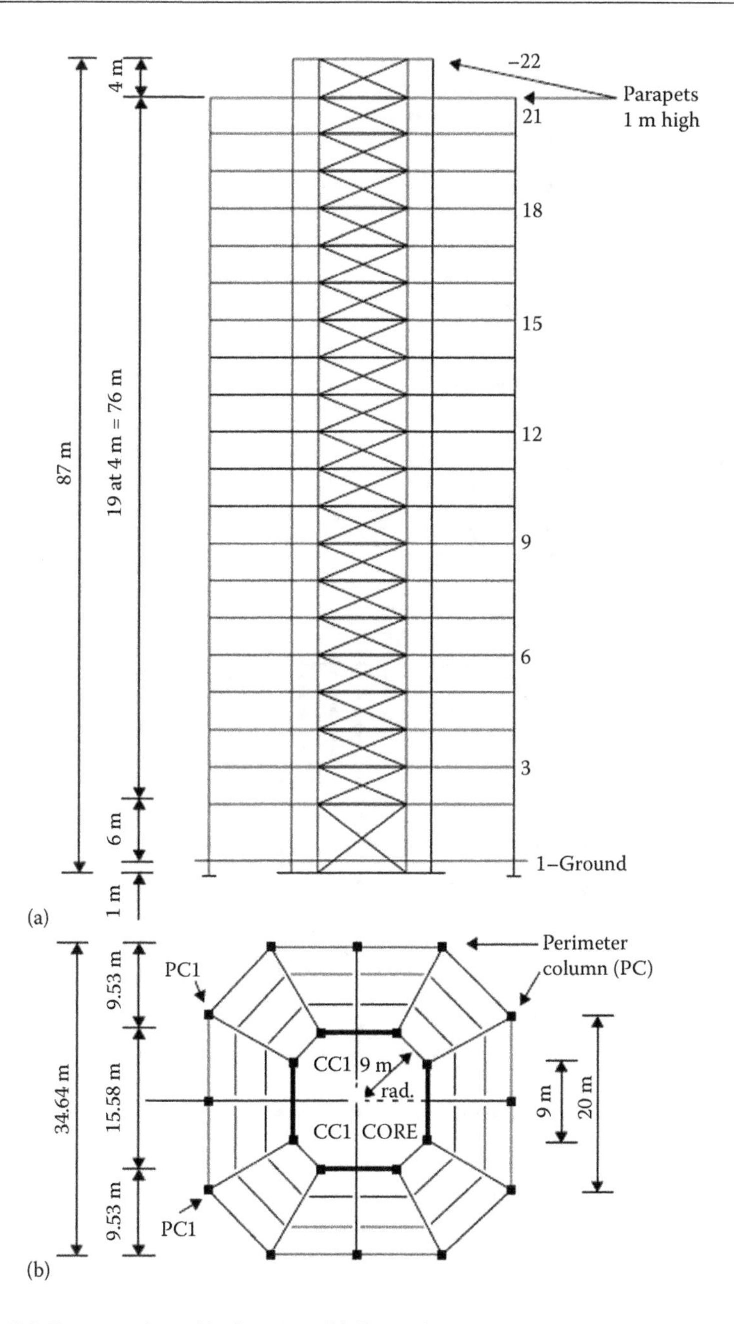

Figure 11.9 Framing plans: (a) elevation; (b) floor plan.

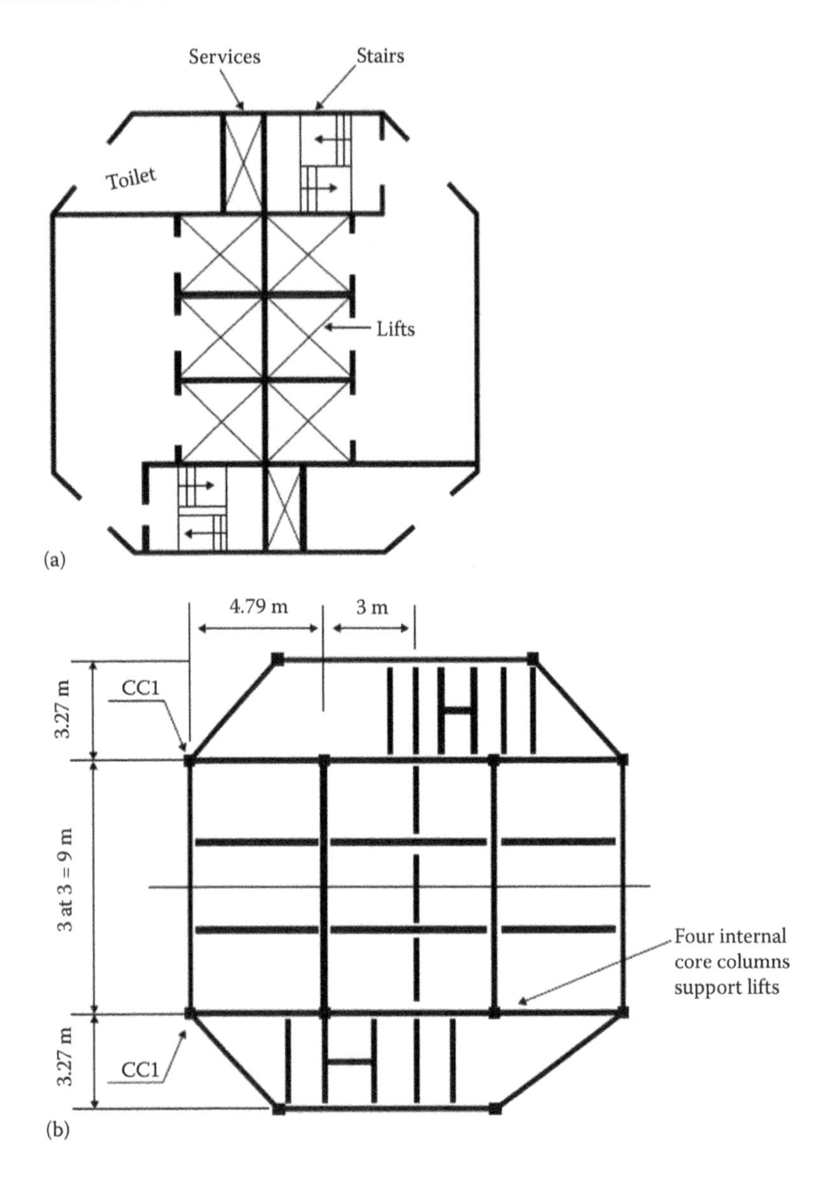

Figure 11.10 Framing plans: (a) core arrangement; (b) core steel beams.

The procedure is as follows:

1. Design floor beams and a perimeter column.
2. Check building stability and design bottom bracing member. The material is to be steel, Grade S355.

(b) *Alternative designs*

The following alternative designs are possible:

1. Steel core with cantilever floor beams;
2. Suspended structure – perimeter hangers with concrete core and four umbrella girders (Figure 11.11).

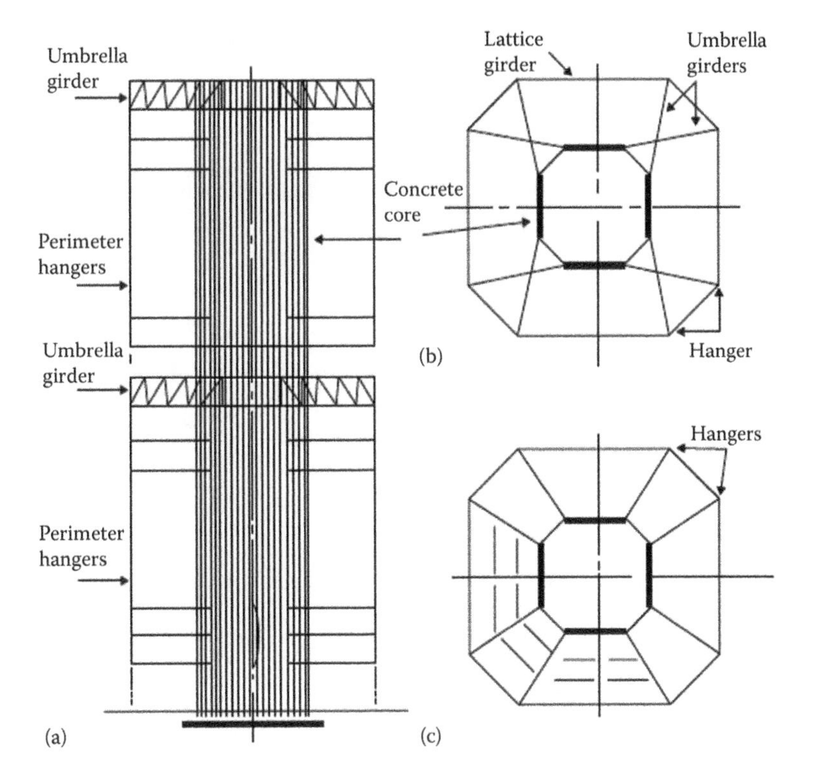

Figure 11.11 Suspended structure: (a) elevation; (b) umbrella girders; (c) floor beams.

11.5.2 Dead and imposed loads

Refer to BS 6399: Part 1 for details.

The loads are

- Roof – dead 5 kN/m^2, imposed 1.5 kN/m^2
- Floors – dead 6 kN/m^2, imposed 3.5 kN/m^2
- Core – dead 6 kN/m^2, imposed 5.0 kN/m^2
- Core (top machine floor) – dead 8 kN/m^2, imposed 7.5 kN/m^2
- Walls
  - Curtain walls, glazing, steel 5 kN/m
  - Internal and access core walls 10 kN/m
  - Braced core walls 15 kN/m
  - Parapet 3 kN/m
- Columns
  - Perimeter with casing 1.4–5 kN/m
  - Core (average) 3 kN/m

The reduction in total distributed imposed load with number of storeys (Table 2 or BS 6399: Part 1) is

- 5–10 floors – 40%
- Over 10 floors – 50%

Note that the lifts and machinery are carried on independent columns within the core.

11.5.3 Beam loads and design

(a) *Office floor beams – design*

The office floor beam layout and beam loads are shown in Figure 11.12.

$$\text{Design load} = (1.4 \times 6) + (1.6 \times 3.5) = 14 \text{ kN/m}^2$$

(i) BEAM FB1 (FB4 SIMILAR)

$$\text{Design load} = (1.77 \times 9.9 \times 14) + (5 \times 10.35 \times 1.4) = 317.8 \text{ kN}$$

$$\text{Imposed load} = 61.3 \text{ kN}$$

$$M = 317.8 \times 10.35/8 = 411.2 \text{ kNm}$$

$$S = 411.2 \times 10^3/355 = 1158 \text{ cm}^3$$

Select 457 × 152 UB 60, with $S = 1290$ cm^3. Deflection satisfactory.

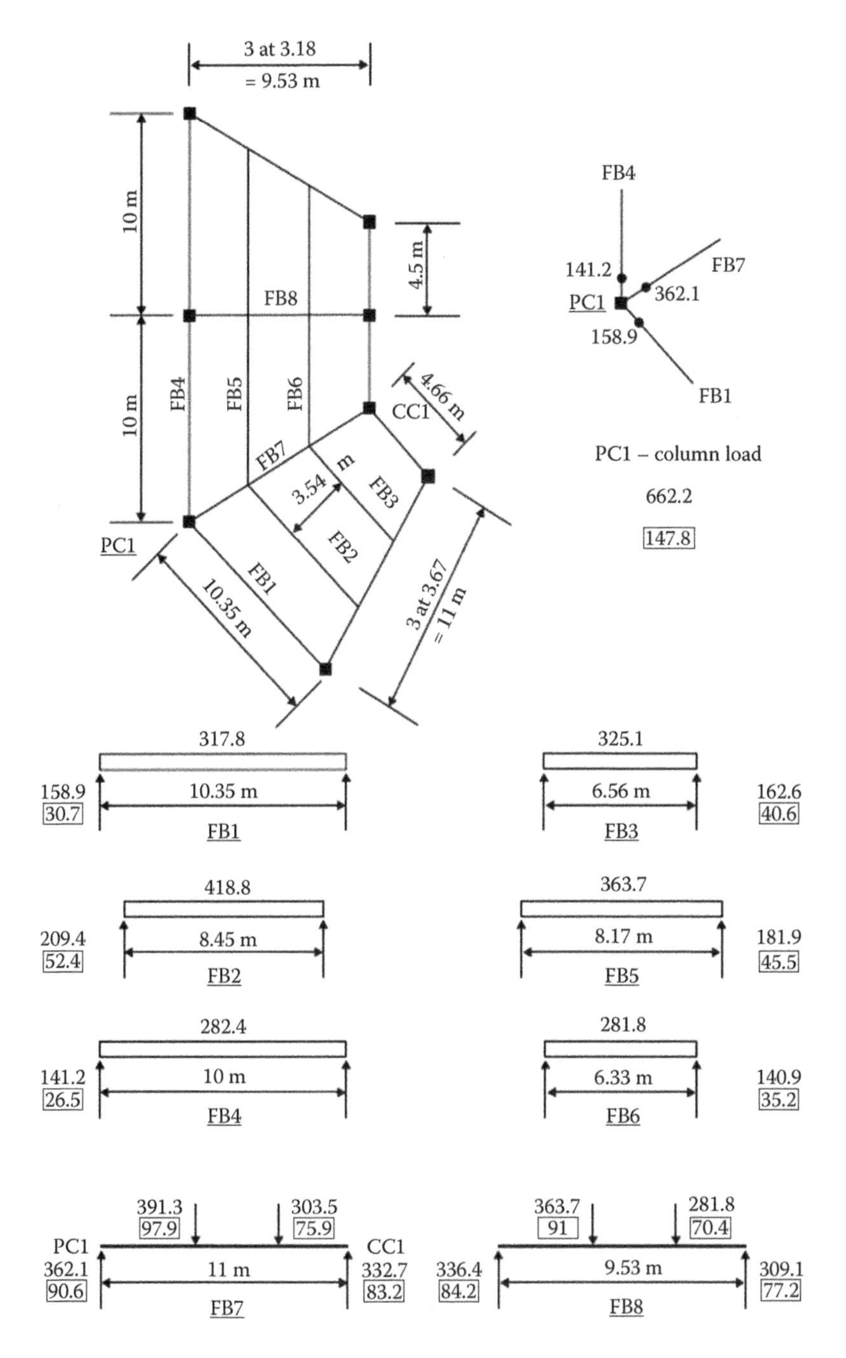

Figure 11.12 Floor beam loads (kN; unfactored imposed loads shown boxed).

(ii) BEAM FB2 (FB5 SIMILAR)

Design load = 418.8 kN

Imposed load = 104.7 kN

M = 442.4 kNm

S = 1246 cm^3

Try 457 × 152 UB 60, with S = 1290 cm^3, I = 25,500 cm^4.

$$\delta = \frac{5 \times 104.7 \times 10^3 \times 8450^3}{384 \times 205 \times 10^3 \times 25{,}500 \times 10^4} = 15.7 = \text{Span}/538$$

This is satisfactory.

(iii) BEAM FB3 (FB6 SIMILAR)

Design load = 325.6 kN

Imposed load = 81.3 kN

Select 406 × 140 UB 46.

(iv) BEAM FB7

$$M = 3.67 \times 362.1 = 1328 \text{ kNm} = \frac{WL}{3}$$

S = 1328 × 10^3/345 = 3849 cm^3

Try 686 × 245 UB 125, with S = 3990 cm^3, T = 16.2 cm, p_y = 345 N/mm^2 (as assumed), I = 118,000 cm^4. Deflection at the centre due to the unfactored imposed loads is

$$\delta = \frac{(97.9 + 75.9)10^3 \times 11{,}000^3}{48 \times 205 \times 10^3 \times 118{,}000 \times 10^4}\left[\frac{3 \times 3.67}{11} - 4\left(\frac{3.67}{11}\right)^3\right]$$
$$= 17 \text{ mm} = \text{span}/647$$

This is satisfactory.

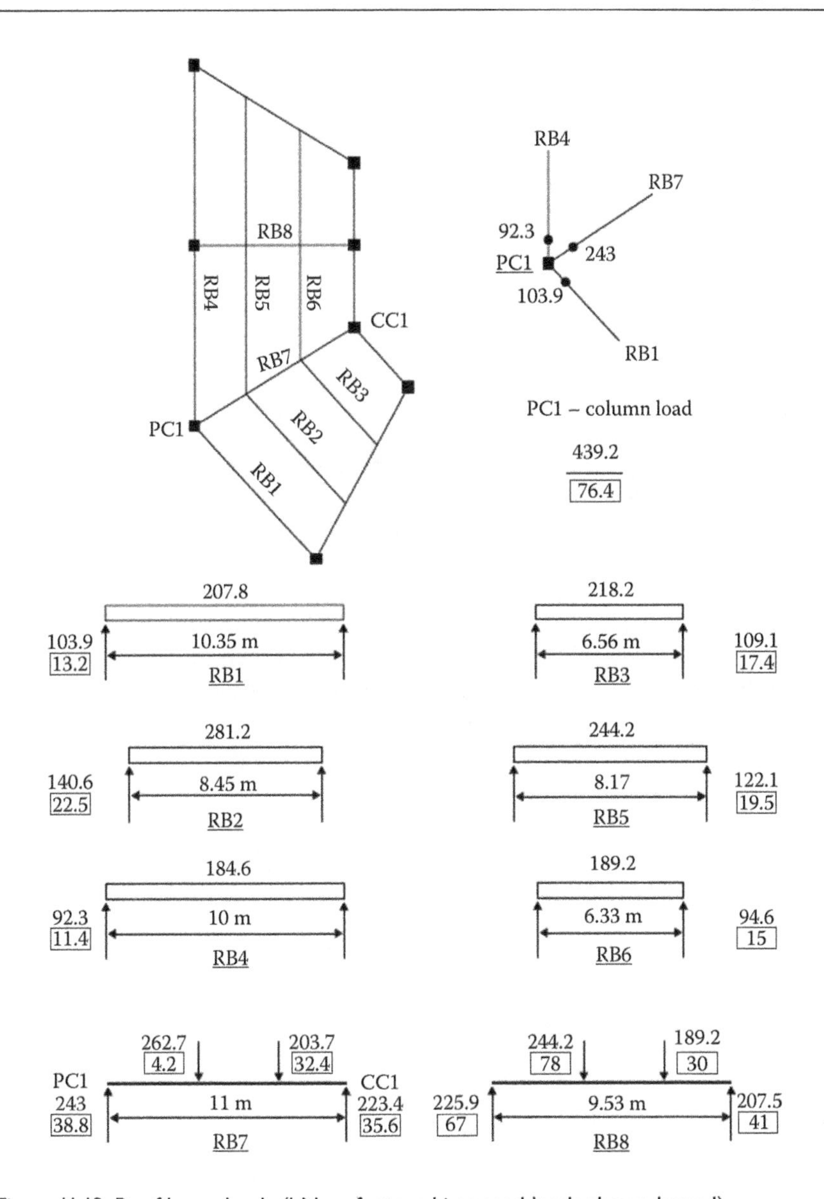

Figure 11.13 Roof beam loads (kN; unfactored imposed loads shown boxed).

(v) BEAM FB8

Select 610 × 229 UB 113.

(b) *Roof beams – loads*

The layout of the roof beams and the beam loads are shown in Figure 11.13.

$$\text{Design load} = (1.4 \times 5) + (1.6 \times 1.5) = 9.4 \text{ kN/m}^2$$

The beams could then be designed.

11.5.4 Design of perimeter column PC1

The column is shown in Figure 11.9. The column sections are changed at every third floor level from the roof down to floor 3 and then from floor 3 to base 1. The column sections are checked at the second level in the three-level length because the splice is made above floor level and the column below is of much heavier section which attracts the bulk of the eccentric moment.

The column loads at the critical sections are shown in Figure 11.14 and the eccentricities for the specimen calculations made are shown in Figure 11.15.

(a) *Roof to floor 18*

Check at floor 19. Reduce imposed load by 10%:

$$F = 1780 - (0.1 \times 224 \times 1.6) = 1758 \text{ kN}$$

Assume the column is 254 × 254 UC:

$$M_x = 0.5[(362.1 \times 0.24) + (158.9 \times 0.03) + (141.2 \times 0.06)] = 50 \text{ kN m}$$

$$M_y = 0.5 \times 0.23(160.3 - 142.2) = 2.0 \text{ kNm (neglect)}$$

Try 254 × 254 UC 73, with r_y = 6.46 cm, S_x = 989 cm^3, A = 92.9 cm^2, T = 14.2 cm, p_y = 355 N mm^2. Also

$$\lambda_y = 0.85 \times 4000/64.6 = 52.6$$

$$p_c = 268.2 \text{ N/mm}^2 \text{ (Table 27(c))}$$

$$p_c A_g = 268.2 \times 92.9/10 = 2491.6 \text{ kN}$$

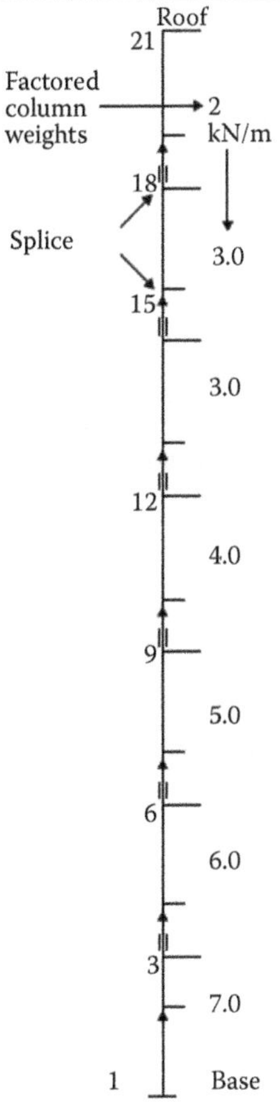

Column	Floor	Total factored load	Unfactored imposed load	Reduction (%)	Design load
Roof 21					
			211		
Factored column weights → 2	20		224	10	
kN/m	19	1780			1758
18					
Splice 3.0	17		667	40	
15	16	3803			3536
3.0	14		1110	40	
	13	5826			5382
12					
4.0	11		1553	50	
	10	7861			7085
9					
5.0	8		1996	50	
	7	9908			8910
6					
6.0	5		2439	50	
	4	11,967			10,748
3	3		2789	50	
7.0	2	13,347			11,953
1 Base					

Figure 11.14 Column PC1 loads.

$$\lambda_{LT} = 0.5 \times 4000/64.6 = 30.9$$

$$P_b = 352.5 \text{ N mm}^2 \text{ (Table 11)}$$

$$M_{bs} = 352.5 \times 989/10^3 = 348.6 \text{ kNm}$$

Then

$$(1758/2491.6) + (50/348.6) = 0.85$$

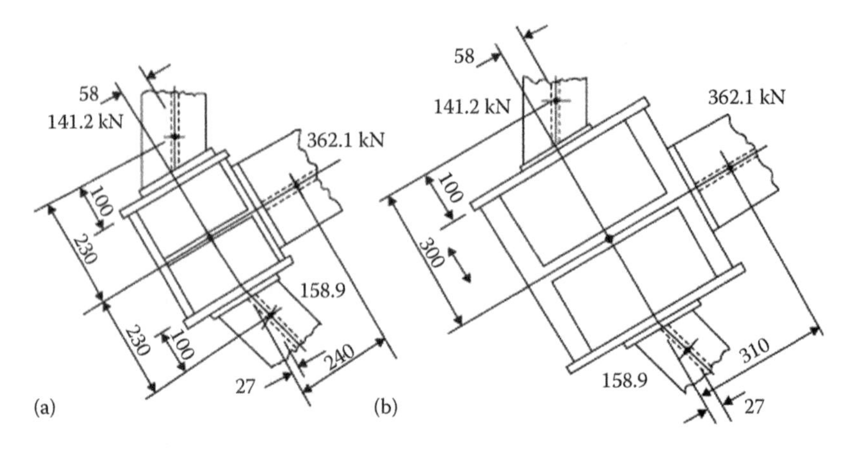

Figure 11.15 Column PCI load eccentricities: (a) 254 × 254 UC 73, floor 19; (b) 356 × 406 UC 393, floor 2.

This is satisfactory.

Note that 203 × 203 UC 71 is too light.

(b) *Floors 18–3*

The following sections are used:

- Floor 18–15 – 305 × 305 UC 118;
- Floor 12–15 – 356 × 368 UC 153;
- Floor 12–9 – 356 × 368 UC 202;
- Floor 9–6 – 356 × 406 UC 235;
- Floor 6–3 – 356 × 406 UC 287.

(c) *Floor 3 – Base 1*

$$F = 13,347 - (0.5 \times 2789 \times 1.6) = 11,953 \text{ kN}$$

Try 356 × 406 UC 393, with $r_y = 10.5$, $S_x = 8220$ cm^3, $A = 501$ cm^2, $T = 49.2$ mm, $p_y = 335$ N/mm^2.

$$M_x = 0.5[(362.1 \times 0.31) + (141.2 \times 0.06) + (158.9 \times 0.03)] = 62.8 \text{ kNm}$$

$$\lambda_y = 0.85 \times 7000/105 = 56.7$$

$$p_c = 244.3 \text{ N/mm}^2, \quad P_c = 12,237 \text{ kN}$$

$$\lambda_{LT} = 0.5 \times 7000/105 = 33.3$$

$$P_b = 327.7 \text{ N/mm}^2, \quad M_{bs} = 2694 \text{ kNm}$$

Then

$$(11{,}953/12{,}237) + (62.8/2694) = 0.99$$

This is satisfactory.

11.5.5 Braced core wall – vertical loads

The building stability is checked by considering the stability of the braced core wall CC1–CC1 (Figure 11.16a). The floor and column loads are calculated using the tributary area method.

The dead loads are calculated for the office and core floors and walls. The other loads are found by proportion (Section 11.5.2 gives load values). The loads on the separate areas and the reactions on core wall CC1–CC1 are shown in Figure 11.16b.

(a) *Office floor – dead load*

Office area A: 9 × 9.53 × 6 = 514.6 kN
+ 5.5 × 9.53 × 6 = 314.5 kN

Office area B: 4.65 × 11 × 6 = 306.9 kN
2.83 × 11 × 6 = 186.5 kN

Core floor area C: 9 × 6 × 4.79/2 = 129.3 kN

Core floor area D: 32 kN (estimate)

(b) *Walls and columns – dead load*

Braced core: (15 × 9 × 20) + (15 × 7.5 × 9/3.5) + (3 × 87 × 2) = 2989 kN

Other walls: 10(4.79 + 4.65)20 + (10 × 9.44 × 7.5/3.5) = 2090 kN

(c) *Separate loads – dead and imposed*

Office roof (dead): (362.1 + 215.7)5/6 = 481.5 kN

Office roof (imposed): 577.8 × 1.5/6 = 144.5 kN

Core roof (dead): (129.3 + 32)5/6 = 134.4 kN

Core roof (imposed): 161.3 × 1.5/6 = 40.3 kN

Machine floor (dead): (129 + 32)8/6 = 215.0 kN

Machine floor (imposed): 161.3 × 7.5/6 = 201.6 kN

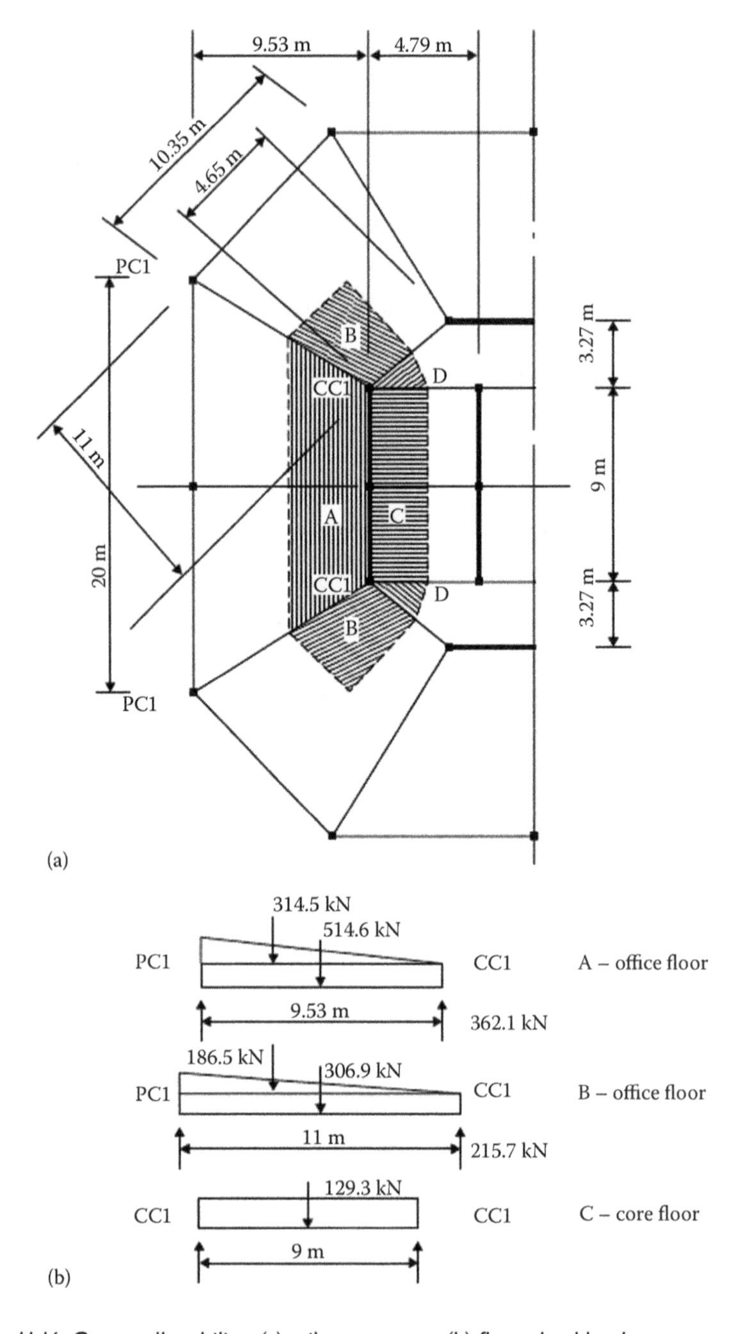

Figure 11.16 Core wall stability: (a) tributary areas; (b) floor dead loads.

Table 11.1 Core wall – vertical loads

Item	*Dead (kN)*	*Imposed (kN)*
Walls and columns		
Braced core	2989	–
Other walls	2090	–
Office roof	482	145
Core roof	134	41
Machine floor	215	202
Office floor × 19	10,984	6407
Core floor × 19	3065	2554
Total	19,959	9349

Office floor (dead): 362.4 + 215.7 = 578.1 kN

Office floor (imposed): 578.1 × 3.5/6 = 337.2 kN

Core floor (dead): 129.3 + 32 = 161.3 kN

Core floor (imposed): 161.3 × 5/6 = 134.4 kN

(d) *Total load at base of braced core wall*

Table 11.1 shows the dead and imposed loads at the core base.

11.5.6 Wind loads

Details are given BS 6399: Part 2.

(a) *Location of building*

The building is located in a city in NE England on a redevelopment site, and stands clear in its own grounds, surrounded by gardens and car park. The site is 10 km from the sea and 75 m above sea level. The building faces west.

(b) *Building dimensions*

The height is 87 m to parapet; the plan is octagonal 36 m × 36 m, corner angle 135° (Figure 11.17).

(c) *Building data (Sections 1.6 and 1.7 of code)*

Building type: framed with structural walls around lift; factor K_b = 1 (Table 1).

Reference height $H_r = H = 87$ m = effective height.

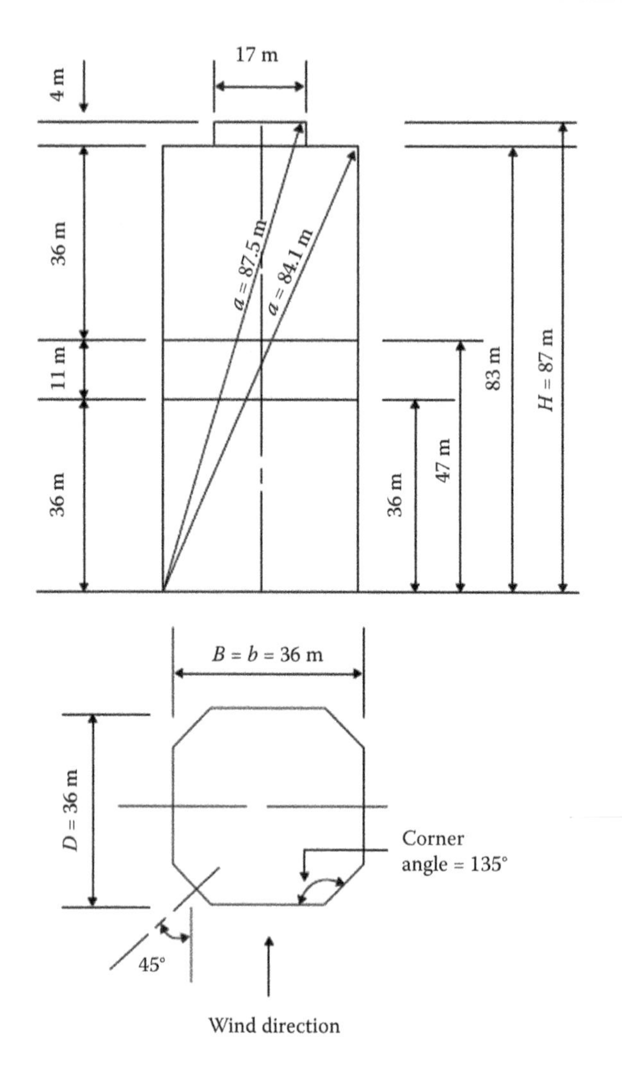

Figure 11.17 **Building dimensions (to top of parapet and outside of building).**

Ground roughness: town, clear of obstruction.

Dynamic augmentation factor $C_r = 0.08$ (Figure 3 in code).

(d) *Wind speeds (Section 2.2 of code)*

Basic wind speed $V_b = 25$ m/s (Figure 6)

Site wind speed $V_s = V_b S_a S_d S_s S_p$

Altitude factor $S_a = 1 + 0.001\ \Delta_s = 1.075 \quad (\Delta_s = 75 \text{ m})$

Direction factor $S_d = 0.99$ (Table 3), $\phi = 270°$ west

Seasonal factor $S_s = 1.0$ (permanent building)

Probability factor $S_p = 1.0$

$V_s = 1.075 \times 0.99 \times 25 = 26.6$ m/s

Effective wind speed $V_e = V_s S_b$ where S_b is the terrain and building factor.

The building is divided into a number of parts as set out in Figure 11 in the code in order to assess the value of *S*b. The four divisions adopted for the building are shown in Figure 11.17. Factors S_b taken from Table 4 are listed in Table 11.2. These depend on the top height of each part.

(e) *Wind loads (Section 2.1 of code)*

$$\text{Dynamic pressure } q_s = 0.613 V_e^2 \text{ N/m}^2$$

Values of q_s for the various building divisions are given in Table 11.2.

The external surface pressure is

$$p_e = q_s C_{pe} C_a$$

where C_{pe} is the external pressure coefficient for the building surface (Table 5) – for $D/H = 36/87 < 1$,

$$C_{pe} = +\ 0.8 \text{ (windward face)} - 0.3 \text{ (leeward face)} = 1.1 \text{ (total)}$$

and C_a is the size effect factor (Figure 4). This depends on the diagonal dimension a (Figure 5 in the code and Figure 11.15). For a = 87.4 m for a town site 10 km from the sea and $H_e > 50$ m, $C_a = 0.82$. Therefore,

$$P_e = (1.1 \times 0.82) q_s = 0.9 q_s$$

Values of p_e are shown in Table 7.2.

$$\text{Net surface load } P = pA$$

where p is the net pressure across the surface and A is the loaded area.

The overall load is found from $P = 0.85(\sum P_{\text{front}} = \sum P_{\text{rear}})(1 + C_r)$ where C_r is the dynamic augmentation factor ((c) above).

$$P = 0.85 \times 1.08 \sum P_{\text{overall}} = 0.918 \sum P_{\text{overall}}$$

Table 11.2 Wind load calculations

Building part	Reference height (m)	Terrain and building factor S_b	Effective wind speed $V_e = V_s S_b$ (m/s)	Dynamic pressure $q_s = 0.613 V_e^2/10^3 (kN/m^2)$	External surface pressure $p_e = q_s C_{pe} C_a$ (kN/m²)
1	87	2.1	55.9	1.92	1.73
2	83	2.09	55.6	1.9	1.7
3	47	1.97	52.4	1.7	1.53
4	36	1.98	52.7	1.7	1.53
Building part	*Area of building part A (m²)*	*Overall load = $0.918 p_e A$ (kN)*	*Lever arm (m)*	*Base moment (kN/m)*	
1	68	108	85	9180	
2	1296	2023	65	131,495	
3	396	556	41.5	23,074	
4	1296	1820	18	32,760	
Total	–	4507	–	196,509	

(f) *Polygonal building* (Figure 11.17)

Referring to Clause 3.3.1.2, from Table 27 the suction coefficients for zone A (Figure 31) can be reduced by 0.4. The corner angle is 135° and the adjacent upward face is 21 m > $b/5$ = 7.2 m. The external pressure coefficients from Table 26 could be used. However, for simplicity the building will be taken as square and coefficients from the Standard Method will be used, giving a conservative result.

(g) *Wind load calculations*

From the code data given above, the calculations for the wind loads are set out in Table 11.2. These give the characteristic wind loads and moments at the base of the braced bay.

11.5.7 Stability, foundations and bracing

(a) *Stability*

The load factors α_f are

- For dead load resisting overturning – 1.0;
- For wind load – 1.4.

The overturning moment is

$$1.4 \times 196{,}509 = 275{,}113 \text{ kNm}$$

The stabilizing moment about the compression leg of bracing is

$$2 \times 19{,}959 \times 4.5 = 179{,}631 \text{ kNm}$$

HD bolts must resist a moment of 95,482 kNm. The bolt tension is

$$95{,}482/(2 \times 9) = 5305 \text{ kN}$$

Use bolts Grade 8.8, strength 560 N/mm^2 with 12 bolts per leg.
The tensile area is

$$5305 \times 10^3/12 \times 560 = 789 \text{ mm}^2$$

Provide 12, 42 mm diameter bolts, tensile area 1120 mm^2.
Under full load with load factor 1.2, the uplift conditions are much less severe.

(b) *Foundations*

The foundations could consist of a thick capping slab under the core supported on say, 12 cylinder piles depending on ground conditions. The two core walls at right angles to those considered would then assist in stabilizing the building. Pits for the lifts would be provided in the centre of the core slab. The outer columns could be supported on separate piled foundations. All foundations are tied together by the ground floor slab.

(c) *Bracing* (Figure 11.9)

Only the section for the bottom bracing member is established. The length of the member is 11.4 m.

From Table 11.2

$$\text{Wind shear} = 4507\ \text{kN}$$

$$\text{Tension} = 4507 \times 11.4/2 \times 9 = 2854\ \text{kN}$$

Provide two channels with four bolt holes per channel. Net tensile area required using Grade 50 steel (Clause 3.33 of BS 5950):

$$A_{net} = 2854 \times 10/(1.1 \times 355 \times 2) = 36.5\ \text{cm}^2$$

Try 300 × 90 × L_{41}, with A = 52.7 cm³, T = 15.5 mm, t = 9.0 mm, p_y = 355 N/mm² and 22 mm diameter. Grade 8.8 bolts, single shear value 114 kN. Then

$$A_{net} = 52.7 - 2 \times 24(15.5 + 9.0)/10^2 = 49.6\ \text{cm}^2$$

This is satisfactory.

The number of bolts required is 2854/114 = 25. Use 26 for the two channels.

Chapter 12

Wide-span buildings

12.1 TYPES AND CHARACTERISTICS

Wide-span buildings may be classified into the following types:

1. One-way-spanning buildings
   - Truss or lattice girder/stanchion frames;
   - Portals and arches;
   - Cable or tie stayed lattice girder roof.
2. Combination portal or cable-stayed spine frame with lateral lattice girders
3. Two-way-spanning trusses, Vierendeel girders or space deck systems
4. Domes springing from circular or polygonal bases
5. One-way-spanning cable girder and two-way-spanning cable net roofs
6. Air-supported roofs

Selected wide-span roof systems are shown in Figure 12.1. The simplest wide-span structure is the flat or sloping roof truss or lattice girder/stanchion frame. Portal frames and arches have been designed to span widths over 60 m. The older sawtooth roof designed to take advantage of natural lighting was once a common type of structure (Figures 2.2 and 2.4).

Various types of spine-supported buildings utilizing portals or cable/tie-stayed frames have been constructed (Figures 12.1 and 12.2). A preliminary design is given for one system (Section 12.2).

Much effort has been expended on the development of the two-way spanning space deck. This system provides a rigid three-dimensional flat roof structure capable of spanning a large distance with a small construction depth. The small depth reduces costs, while the exposed framing forms a pleasing structure. Commercial space deck systems are described. A preliminary design for a space deck is given.

The large framed dome is one of the most spectacular structures. Domes are constructed to cover sports arenas, auditoria, exhibition pavilions, churches, etc. The masonry dome shell is an ancient structural form.

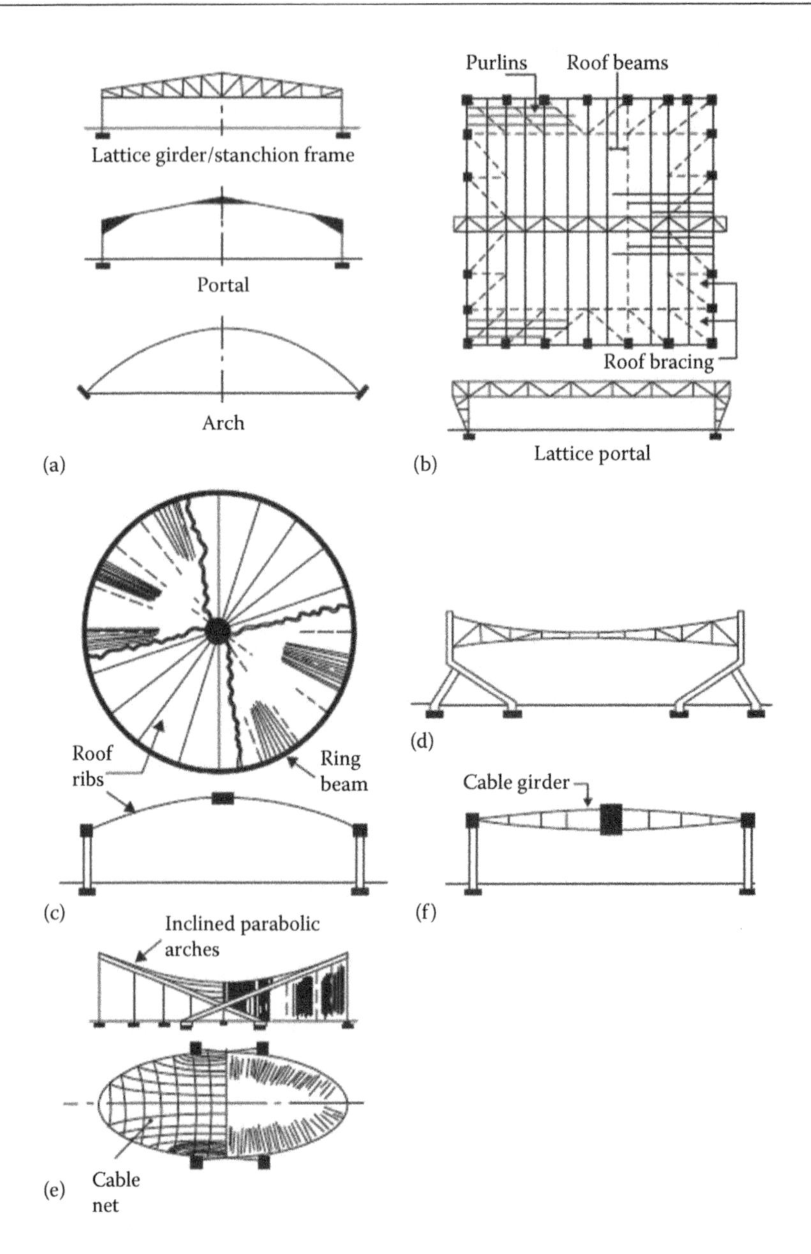

Figure 12.1 Roof systems for wide-span buildings: (a) one-way spanning systems; (b) spine portal/roof beams; (c) dome roof; (d) one-way cable girder; (e) cable net roof; (f) cable roof, circular building.

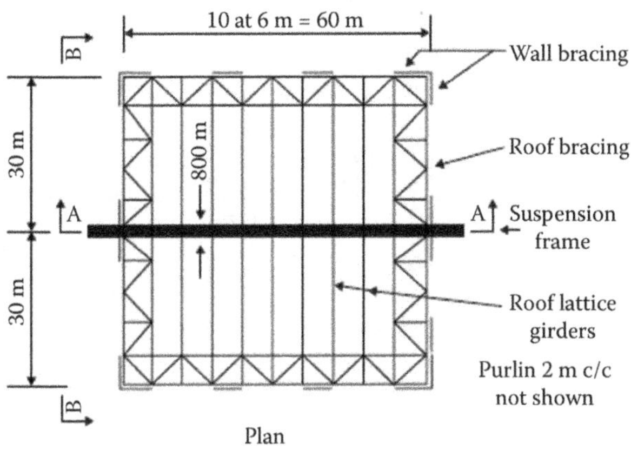

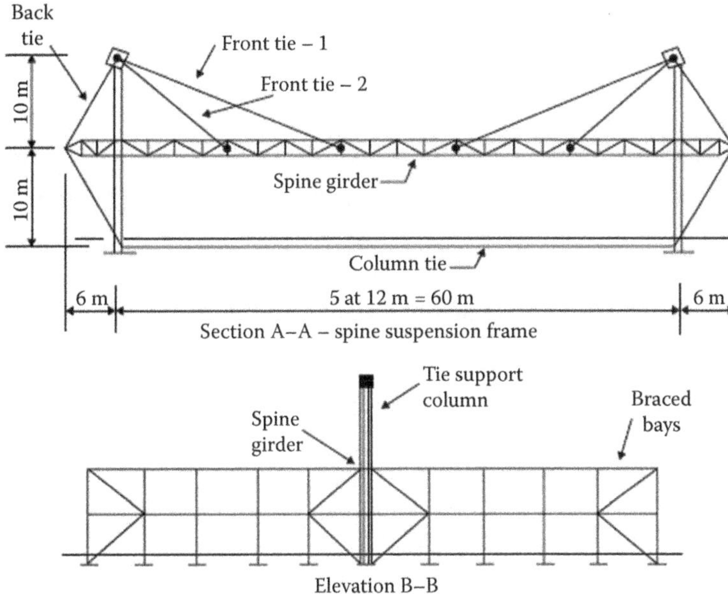

Figure 12.2 Roof and spine suspension frame.

The cable girder and cable net roof with appropriate supporting structure are architectural forms in great demand for sports and exhibition buildings. Very large spans have been covered with air-supported, cable-stiffened membranes. These structures are outside the scope of the book.

One further modern development, the retractable roof structure, is discussed. Such structures, for sports arenas, become attractive where disruption due to adverse weather causes heavy financial losses. Some outline proposals are given for a football stadium with retractable roof.

12.2 TIE-STAYED ROOF – PRELIMINARY DESIGN

12.2.1 Specification

Make a preliminary design for a wide span roof with a tie-stayed spine support structure as shown in Figure 12.2. The roof construction is to consist of steel decking supported on purlins carried on lattice girders. The covering is three layers of felt on insulation board. The total dead load of decking and girders is taken as 1.0 kN/m^2 and of the spine girder 5 kN/m. The imposed load is 0.75 kN/m^2. The steel is Grade 50 for the lattice girders and columns and Grade S355 for the stays.

Preliminary designs are to be made for:

- Roof lattice girders;
- Spine girder;
- Front and back stays and column tie;
- Spine frame column.

The design for stability and wind load is discussed.

The analysis carried out does not take into account the sinking supports of the girder. Secondary analysis tends to show that the girder section is satisfactory.

The aim of the analysis is to give member sections for a rigorous computer analysis.

Asymmetrical load cases should also be considered.

12.2.2 Preliminary design

$$\text{Design load} = (1.4 \times 1.0) + (1.6 \times 0.75) = 2.6 \text{ kN/m}^2$$

(a) *Roof lattice girders* (Figure 12.3a)

The span is 29.6 m, depth is 1.5 m and spacing 6 m. Thus,

$$\text{Load} = 2.6 \times 29.6 \times 6 = 462 \text{ kN}$$

$$\text{Chord force} = 462 \times 29.6/(8 \times 1.5) = 1140 \text{ kN}$$

The top chord is supported at 2 m centres by purlins (Steel Construction Institute, 2000). Select 120 × 120 × 8 SHS, with P_C = 1150 kN, P_t = 1260 kN. Reduce to 120 × 120 × 6 RHS at quarter points. For the web at support,

$$F = 2.5 \times 231/1.5 = 385 \text{ kN}$$

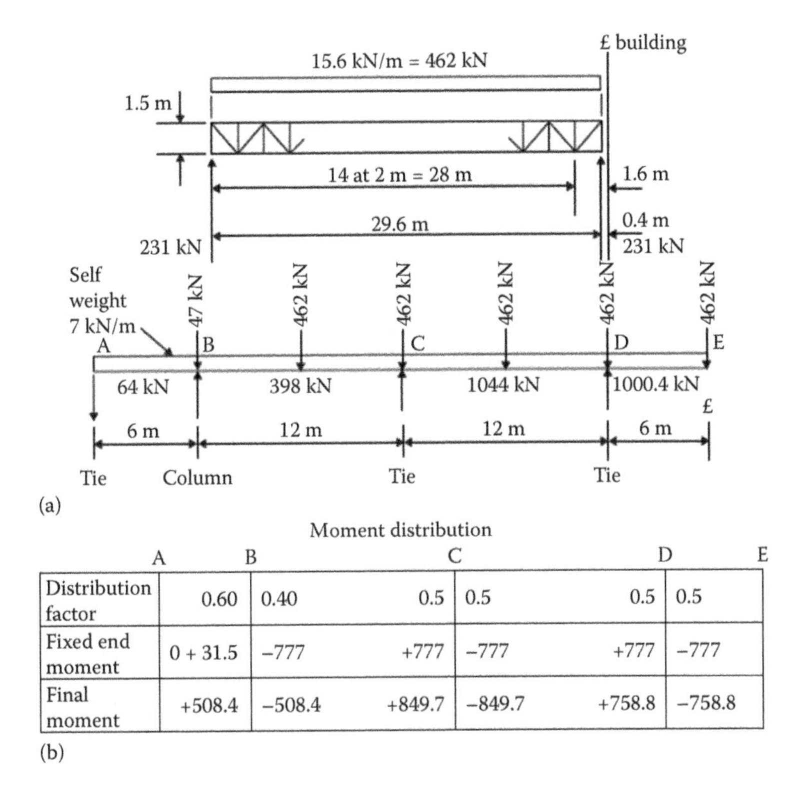

	A–B	B		C		D	
Distribution factor		0.60	0.40	0.5	0.5	0.5	0.5
Fixed end moment		0 + 31.5	−777	+777	−777	+777	−777
Final moment		+508.4	−508.4	+849.7	−849.7	+758.8	−758.8

Figure 12.3 Roof and spine girder analysis: (a) roof lattice girder; (b) spine girder.

Use 90 × 90 × 5 SHS on outer quarter lengths. Reduce to 70 × 70 × 3.5 SHS over centre half.

(b) *Spine structure – analysis*

The loads on one-half of the spine girder are shown in Figure 12.3b.

Self-weight = 1.4 × 5 = 7 kN/m

The fixed end moments are

$$\text{Span AB: } M^{F}_{BA} = 7 \times 6^2/8 = 31.5 \text{ kNm}$$

$$\text{Spans BC, CD, DE: } M^{F} = \frac{462 \times 12}{8} + \frac{7 \times 12^2}{12} = 777 \text{ kNm}$$

The distribution factors are

$$DF_{BA} : DF_{BC}, \text{from} \frac{0.75}{6} \Big/ \left(\frac{0.75}{6} + \frac{1}{12}\right) = 0.60 : 0.40$$

$$DF_{CB} : DF_{CD}, \text{and } DF_{DC} : DF_{DE} = 0.5 : 0.5$$

Results of the moment distribution are shown in the figure, from which the spine beam shears and reactions are calculated.

The forces in the ties, column and column tie are shown in Figure 12.4. The self-weight of the column is taken to be 40 kN/m. The forces are found

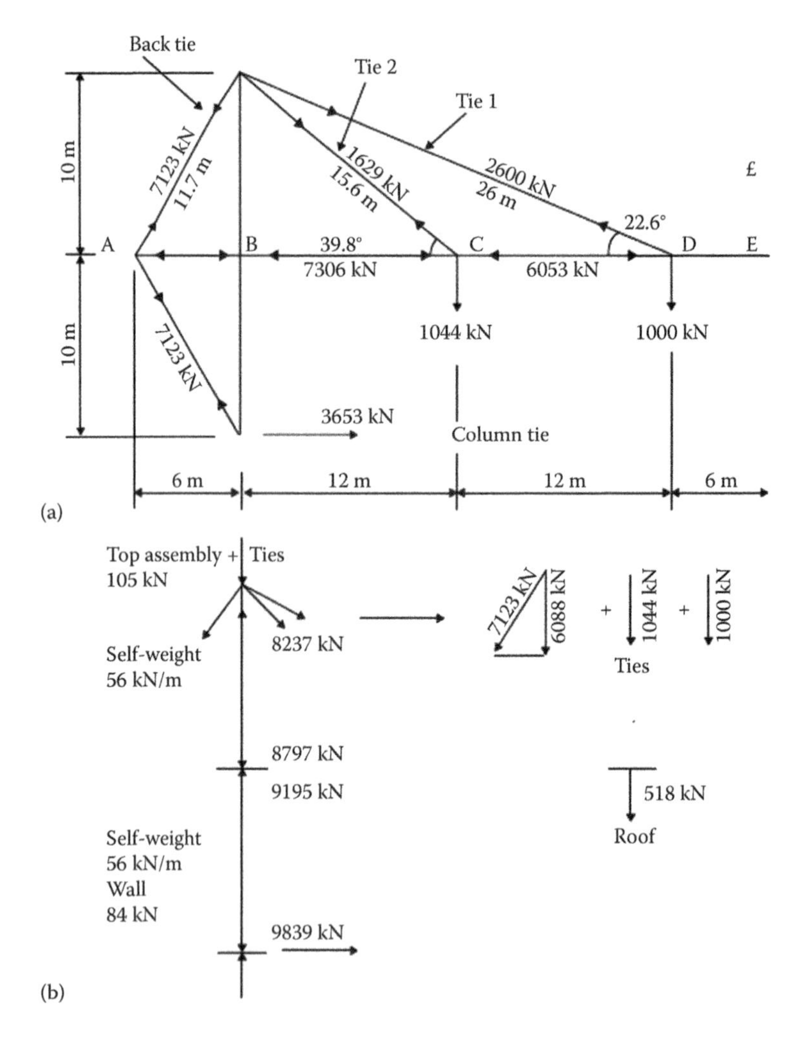

Figure 12.4 Forces: (a) stays and column tie; (b) column.

using statics. The tie forces are slightly changed by the end reaction at A in the spine girder. This is neglected.

(c) *Spine girder design*

The maximum design conditions are

$$M = 850 \text{ kNm}, F = 6909 \text{ kN}, V = 259 \text{ kN}$$

The trial section shown in Figure 12.5a consists of 4 Wo. 200 × 200 × 12.5 SHS chords, with

$$A = 4 \times 92.1 = 368 \text{ cm}^2$$

$$I_X = 4 \times 92.1 \times (75)^2 + (4 \times 5340) = 2{,}093{,}610 \text{ cm}^4$$

$$S = 4 \times 92.1 \times 75 = 27{,}630 \text{ cm}^3$$

$$r_x = (2{,}093{,}610/368)^{0.5} = 75.4 \text{ cm}$$

$$r = 7.61 \text{ cm for SHS}$$

Assume that the effective length for buckling about the XX axis is 0.9 times the total girder length of 60 m:

$$\lambda_X = 0.9 \times 6000/75.4 = 71.6$$

For one chord,

$$\lambda_Y = 600 \times 0.9/7.61 = 70.9$$

$$P_c = 265.2 \text{ N/mm}^2 \text{ (Table 24(a))}$$

$$P_c = 265.2 \times 368/10 = 9759 \text{ kN}$$

$$M_c = 27{,}630 \times 355/10^3 = 9808 \text{ kNm}$$

Combined: $(7306/9759) + (850/9808) = 0.83$

For the web member,

$$\lambda = 0.9 \times 3.36 = 3 \text{ m}$$

Select 100 × 100 × 5 RHS (Steel Construction Institute, 2000). Adopt this section throughout the girder length.

For cross members, use 80 × 80 × 5 RHS.

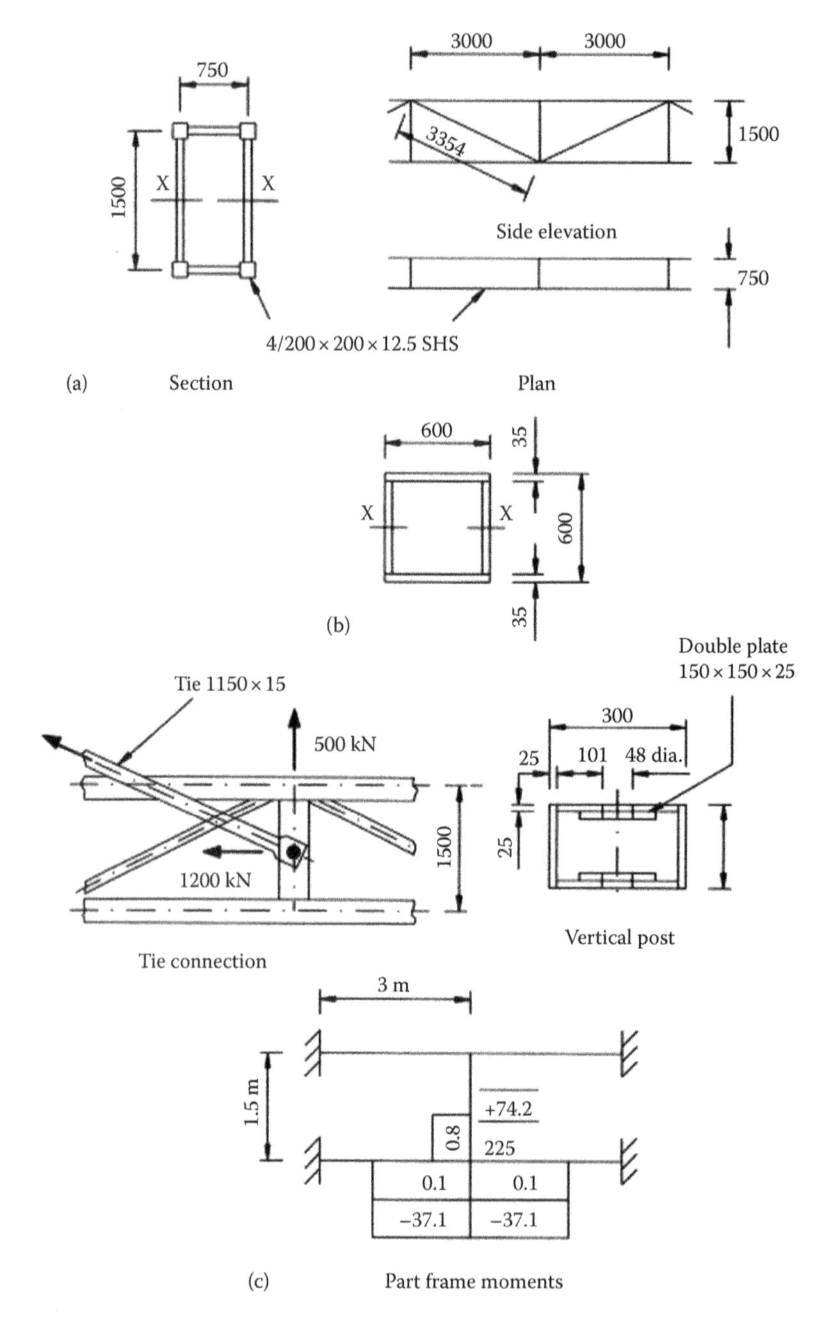

Figure 12.5 Suspension frame: (a) spine lattice girder; (b) column; (c) tie connection to post.

(d) *Column design*

The maximum design conditions are

Top length, $F = 8797$ kN

Bottom length, $F = 9839$ kN

The trial section is a 600 mm × 600 mm box × 35 mm plate (Figure 12.5b), with $P_y = 345$ N/mm^2 and

$b/t = 530/35 = 15.1$

$< (275/345)^{0.5} \times 28 = 24.99$ (plastic)

$A = 791$ cm^2, $r_x = 23.1$ cm

$\lambda = 2.5 \times 1000/23.1 = 108.2$ (Table 14, BS 5950)

$P_c = 136.3$ N/mm^2 (Table 24(b))

$P_c = 10{,}779$ kN

Refer to stability assessment in Section 12.2.3.

(e) *Front tie 1* (Figures 12.4a and 12.5c)

The stay material is Grade S480 Steel.

Tie force = 2600 kN + (self-weight ≅ 30 kN) ≃ 2630 kN

Provide four flats, with $T = 658$ kN per flat. Allow for splices – 4 No. 20 mm diameter Grade 8.8 bolts, double shear capacity 184 kN. Try flat 150 mm × 15 mm with 2 No. 22 mm diameter holes:

Capacity $P_t = (150 \times 15 - 2 \times 15 \times 22)460/10^3 = 731.4$ kN

The tie is connected by pin to the centre vertical of the spine girder. For Grade 8.8 material, shear strength $P_s = 375$ N/mm^2.

Pin diameter $= (658 \times 10^3 \times 4/375\pi)^{0.5} = 47.2$ mm

Provide 50 mm diameter pin. For post plate use Grade S355 steel, P_y = 345 N/mm^2.

Design for bearing (Clause 6.5.3.3 of BS 5950):

$t = 658 \times 10^3/1.5 \times 345 \times 50 = 25.4$ mm

If rotation is not required (4 pin is not intended to be removable). Or $t = 658 \times 10^3/0.8 \times 345 \times 50 = 47.7$ mm (for rotation and removable pin). Provide 50 mm thickness – use 25 mm doubler plate as shown in Figure 12.4c.

For tie plate *at pin* use Grade S460 steel, $P_y = 430$ N/mm^2. Thickness for bearing;

$$t = 658 \times 10^3/1.5 \times 430 \times 50 = 20.4 \text{ mm or}$$

$$t = 658 \times 10^3/0.8 \times 430 \times 50 = 38.3 \text{ mm}$$

Provide 40 mm plate and splice to 20 mm plate. The tie end can be designed to conform to Figure 26 of BS 5950.

The vertical post section is shown in Figure 12.5c.

$$I_X = 25{,}468 \text{ cm}^4 \text{ (gross)}$$

$$A = 225 \text{ cm}^2, S = 2960 \text{ cm}^3 \text{ (at centre)}$$

The pin load causes bending and axial load in the post and bending in the chords. Consider the part frame in the figure. Neglect diagonal members.

For the chords (200 × 200 × 12.5 SHS),

$$I/L = 5340/300 = 18$$

For the post (300 × 200 box),

$$I/L = 25{,}468/150 = 170$$

The distribution factors are

$$\text{Chord} : \text{Post} : \text{Chord} = \frac{18:170:18}{206} = 0.1:0.8:0.1$$

The fixed end moment is

$$1200 \times 1.5/8 = 225 \text{ kNm}$$

After distribution the final moments are

$$\text{Chord } M = 37.1 \text{ kNm}$$

$$\text{Post-support } M = 74.2 \text{ kNm}$$

$$\text{Centre } M = 375.8 \text{ kNm}$$

For the chord the actions at D from Figures 12.3b and 12.4a are

$F = 6053$ kN

$M_{girder} = 758.8$ kNm

$M_{chord} = 37.1$ kNm

$M_c = 228$ kNm

Combined ((c) above):

$$\frac{6053}{9759} + \frac{758.8}{9808} + \frac{37.1}{228} = 0.86$$

For the post,

$F = 500$ kN, $M = 375.8$ kNm

$M_c = 1021$ kNm, $P_c = 7763$ kN

Combined = 0.43

The post side plate thicknesses are determined by the bearing of the pin.

(f) *Front tie 2*

Tie force $T = 1629$ kN + 20 kN factored self-weight

Use 4 No. 130 mm × 15 mm flats for 412 kN/flat.

Splice = 3 No. 20 mm dia. Grade 8.8 bolts

Pin diameter = 40 mm

Tie end thickness for bearing = 22 mm

Column plate thickness for bearing = 25 mm

(g) *Back ties*

Tie force $T = 7123$ kN + 40 kN factored self-weight

Use 4 No. 260 mm × 25 mm flats for 1790 kN/flat.

Splice = 6 No. 27 mm dia. Grade 8.8 bolts

Pin diameter = 80 mm

Tie end thickness for bearing = 45 mm

Column plate thickness for bearing = 55 mm

(h) *Column ties*

Tie force T = 3653 kN

Use 2 No. 275 mm × 25 mm flats. For bolts to column plates, use 9 No. 30 mm diameter Grade 8.8 m single shear for each flat.

(i) TIE CONNECTION ARRANGEMENTS

The connection arrangements for the stays are shown in Figure 12.6.

12.2.3 Stability and wind load

The spine column must be restrained laterally and longitudinally in position at roof level. Restraint is provided by the roof and wall bracing systems. These systems, shown in Figure 12.6d also resist wind loads on the building.

(a) *Column restraint force*

Refer to Clauses 4.7.3, 4.7.9 and 4.7.12 of BS 5950. The code states that the restraint system must be capable of resisting 2.5% of the factored load in the column. Also for comparison see the relevant clauses of EC3.

The restraint loads are as follows.

(i) DEAD AND IMPOSED LOAD CASE

Factored column load = 9195 kN (Figure 12.4b)

Restraint lateral load = 0.025 × 9195 = 230 kN

(ii) DEAD, IMPOSED AND WIND LOAD CASE

The frame is reanalysed to give

Dead load = 4002 kN

Imposed load = 2244 kN

Factored column load = 1.2(4002 + 2244) = 7495 kN

Restraint lateral load = 187.4 kN

This load is to be added to the factored wind load.

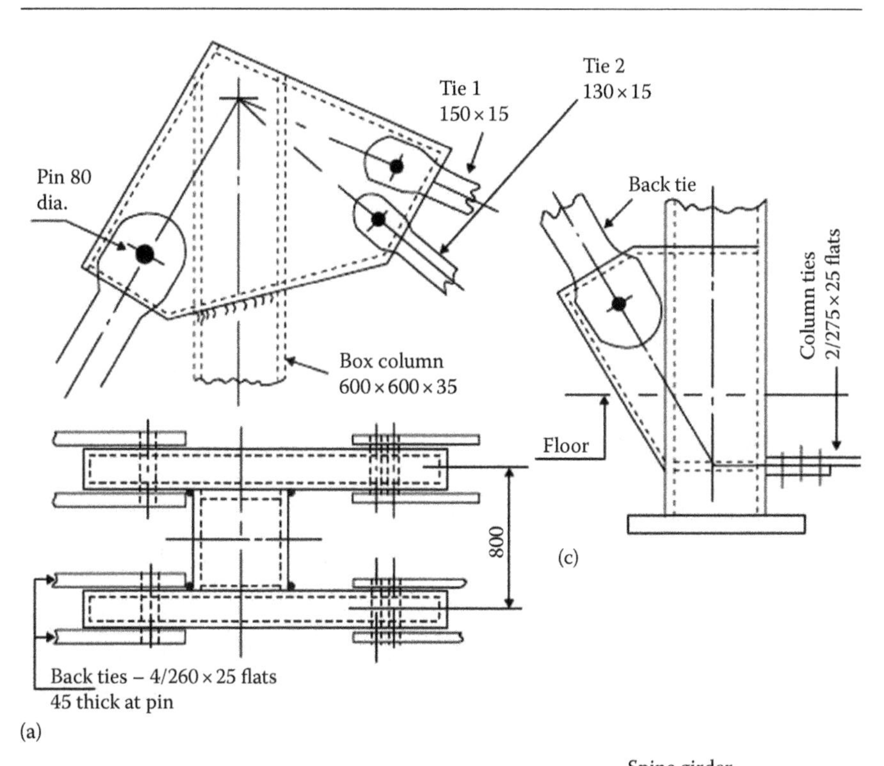

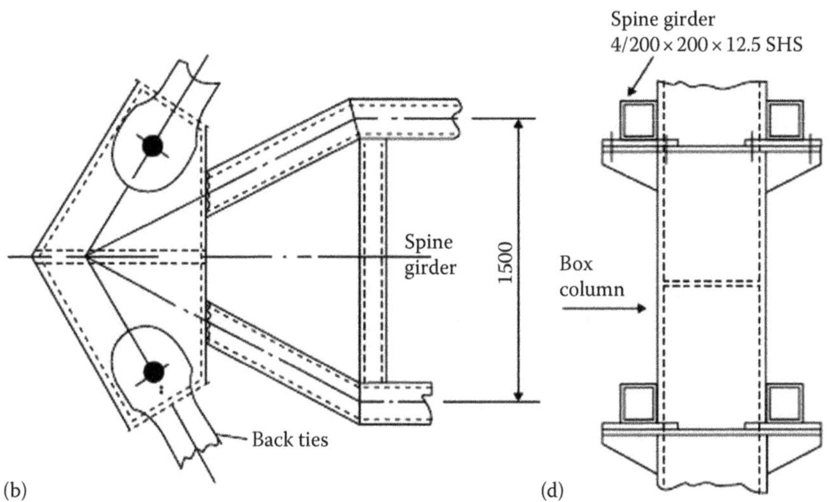

Figure 12.6 Tie connections: (a) top, stays–column; (b) back, stays–spine girder; (c) base, stays–column tie–column; (d) spine girder–column. (Not all stiffeners are shown.)

(b) *Wind load*

The following components of wind load must be calculated for wind blowing in the transverse and parallel directions.

- Wind on the building face;
- Wind on the suspension frame and column;
- Frictional drag on the roof and walls.

The various components of wind load are shown in Figure 12.7b.

12.3 SPACE DECKS

12.3.1 Two-way spanning roofs

If the length of the area to be roofed is more than twice the breadth, it is more economical to span one way. If the area is nearer square, the more economical solution, theoretically, is to span two ways. A rectangular area can be divided into square or near-square areas with lattice girders and then two-way-spanning structures can be installed in the subdivided roof (Figure 12.8a).

The two-way-spanning grid may be a single or double layer. The single-layer grid of intersecting rigid jointed members is expensive to make and more easily constructed in reinforced concrete than steel.

The double-layer grid can be constructed in a number of ways. Lattice or Vierendeel girders intersecting at right angles can be used to form two-way grids where the bottom chord lies below the top chord (Figure 12.8b). The roof is divided into squares which can be covered with plastic roof units. Three-way grids, where the surface is divided into equilateral triangles, are shown in Figure 12.8c.

Double-layer grids, where the bottom chords do not lie in the same vertical plane or in some cases do not have the same geometrical pattern as the top chords, are termed space grids or decks.

12.3.2 Space decks

The basic form of a space deck shown in Figure 12.8d is square on square offset with cornice or mansard edge, where the top and bottom chords form squares of equal area. The basic unit is the inverted square-based pyramid shown in Figure 12.8e. Other variations are possible such as that shown in Figure 12.8f, termed square on larger square set diagonally.

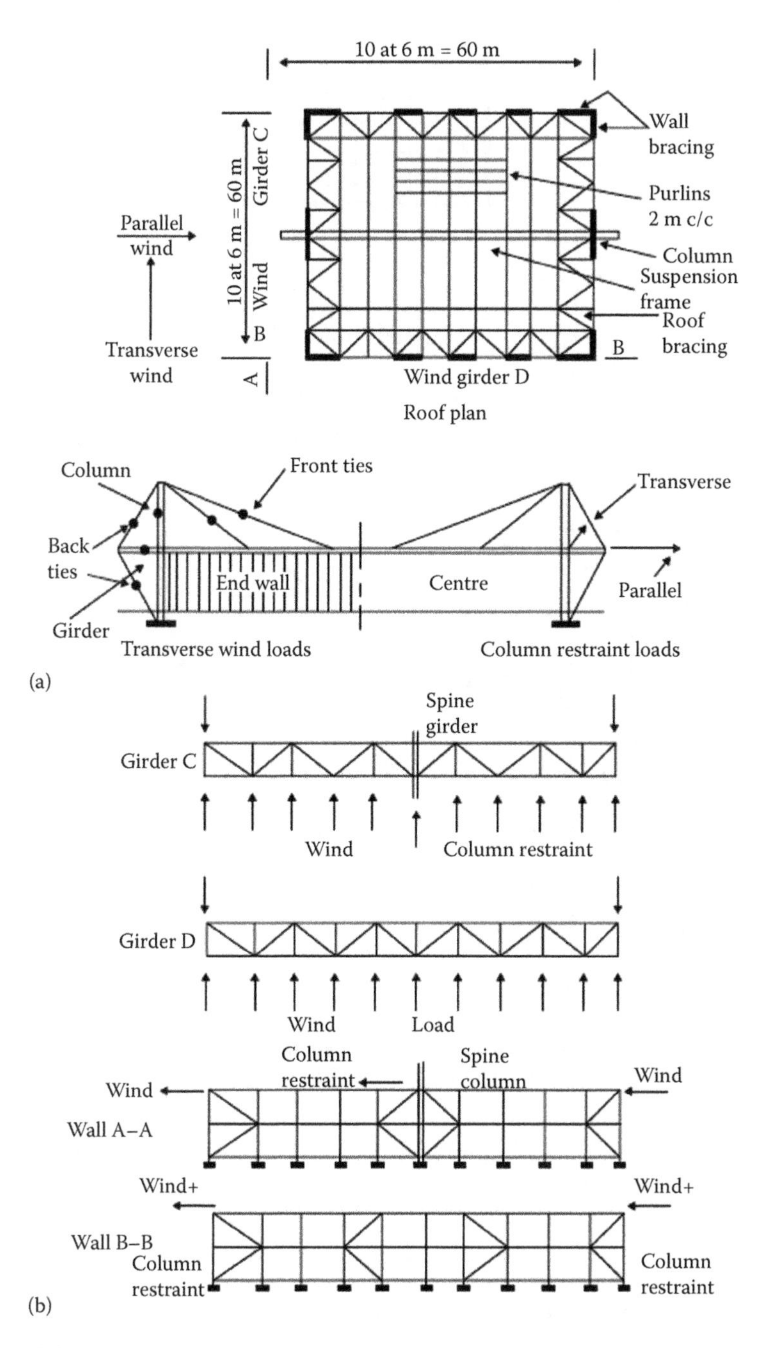

Figure 12.7 Wind and column restraint loads: (a) roof and wall bracing systems; (b) load components.

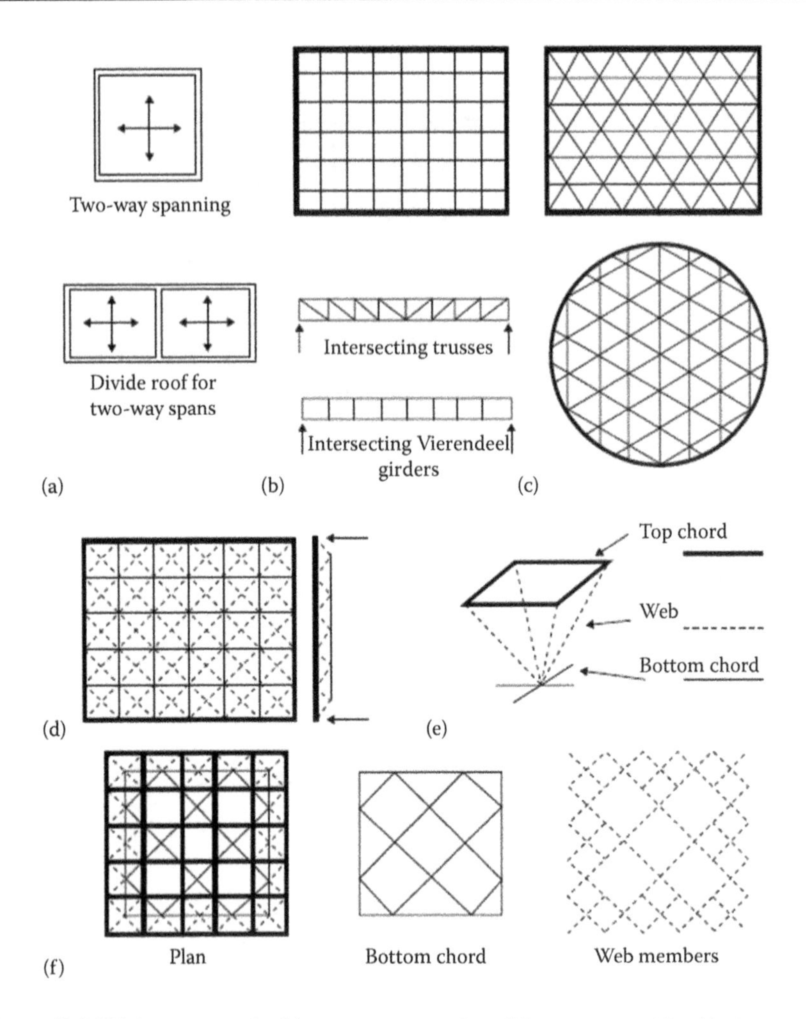

Figure 12.8 Wide-span roofs: (a) two-way spanning; (b) two-way grids; (c) three-way grids; (d) space deck with cornice edge; (e) basic pyramid unit; (f) square on larger square set diagonally, cornice edge.

A great effort in inventiveness, research and testing has gone into perfecting systems for constructing space decks. Two main construction methods are used:

- Division into basic pyramid units;
- Joint systems to connect deck members.

These systems are discussed below. Triangular trusses have also been used as the shop fabricated unit.

(a) *Basic pyramid unit*

As stated above, the inverted pyramid forms the basic unit of the square on square space deck. The units are prefabricated with top chords as angles or channels, tubular section web members and bottom chords consisting of high tensile bars with screwed ends. Top chord lengths vary from 1.2 to 2.5 m, while depths vary from 0.8 to 1.5 m. The pyramids conveniently stack together for transport.

The grid is assembled by bolting the top chords together and connecting the bottom chords through the screwed joints. The grid is constructed at ground level and hoisted into position as a complete structure.

(b) *Joint system*

The joint system gives greater flexibility in space deck construction than the fixed component system. The joint connects eight members coming from various directions in space to meet at a point. The finished structure is assembled from straight members, usually hollow sections and joints.

Three commercial joint systems are briefly described below.

- Nodus System – This was developed by the British Steel Corporation and is now manufactured and marketed by Space Deck Limited. It consists of two half castings clamped together by a bolt. The chords lock into grooves in the castings, while the webs have forked ends for pin connections to the lugs on the castings.
- Mero Joint System – This consists of cast steel balls into which the ends of members are screwed through a special end connector. The Mero balls are made in a range of sizes to accommodate different member sizes and web member inclinations. There is no eccentricity at the joint.
- Nippon Steel Corporation NS Space Truss System – The joint consists of a steel bowl node to which the space deck members are connected by special bolted joints. There is no eccentricity at the joint.

Very large and heavy space deck systems have been constructed for aircraft hangers, wide conference halls. etc. These have heavy tubular or box members and site-welded joints.

12.3.3 Space deck analyses and design

Space decks are highly redundant and analysis is carried out using a space frame program. Joints are normally taken as pinned. The space deck dead load from the flat roof decking and self-weight of grid is greater than uplift due to wind. Only the dead and imposed load case need be considered.

The space deck can be considered as a plate supported on four sides to obtain member forces for preliminary design. British Steel Corporation Tubes Division (n.d.) gives approximate formulae for chord forces and reactions. These can be used to obtain sections for computer analysis.

For a square or rectangular grid only one-quarter of the frame need be considered in the analysis. Rotational and linear restraints are applied as required to members cut by the section planes. If members are sectioned lengthwise, one-half the properties are used in the analysis.

For a grid considered as pin jointed, member design is as follows.

- Top chord – Design for axial compression and bending due to roof load applied through purlins or roof units. Effective length = 0.9 × member length.
- Bottom chord – Design for axial tension and any bending from ceiling or service loads.
- Webs – Design for tension or compression as applicable. Effective length = 0.9 × member length.

12.4 PRELIMINARY DESIGN FOR A SPACE DECK

12.4.1 Specification

A preliminary design is made for a space deck roof for a building 60 m square. The roof construction consists of steel decking supported on the space frame and purlins, insulation board and three layers of felt. The total dead load including an allowance of 0.3 kN/m^2 for the space deck is 1.0 kN/m^2. The imposed load is 0.75 kN/m^2. The deck steel is Grade S355.

12.4.2 Arrangement of space deck

The arrangement for the space deck is shown in Figure 12.9a. The deck is square on square offset with mansard edge. The square module is 5 m × 5 m and the deck depth is 3 m. The deck is supported around the perimeter on 8 m high columns located at the bottom chord node points. The side walls are braced to resist wind loading.

12.4.3 Approximate analysis and design

An approximate analysis as a plate is made to obtain the maximum forces in the space deck members from which the sizes to be used in the final analysis can be determined.

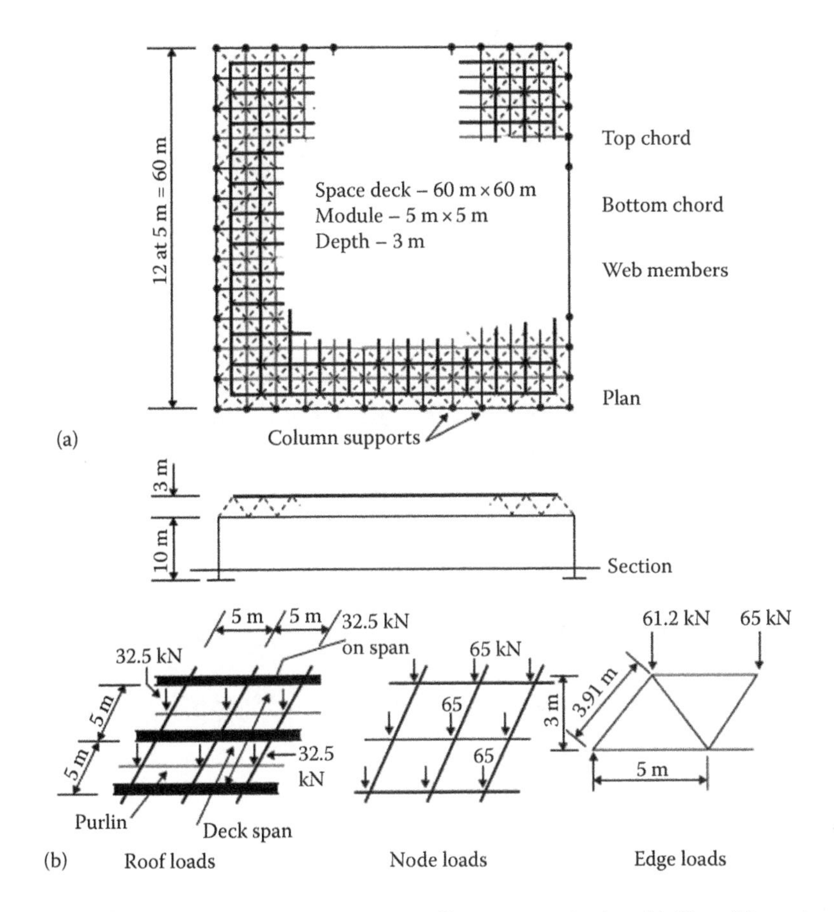

Figure 12.9 Space deck – square on square offset, mansard edge. (a) Plan; (b) section: roof loads, node loads and edge loads.

The approximate analysis is from British Steel Corporation Tubes Division (n.d.).

Design load = $1.4 \times 1.0 + 1.6 \times 0.75 = 2.6$ kN/m^2

The load arrangement on the top chord members is shown in Figure 12.9b where the purlin applies a point load and the decking a uniform load on the top chord members.

Total load on the deck $T = 2.6 \times 60^2 = 9360$ kN

Deck module $n = 5$ m

Depth of deck $D = 3$ m

Chord force factor $F = 0.08$

Maximum chord force = 9360 × 5 × 0.08/3 = 1248 kN

Maximum reaction = 9360 × 5 × 0.08 × 8/60 = 500 kN

The initial design of the grid members is as follows.

(a) *Bottom chord*

$F = 1248$ kN

Try 150 × 150 × 6.3 RHS, with $P_t = 1271$ kN or 168.3 × 8 CHS, P_t = 1430 kN.

(b) *Top chord*

$F = 1248$ kN

$M = 2.6 \times 2.5 \times 5^2/4 = 40.6$ kNm

Try 200 × 200 × 8 RHS, with $r = 7.81$ cm, $A = 60.8$ cm^2, $S = 436$ cm^3.

$\lambda = 0.9 \times 5000/78.1 = 57.5$

$P_c = 304$ N/mm^2 (Table 24(a))

$M_c = 436 \times 355/10^3 = 154.8$ kNm

Combined:

(1248 × 10/304 × 60.8) + (40.6/154.8) = 0.93

An alternative section is 219.1 × 10 CHS.

(c) *Web members*

Length = (2 × 2.5^2 + 32)0.5 = 4.64 m

Load = 500 × 4.64/(2 × 3) = 386 kN

Effective length = 0.9 × 4.64 = 4.18 m

Select 139.7 × 5 CHS, with $P_c = 441$ kN.

12.4.4 Computer analysis

The computer analysis is carried out for a pin-jointed frame for dead and imposed loads only. Wind uplift is less than the dead load. Due to symmetry, only one-quarter of the frame need be considered for analysis. The data for the space frame program are discussed and set out below.

(a) *Joint coordinates*

The quarter frame and joint numbering are shown in Figure 12.10. The joint coordinates and member connection can be taken off the figure and member numbers assigned.

Number of joints = 97

Number of members = bottom chords (84) + top chords (72)
+ web (144) = 300

The joints are taken as pinned.

(b) *Member properties*

The properties of the various member types are taken from the approximate design. Section properties are reduced for the two outer modules of the top and bottom chords and for the inner 4.5 modules of the web, that is:

- Top chord – Outside lines 38–94 and 38–42;
- Bottom chord – Outside lines 31–87 and 31–35;
- Web – Outside lines 31–87 and 31–35.

One-half of the properties of the bottom chord members lying on the section planes are used in the analysis. The member properties are given in Table 12.1.

(c) *Restraints*

The restraints at supports and where members are cut by section planes are listed in Table 12.2.

The maximum member forces are given for the inner and outer modules as specified in Section 12.4.4(b) (Figure 12.9 and Table 12.1). The top chord moment due to the purlin load is 40.6 kNm in members carrying purlins (Figure 12.9b and Section 12.4.3(b)). In members at right angles the moment is 20.8 kNm.

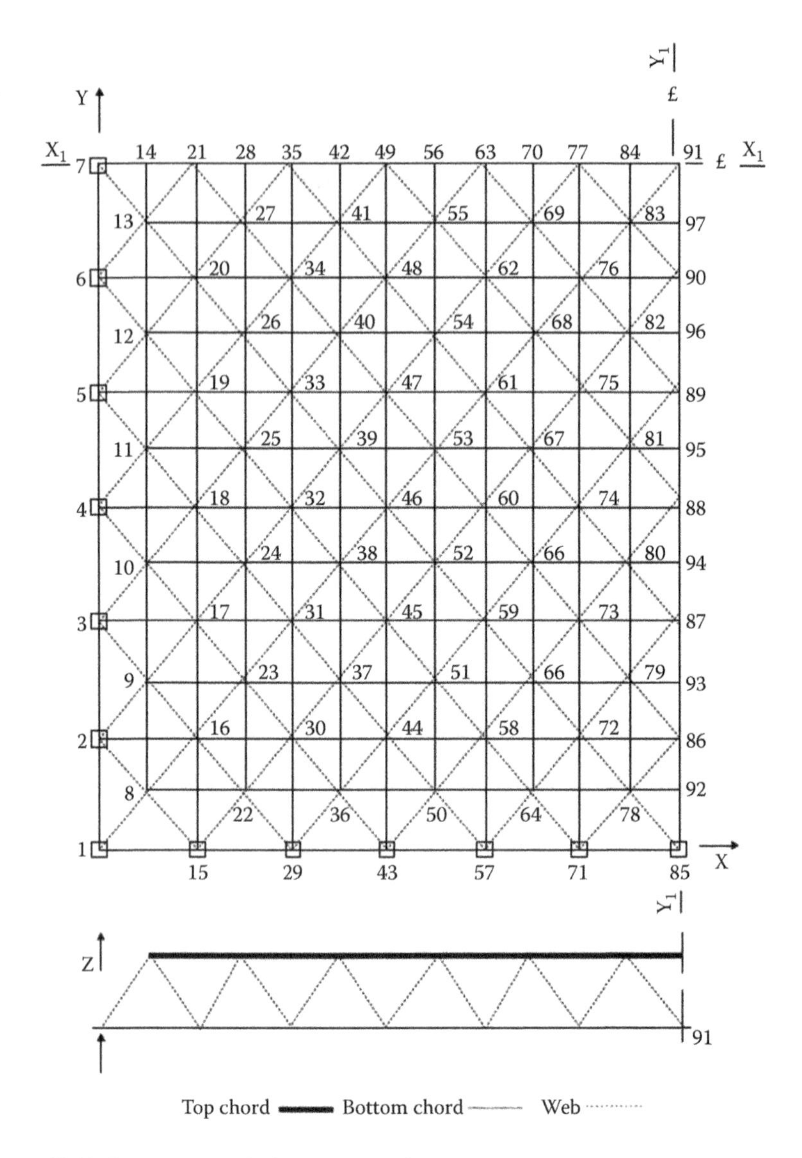

Figure 12.10 One-quarter deck – joint members.

(d) *Loading*

All loads are applied to the top chord nodes as follows (Figures 12.9b and 12.10):

- Node 8

$$\text{Load} = [2.5^2 + (2.5 + 0.625)]3.91 \times 1.4/2 + (3.75^2 \times 1.2) = 42.8 \text{ kN}$$

Table 12.1 Member properties for analyses

Members	*Section*	*Area* (cm^2)	*Moment of inertia* (cm^4) I_X	I_Y	*Torsion constant* J
Top chord					
Outer modules	180 × 180 × 6.3 RHS	43.6	2170	2170	3360
Inner modules	200 × 200 × 8 SHS	60.8	3710	3710	5780
Bottom chord					
Outer modules	139.7 × 8 CHS	33.1	720	720	1440
Inner modules	168.3 × 8 CHS	40.3	1300	1300	2600
Web					
Outer modules	139.7 × 5 CHS	21.2	481	481	961
Inner modules	114.3 × 5 CHS	17.2	257	257	514

Table 12.2 Restraints

Joint no.	*Restraint*
Supports:	
1, 2, 3, 4, 5, 6	LZ
15, 29, 43, 57, 71	LZ
7	LY, LZ, RX
85	LX, LZ, RY
Section plane X1–X1:	
14, 21, 28,..., 70, 77, 84	LY, RX
Section plane Y1–Y1:	
92, 86, 93,..., 90, 97, 91	LX,RY
Centre: 91	LX, LY, RX, RY, RZ

Note: L = linear restraint and direction axis; R = rotational restraint and axis about which it acts.

- Nodes 9, 10, 11, 12, 13, 22, 36, 50, 64, 78 –

$$\text{Load} = [(3.91 + 5)5 \times 1.4/2] + [(5 + 2.5)5 \times 1.2/2] = 53.7 \text{ kN}$$

- Nodes 23, 24, 25, 26, 27, 37, 38, 39, 40, 41, 51, 52, 53, 54, 55, 65, 66, 67, 68, 69, 79, 80, 81, 82, 83 –

$$\text{Load} = 2.6 \times 25 = 65 \text{ kN}$$

The maximum moment in top chord members is

$$M = 32.5 \times 5/4 = 40.63 \text{ kNm}$$

The loads over supports are not listed.

Table 12.3 Computer results for critical members

Location	*Member*	*Force[a] (kN)*	*Maximum moment (kNm)[b]*
Top chord			
Inner modules	83–84	−1587	40.6
Outer modules	79–80	−936	40.6
Bottom chord	76–77	+1225	–
Inner modules	90–91	+1126	–
Outer modules	17–33	+586	–
Web	7–13	−310	–
Outer modules	13–21	+333	–
Inner modules	35–41	−211	–
	41–49	+197	–

[a] + = tension; − = compression.
[b] Moments for top chord calculated by statics.

12.4.5 Computer results

The computer output for critical members in top and bottom chords and web members is given in Table 12.3.

12.4.6 Member design

(a) *Top chord – inner modules*

$$C = -1587\ \text{kN},\ M = 40.6\ \text{kNm}$$

Try 200 × 200 × 10 SHS, with $r = 7.72$ cm, $A = 74.9$ cm^2, $S = 531$ cm^3.

$$\lambda = 0.9 \times 5000/77.2 = 58.1$$

$$P_c = 302.8\ \text{N/mm}^2$$

$$P_c = 302.8 \times 74.9/10 = 2268\ \text{kN}$$

$$M_c = 355 \times 531/10^3 = 188.6\ \text{kNm}$$

Combined: (1587/2268) + (40.6/188.6) = 0.91
This is satisfactory.

(b) *Top chord – outer modules*

$$C = -936\ \text{kN},\ M = 40.6\ \text{kNm}$$

Select 180 × 180 × 8 SHS.

(c) *Bottom chord – inner modules*

$T = 1225$ kN

Select section from capacity tables. Select 168.3 × 8 CHS, with $P_t = 1430$ kN from capacity tables (Steel Construction Institute, 2000).

(d) *Bottom chord – outer modules*

$T = 586$ kN

Select 114.3 × 5 CHS, with $P_t = 611$ kN.

(e) *Web – outer modules*

$C = -310$ kN, $T = 333$ kN

Effective length $l = 4.64 \times 0.9 = 4.2$ m

Select 139.7 × 5 CHS, with $P_c = 441$ kN for $l = 4$ m, $P_t = 753$ kN.

(f) *Web – inner modules*

$C = -211$ kN, $T = 197$ kN

Select 114.3 × 5 CHS.

12.5 FRAMED DOMES

12.5.1 Types

Domes are most usually generated by rotating a plane curve, often a sector of a circle, about the vertical axis. Other curves, such as a parabola or ellipse, could be used, or the dome could be formed from intersecting cones. Again, domes are usually constructed on circular or regular polygonal bases where apexes touch the circumscribing circle. Other base shapes can be used.

From classical times masonry domes have been constructed. Another example is the Eskimo's igloo. The large-span skeletal or braced dome dates from the last century. Members may be curved or straight to meet at joints lying on the shell surface. Domes are doubly curved synclastic surfaces, that is, the curvature is of the same sign in each direction.

Braced domes are classified according to the way in which the surface is framed. Many different patterns have been devised. The main types of spherical dome (Figure 12.11) are as follows.

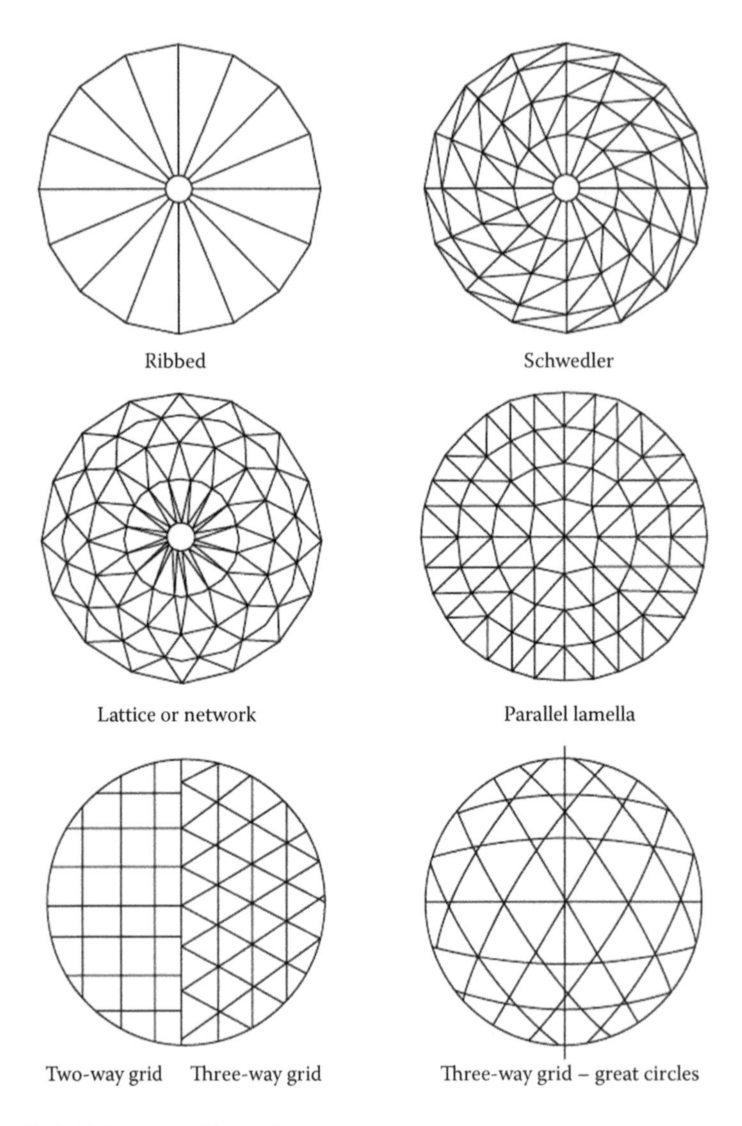

Figure 12.11 Main types of framed domes.

(a) *Ribbed dome*

This dome consists of equally spaced radial ribs on arches supported by a compression ring at the top and a tension ring or separate bases at ground level. The ribs carry triangular loading and the dome can be designed as a series of two- or three-pinned arches.

(b) *Schwedler dome*

This consists of ribs or meriodinal members and parallel rings or hoops which support the ribs. The two member systems divide the surface into trapezoidal panels, which are braced diagonally to resist shear due to asymmetrical loading. The joints between ribs and rings may be made rigid as an alternative method of resisting shear. If the dome is loaded symmetrically and the joints are taken as pinned the structure is statically determinate. An analysis can also be based on spherical thin-shell theory.

(c) *Lattice or network dome*

In this type, parallel rings are spaced equidistantly. The annular spaces are then subdivided by triangular networks of bars. Members between any two adjacent rings are equal in length.

(d) *Lamella domes*

Two types, the curved and parallel lamella, are defined. In the curved type the surface is divided into diamond-shaped areas, while the parallel lamella type consists of stable triangular divisions. The world's two largest domes are of the parallel lamella type. The curved lamella type was a development for timber domes, where the timber cladding provided stability. Note that in the network dome the horizontal rings cover this requirement.

(e) *Grid dome*

This type of dome is formed by a two- or three-way intersecting grid of arcs. Where the arcs are great circles, the geodesic dome is one special case of the grid dome.

(f) *Geodesic dome*

This system was developed by Buckminster Fuller. Most geodesic dome construction is based on the icosohedron, the regular 20-sided solid whose apexes touch the surface of the circumscribing sphere. The dome is formed from part of the sphere. Each primary spherical triangle may be subdivided further to make the framing of large domes possible. The main advantage of this type of dome is that all members are of approximately equal length and the dome surface is subdivided into approximately equal areas.

12.5.2 Dome construction

(a) *Framing*

Dome framing may be single or double layer. Large domes must be double layer to prevent buckling. All types of members have been used. Hollow sections with welded joints are attractive where the steelwork is exposed. Members are usually straight between nodes. The dome must be broken down into suitable sections for shop fabrication. Lattice double-layer domes, can be assembled on site using bolted joints.

(b) *Proprietary jointing systems*

Domes systems using the Mero and Nippon NS Space Truss joints are available. The domes can be single or double layer. Network and geodesic systems are available.

(c) *Cladding*

Cladding causes problems because panel dimensions vary in most domes and twisted surface units are often needed. The systems used are

- Roof units, triangular or trapezoidal in shape, supported on the dome frame – These may be in transparent or translucent plastic or a double-skinned metal sandwich construction;
- Timber decking on joists with metal sheet or roofing felt covering;
- Steel decking on purlins and dome members with insulation board and roofing felt – This can only be used on flat surfaces.

12.5.3 Loading

(a) *Dead load*

The dead load varies from about 0.5 to say 1.2 kN/m^2 of the dome surface depending on the type of roof construction and cladding used and whether a ceiling is provided over the inner surface. The load acts uniformly over the roof surface area.

(b) *Imposed load*

The imposed load is 0.75 kN/m^2 as specified in BS 6399: Part 1. This load acts on the plan area of the dome. It is necessary to consider cases where the load covers part only of the roof.

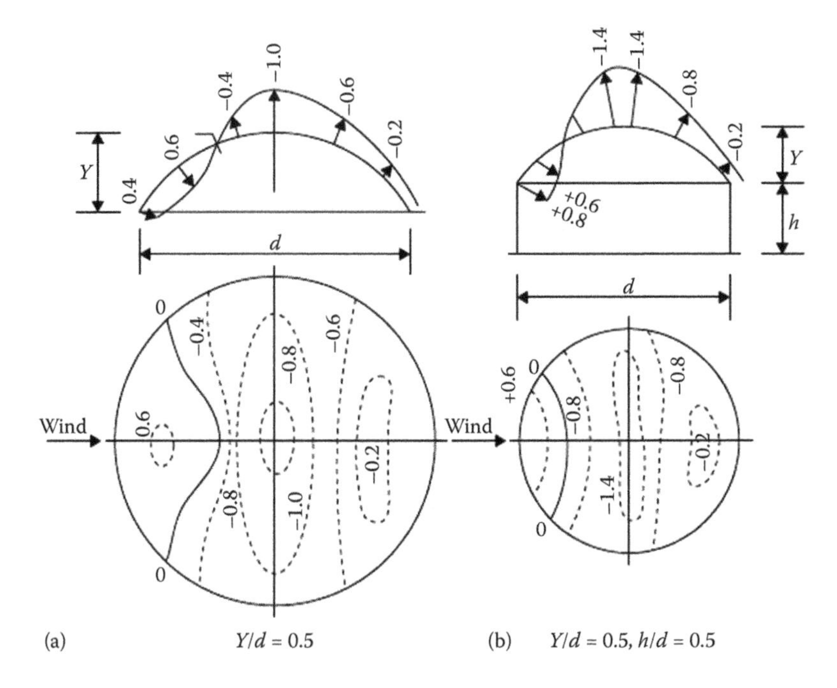

Figure 12.12 Pressure distribution – C_{pe} values for: (a) dome rising from ground; (b) dome on cylindrical base.

(c) *Wind load*

The distribution of wind pressures on domes determined by testing is given in various references (e.g., Newberry and Eaton, 1974; Makowski, 1984).

The external pressure and suction distribution depend on the ratio of dome height to diameter for a dome rising from the ground. If the dome rests on a cylindrical base, the cylinder height as well as the dome rise affects the values. In general there is a small area directly normal to the wind in pressure, while the major part of the dome is under suction. The wind pressure distributions for the two cases are shown in Figure 12.12. The distributions are simplified for use in analysis.

12.5.4 Analysis

The ribbed dome with ribs hinged at the base and crown and the pin-jointed Schwedler dome subjected to uniform load are statically determinate. The Schwedler dome under non-uniform load and other types of domes are highly redundant.

Shell membrane theories can be used in the analysis of Schwedler domes under uniform load. Standard matrix stiffness space frame programs can be used with accuracy to analyse stiff or double-layer domes. The behaviour of flexible domes may be markedly nonlinear and the effect of deflection must be considered. Dome stability must be investigated through nonlinear analysis. A linear analysis will be sufficiently accurate for design purposes in many cases.

12.5.5 Stability

Flexible domes, that is, shallow or single-layer large-span domes present a stability problem. Three distinct types of buckling are discussed in Galambos (1988). These are as follows:

(a) General buckling – A large part of the surface becomes unstable and buckles. Failures of this type have occurred where snow loads have covered part of the dome surface. Shell theory is extended to predict the critical pressure causing buckling. The critical pressure on a thin shell is

$$p_{cr} = CE(t/R)^2$$

where R is the radius, E the modulus of elasticity, t the shell thickness and C the shell coefficient.

Modified forms are given for the thickness t to allow for membrane and ribs in a braced dome. Various values are given for C.

(b) Snap through or local buckling – One loaded node deflects or snaps through, reversing the curvature between adjacent nodes in that area. Expressions are given for checking this condition.

(c) Member buckling – An individual member buckles as a strut under axial compression. This is considered in member design.

Dome stability can be studied using nonlinear matrix analysis.

12.6 SCHWEDLER DOME

12.6.1 Specification

A circular arts pavilion is required for a city centre cultural development. The building is to be 50 m diameter with 4 m height at the perimeter walls. It is proposed to construct a Schwedler dome, diameter 59.82 m, spherical radius 43.06 m and height 12 m at the crown to meet these requirements. The dome is to have 20 radial ribs to give a 20-sided polygonal plan shape. The arrangement and framing for the dome are shown in Figure 12.13.

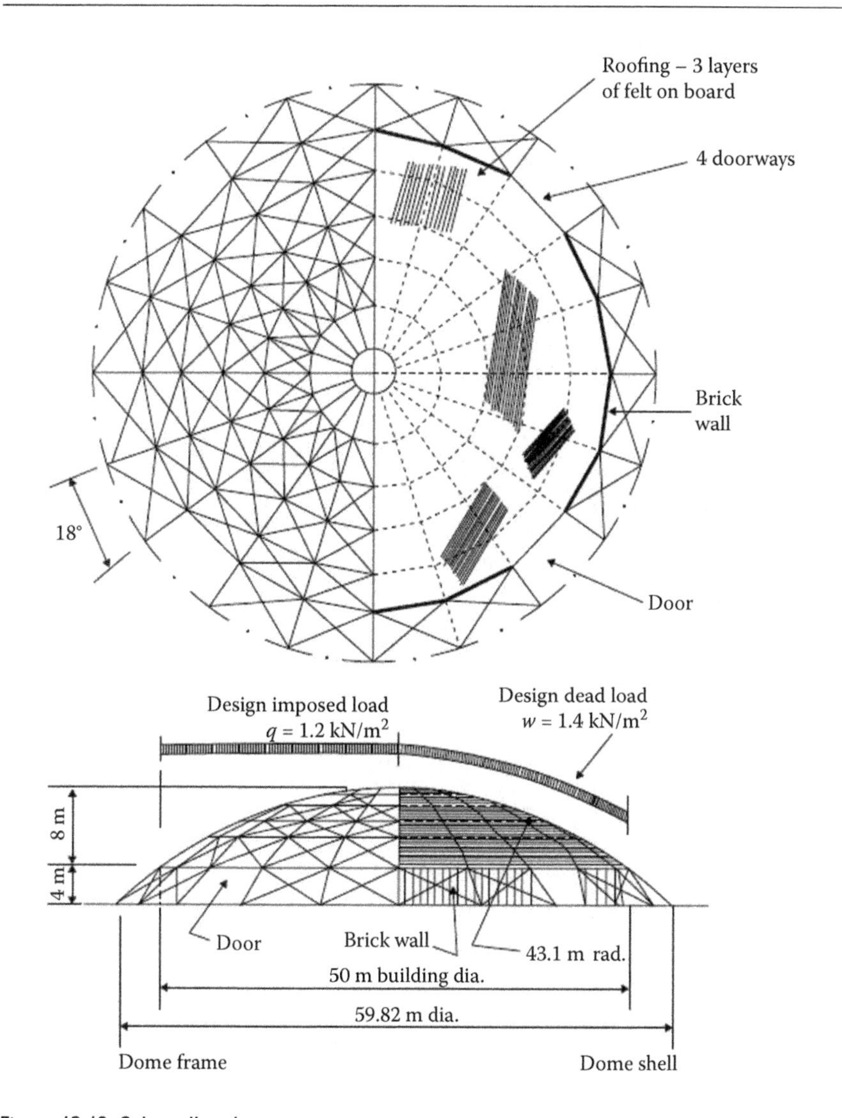

Figure 12.13 Schwedler dome.

The roofing material is to be timber supported on purlins at about 1.5 m centres spanning up the roof slope and covered with three layers of felt. The ceiling inside is plasterboard on joists. The roof dead load is 1.0 kN/m^2 and the imposed load is 0.75 kN/m^2.

Make a preliminary design to establish sections for the ribs and rings when the dome is subjected to dead and imposed load over the whole roof. These sections would then be used as the basis for detailed computer analyses. Compare the solution from the statical analysis in the preliminary design with that from a shell membrane analysis.

12.6.2 Loading for statical analysis

The design loads are

Dead load on slope = 1.4 × 1.0 = 1.4 kN/m^2

Imposed load on plan = 1.6 × 0.75 = 1.2 kN/m^2

Dome steel on slope = 1.4 × 0.25 = 0.35 kN/m^2

The dimensions for calculating the rib loads are shown in Figure 12.14. The loads at the ring levels are

$$F_1 = 0.35 \times 6.18 \times 8.58/2 = 9.3 \text{ kN}$$

$$F_2 = 9.3 + (1.4 \times 7.5 \times 2.44) + (1.2 \times 7.5 \times 2.06) = 53.4 \text{ kN}$$

$$F_3 = (1.4 \times 6.51 \times 4.88) + (1.2 \times 6.51 \times 4.27) = 77.9 \text{ kN}$$

$$F_4 = (1.4 \times 5.12 \times 4.88) + (1.2 \times 5.12 \times 4.5) = 62.6 \text{ kN}$$

$$F_5 = (1.4 \times 3.66 \times 4.88) + (1.2 \times 3.66 \times 4.68) = 45.6 \text{ kN}$$

$$F_6 = (1.4 \times 2.16 \times 4.88) + (1.2 \times 2.16 \times 4.8) = 27.0 \text{ kN}$$

$$F_7 = (1.4 \times 1.24 \times 3.41) + (1.2 \times 1.24 \times 3.41) = 11.0 \text{ kN}$$

$$F_8 = 2.6 \times 0.32 \times 1/2 = 0.4 \text{ kN}$$

The rib loads are shown in the figure.

12.6.3 Statical analysis

The axial forces in rib and ring members are calculated using statics. To calculate moments the ribs are taken as continuous and the rings as simply supported members.

(a) *Rib members – axial loads* (Figure 12.15a)

For joint 2, the total axial load above joint is 277.9 kN.

For rib 1–2, $F = 277.9 \times 6.24/4 = 434.9$ kN (compression).

For joint 3, rib 2–3, $F = 421.7$ kN (compression).

For joint 4, rib 3–4, $F = 338.4$ kN (compression).

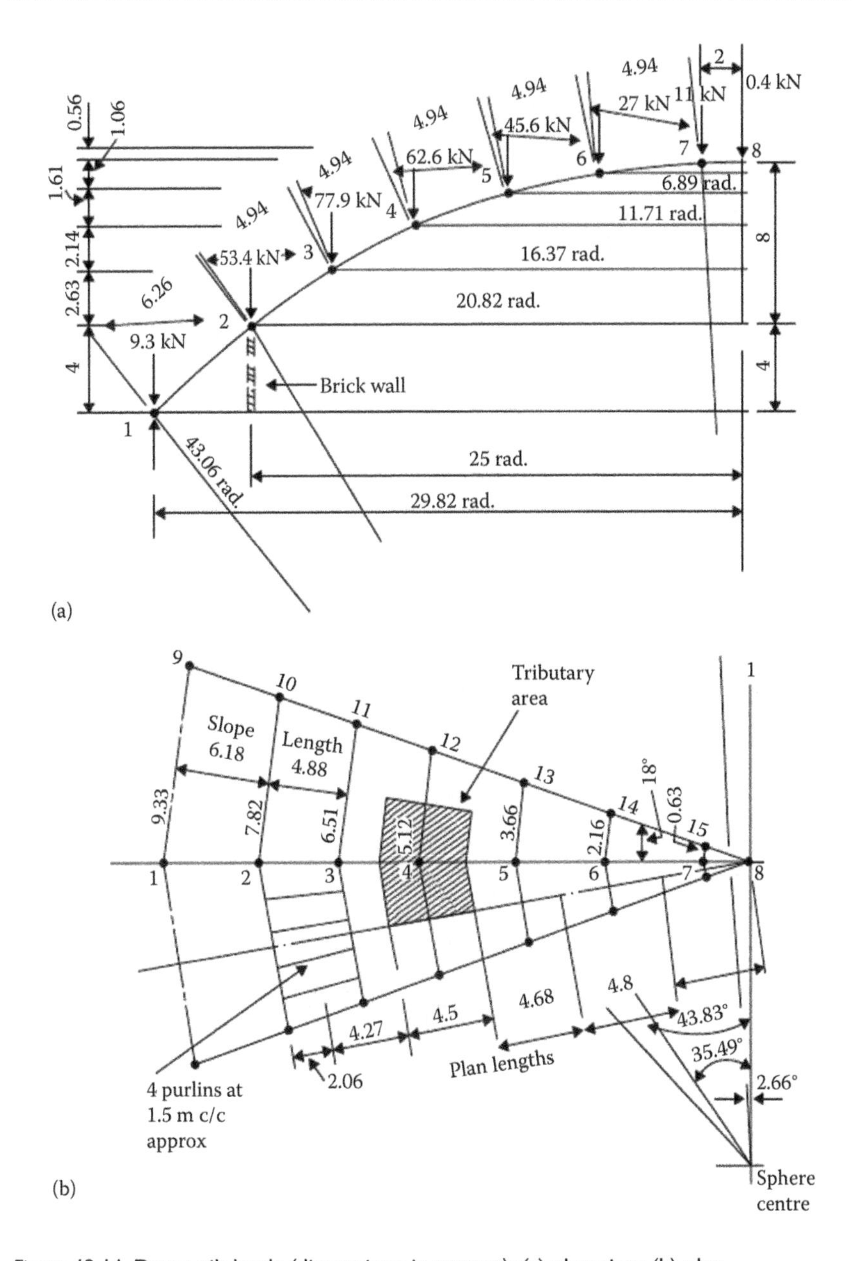

Figure 12.14 Dome rib loads (dimensions in metres): (a) elevation; (b) plan.

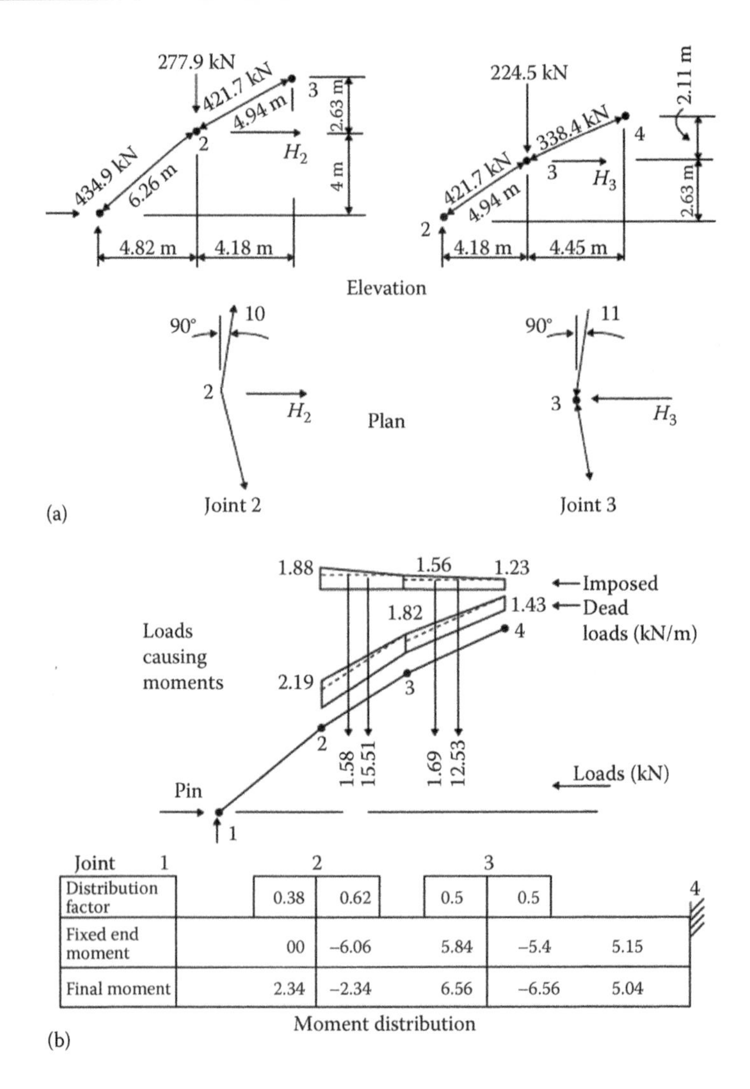

Joint	1	2	2	3	3	4
Distribution factor		0.38	0.62	0.5	0.5	
Fixed end moment		00	−6.06	5.84	−5.4	5.15
Final moment		2.34	−2.34	6.56	−6.56	5.04

Figure 12.15 Statical analysis: (a) axial forces in ribs; (b) moments in arch rib.

(b) *Rib members – moments*

The lateral loads on the ribs between rings 2–3 and 3–4 with four purlins are shown in Figure 12.15b.

The distributed loads at the joints are

- Imposed – joint 2, 1.2 × 7.82/5 = 1.88 kN/m; joint 3, 1.56 kN/m; joint 4, 1.23 kN/m;
- Dead – joint 2, 1.4 × 7.82/5 = 2.19 kN/m; joint 3, 1.82 kN/m; joint 4, 1.43 kN/m.

The member loads are then calculated. For rib 2–3, the uniform load is

$$(1.56 \times 4.18) + (1.82 \times 4.94) = 15.51 \text{ kN}$$

The triangular load is

$$(0.32 \times 4.18/2) + (0.37 \times 4.94/2) = 1.58 \text{ kN}$$

For rib 3–4 the uniform load is 12.53 kN and the triangular was 1.69 kN. The self-weight or member 1–2 is neglected.

The fixed end moments are

$$M_{2\text{-}3} = (15.51 \times 4.18/12) + (1.58 \times 4.18/10) = 6.06 \text{ kNm}$$

$$M_{3\text{-}2} = 5.4 + (1.58 \times 4.18/15) = 5.84 \text{ kNm}$$

$$M_{3\text{-}4} = 5.4 \text{ kNm}; \; M_{4\text{-}3} = 5.15 \text{ kNm}$$

The distribution factors are, for joint 2:

$$(\text{DF})_{2\text{-}1}\text{: } (\text{DF})_{2\text{-}3} = (0.75/6.25\text{: } 1/4.94)/\Sigma(0.12 + 0.202) = 0.38\text{: } 0.62$$

Joint 3:

$$(\text{DF})_{3\text{-}2}\text{: } (\text{DF})_{3\text{-}4} = 0.5\text{: } 0.5$$

The results of the moment distribution are shown in Figure 12.15b. For joint 3, $M = 6.56\text{k Nm}$.

(c) *Ring members – axial loads* (Figure 12.15a)

For joint 2, $\Sigma H = 0$, and

$$h_2 = (434.9 \times 4.82/6.26) - (421.7 \times 4.18/4.94) = -21.9 \text{ kN}$$

For ring 2–10:

$$T = -21.9/(2 \sin 9°) = -70 \text{ kN (tension)}$$

For joint 3:

$$H_3 = +52 \text{ kN}$$

For ring 3–11:

$$F = 166.2 \text{ kN (compression)}$$

(d) *Ring members – moments* (Figure 12.14)

For joint 2, ring 2–10:

$$M = 53.4 \times 7.82/8 = 52.2 \text{ kNm}$$

For joint 3, ring 3–11:

$$M = 63.4 \text{ kNm}$$

12.6.4 Member design

(a) *Ribs*

For rib 1–2:

$$F = 434.9 \text{ kN (compression)}, M = 2.34 \text{ kNm}, L = 6.26 \text{ m}$$

For rib 2–3:

$$F = 421.7 \text{ kN (compression)}, M = 6.56 \text{ kNm}, L = 4.94 \text{ m}$$

Try 150 × 150 × 6.3 SHS with $r = 5.85$ cm, $A = 35.8$ cm², $S = 192$ cm³, $p_y = 355$ N/mm².

For rib 1–2:

$$\lambda = 6260/58.5 = 107$$

$$P_c = 153 \text{ N/mm}^2 \text{ (Table 24(a))}$$

$$P_c = 153 \times 36/10 = 550.8 \text{ kN}$$

$$M_c = 68.9 \text{ kNm}$$

Combined:

$$(434.9/550.8) + (2.34/68.9) = 0.82$$

This is satisfactory. Rib 2–3 is also satisfactory.

(b) *Rings*

For ring 2–10:

$T = 70$ kN (tension), $M = 52.2$ kNm

For ring 3–11:

$F = 166.2$ kN (compression), $M = 63.4$ kNm, $L = 6.51$ m

Try 150 × 150 × 10 SHS, with $r = 5.7$ cm, $A = 54.9$ cm^2, $S = 286$ cm^3.
For ring 3–11:

$\lambda = 6510/57 = 114.2$

$P_c = 135.6$ N/mm^2 (Table 24(a))

$P_c = 744.4$ kN

$M_c = 101.5$ kNm

Combined = 0.84

This is satisfactory. Ring sections nearer the crown could be reduced.

12.6.5 Membrane analysis

The forces in the members of a Schwedler dome can be determined approximately using membrane theory for spherical shells (Makowski, 1984; Schueller, 1977).

Membrane theory gives the following expressions for forces at P (Figure 12.16).

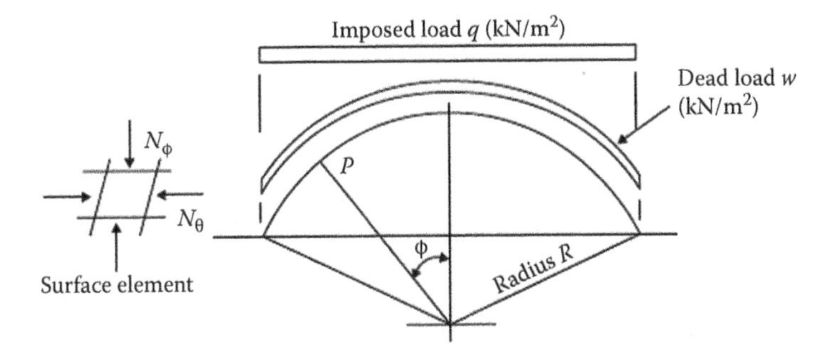

Figure 12.16 Membrane analysis – Schwedler dome.

The meridional or rib force (kN/m) is

$$N_\phi = wR/(1 + \cos\phi) + qR/2$$

The hoop on ring force is

$$N_\theta = wR[\cos\phi - 1/(1 + \cos\phi)] + \frac{1}{2}qR\cos 2\phi$$

where

w = dead load = 1.4 kN/m²;
q = imposed load = 1.2 kN/m²;
R = shell radius = 43.06 m;
ϕ = angle at point of force P.

These forces are calculated at joint 3 in Figure 12.16, where $\phi = 28.92°$. The ribs are spaced at 6.51 m and the rings at 4.91 m.

$$N_\phi = (1.4 \times 43.06)/1.88 + (1.2 \times 43.06/2) = 57.9 \text{ kN/m}$$

For rib 2–3:

$$F = 57.9 \times 6.51 = 376.9 \text{ kN}$$

This compares with average for ribs 2–3, 3–4 of 380.1 kN.

$$N_\theta = 1.4 \times 43.06(0.88 - 1/1.88) + (1/2 \times 1.2 \times 43.06 \times 0.53)$$
$$= 34.7 \text{ kN/m}$$

For ring 3–11:

$$F = 34.7 \times 4.94 = 171.4 \text{ kN}$$

This compares with 166.2 kN.

12.7 RETRACTABLE ROOF STADIUM

12.7.1 Introduction

Conventional stadium structures consist of single or multitiered grandstands grouped around an open rectangular or oval games area. Performers are in the open, while spectators may or may not be under cover. Such a state is ideal in good weather, but bad weather can cause heavy financial losses.

The normal solution is to enclose the entire area with a large fixed roof, or cable- or air-supported roof. However, a case can be made for large retractable roof structures with rigid moving parts, which combine the advantages of the traditional grandstand with the fixed roof arch or dome building. A number of such structures have already been built. An example is the National Tennis Centre in Melbourne, Australia.

Retractable roof structures pose a number of problems not encountered with normal fixed structures. Heavily loaded long-span lattice girders and cantilevers feature prominently. Loss of continuity increases weight. Further problems are discussed in Section 12.7.4.

12.7.2 Proposed structure

Framing plans are set out in Figure 12.17 for a stadium with retractable roof to cover a games area 140 m × 90 m with 40 m clear height under the

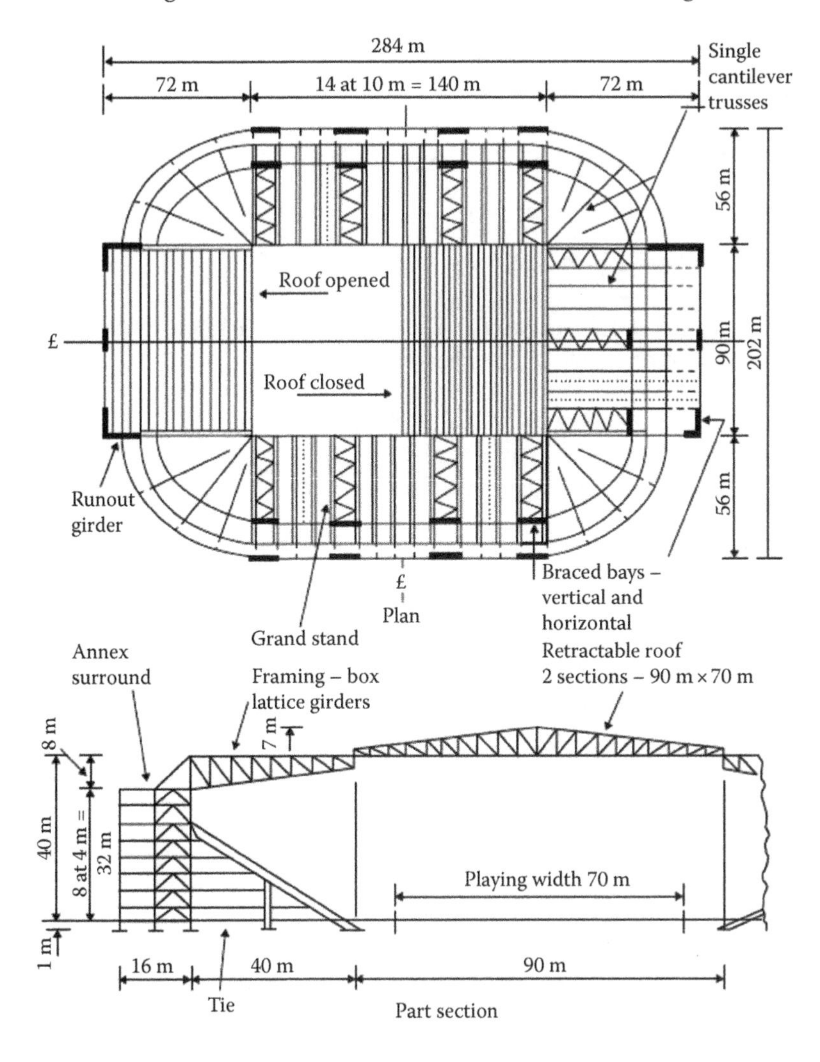

Figure 12.17 Stadium with retractable roof.

roof. All-round seating on a single tier stand is provided with the whole area visible from all seats.

The grandstand area is surrounded by a 16 m wide multistorey annex which contains administration offices, changing rooms, executive viewing boxes, gymnasia, health spas, shops, etc. This structure is a normal multi-storey steel-framed concrete floor slab construction.

The structure proposed consists of two movable roof sections carried on the cantilever grandstand frames, spaced at 10 m centres. The cantilevers are propped by the inclined seating girders. The preliminary sizes of the main sections for the roof and grandstand structures are given. Detail calculations are omitted.

The alternative proposal would be to adopt lattice girders 180 m long on each side to carry the roof. These girders would need to be very deep and would be very heavy.

12.7.3 Preliminary section sizes

The material is Grade S355 steel.

(a) *Movable roof* (Figure 12.18a)

The purlins are at 2.5 m centres – use 100 × 50 × 5 RHS.

The lattice girders are at 5 m centres:

- Top chord – Use 180 × 180 × 10 SHS;
- Bottom chord – Use 150 × 150 × 8 SHS;
- Web – Use 120 × 120 × 6.3 SHS.

The end lattice girder is supported on wheels at 10 m centres:

- Chords – Use 120 × 120 × 6.3 SHS;
- Web – Use 120 × 120 × 5 SHS.

(b) *Runway girder* (Figure 12.18b)

This is a plate girder 1400 mm × 500 mm, with flanges 40 mm thick and web 15 mm thick.

(c) *Grandstand roof cantilever* (Figure 12.18b)

- Diagonal 1–3 – Use 350 × 350 × 16 SHS;
- Top chord 3–5 – Use 300 × 300 × 10 SHS;
- Bottom chord 2–4 – Use 300 × 300 × 12.5 SHS.

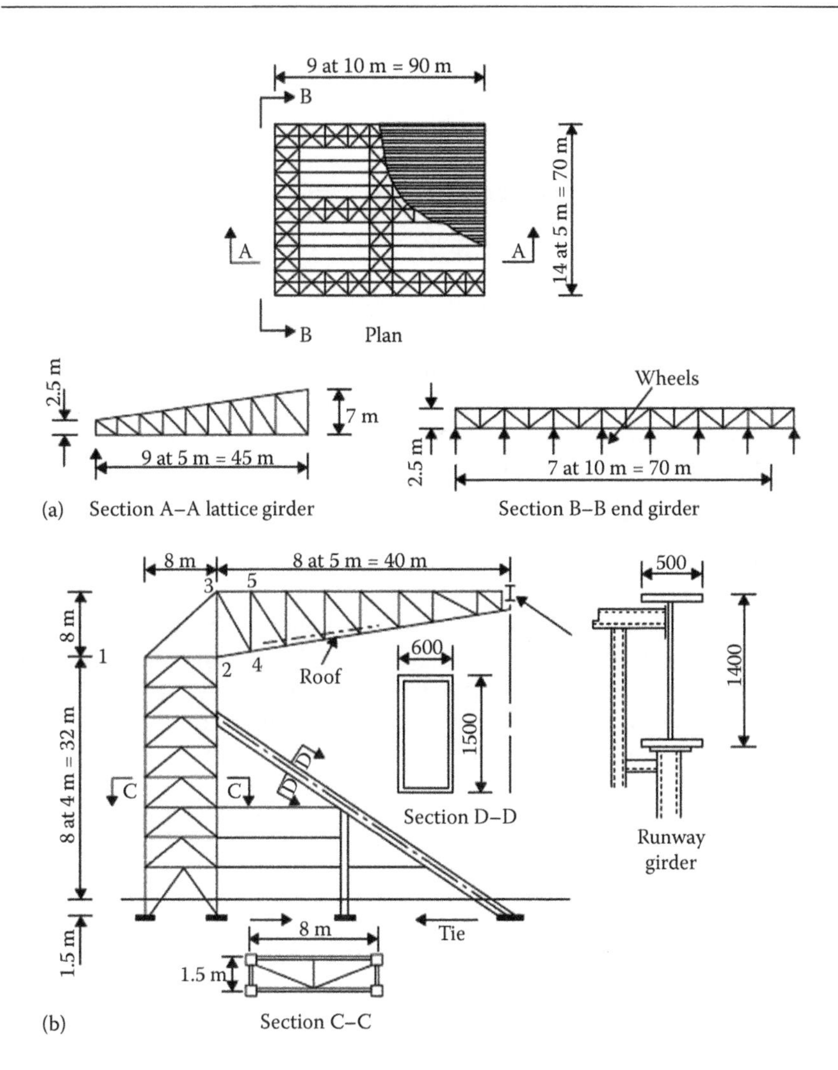

Figure 12.18 Dimensions for: (a) retractable roof unit; (b) grandstand support structure.

(d) *Seating girder* (Figure 12.18b)

This is a box girder 1500 mm deep × 600 mm wide with flanges 40 mm thick and webs 10 mm thick.

(e) *Lattice box column* (Figure 12.18b)

There are four column legs at the base – use 400 × 400 × 20 SHS.

(f) *Column tie* (Figure 12.18b)

Use two 250 × 250 × 20 mm plates.

12.7.4 Problems in design and operation

The following problems must be given consideration.

1. Deflections at the ends of the cantilever roof must be calculated. Differential deflection must not be excessive, but precise limits acceptable would need to be determined from mechanical engineering criteria. The factored design load for the retractable roof is 1.9 kN/m^2 while the unfactored dead load is 0.5 kN/m^2. The latter load would apply when moving the roof.
2. The retractable roof must be tied down against wind uplift and could not be moved under high wind speeds. Wind tunnel tests would be required to determine forces for the open roof state.
3. Temperature effects create problems. Expansion away from the centre line on 100 m width is 60 mm for 50°C temperature change. The roof could be jacked to rest on transverse rollers when it is in the open or closed position. Note also the need to secure against uplift.
4. Problems could arise with tracking and possible jamming. Some measure of steering adjustment to the wheel bogies may be necessary. Model studies could be made.
5. The movable roof should be made as light as possible. Some alternative designs could be:

    - Tied barrel vault construction;
    - Design using structural aluminium;
    - Air-inflated, internally framed pneumatic roof structure.

It is considered that the above problems could be successfully overcome.

Chapter 13

Composite steel columns to EC3 and EC4

Theory, uses and practical design studies

13.1 THEORY AND GENERAL REQUIREMENTS

Concrete cover to steel compression members acts as a fire protection and assists in carrying the load and preventing the members from buckling about the weak axis. The two main types of steel-cased columns are

1. Rolled hollow sections (RHSs) filled with normal-weight concrete.
2. Totally or partially encased universal column (UC) or universal beam (UB) sections with equal flanges (Figure 13.1). Other types of sections can be cased with concrete; see Figure 6.7.1 and Clause 6.7.1 of EC4.

13.2 DESIGN NOTES, PARTIALLY OR TOTALLY ENCASED UNIVERSAL COLUMNS

1. Fire protection; to decide minimum concrete cover for steel column; see Table 4.7 BS EN 1994-1-2, Clause 4.2.3.4 of Eurocode. Note: before considering using the tables and the information, check the rules, steel grade, concrete grade, etc.

   For more information and typical cross sections of composite column, see Table 4.7 of Eurocode. For example, for standard fire resistance for 90 minutes (R90) concrete cover, C_y = 30 mm and C_z = 30 mm. Also, see Clause 6.7.3.1 of EN 1994-1-1:2004(E).
2. See Table 4.7 of Eurocode and relevant literature for deciding the minimum: (i) steel column cross-sectional dimensions, (ii) minimum axis distance of reinforcing bars and (iii) minimum reinforcement rations. This table is applicable for composite columns made from partially encased steel sections.
3. See Table 4.7 of Eurocode and relevant literature for deciding the minimum: (i) steel column cross-sectional dimensions, (ii) minimum reinforcement rations and (iii) minimum axis distance of reinforcing bars.

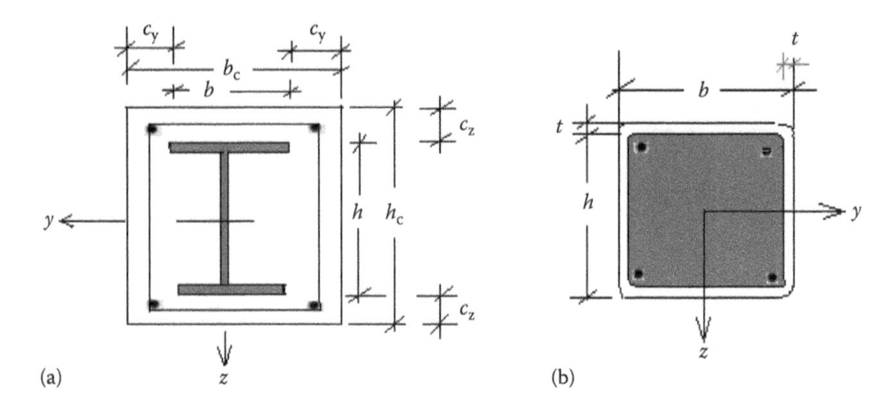

Figure 13.1 Typical cross section of composite column and notation. (a) Composite UC or UB; (b) composite RHS filled with normal-weight concrete.

4. Based on practical experience and the code of practices, the following are normally recommended for the construction and design of composite column sections:

   - The steel surface is clean and unpainted.
   - For steel-grade S235 to S460, use ordinary nominal dense structural concrete of at least grade C25/30 to C50/60.
   - Consult Clause 4.2.3.4 and Table 4.7 of EC4 and up-to-date literature and code of practices for the requirements of concrete reinforcement and cover.
   - For steel reinforcement of a totally encased steel section, ensure that the following are satisfied:

     - Normally, select at least four main longitudinal steel reinforcement bars with bar diameter = 12 mm each. In all cases, the requirements of code of practices should be satisfied.
     - Consult the code of practices to keep the maximum percentage of longitudinal reinforcement bars within the specified limits.

5. For main steel reinforcement bars:

   - $A_{s(min)} = 0.1\ N_{Ed}/f_{yd}$; or
   - $A_{s(min)} = 0.002A_c$, whichever is the greatest; also $0.002A_c < A_s < 0.04A_c$.
   - Mesh reinforcements Ø > 5 mm.
   - Stirrups (Links): 6 mm < Ø (links) < Ø (main bars)/4.
   - See EN 1992-1-1 for further information on stirrups (links).

- Spacing of main bars is
    - o The smallest of column dimensions.
    - o 20 Ø main bars, where Ø is the bar diameter.
    - o 400 mm.

- If fire resistance is not required. Normally no A_s *in hollow filled section.*
- A_s contribution is neglected for Class XO environmental condition, see Clause 6.7.5.2 (4) of EN 1994-2 2005(E).
- Additional steel normally used as a reinforcement for cracking prevention.

13.3 COLUMN SYMMETRIC ABOUT BOTH AXES. GENERAL REQUIREMENT OF BS EC EN 1994-1-1

13.3.1 The steel contribution ratio δ satisfies the following limits

$0.2 \leq \delta \leq 0.9$

where

$$\delta = \frac{A_a f_{yd}}{N_{pl,Rd}} \tag{13.1}$$

A_a = area of steel column section
f_{yd} = steel design strength; $f_{yd} = f_{ys}/\gamma_m$
f_{ys} = steel yield strength
γ_m = steel partial safety factor; see relevant table in EC3
$N_{pl,Rd}$ = characteristic plastic compression capacity, where

$$N_{pl,Rd} = A_a(f_y/\gamma_a) + A_c\alpha_c(f_{ck}/\gamma_c) + A_s(f_{sk}/\gamma_s) \tag{13.2}$$

where

A_a = area of steel cross section
A_c = area of concrete
A_s = steel reinforcement area
f_{ck} = concrete cylinder characteristic strength
f_y = steel yield strength
γ_a, γ_c and γ_s are material partial safety factor for steel column, concrete and steel reinforcement, respectively.

$\alpha_c = 1.0$ for steel hollow section filled with normal weight concrete

$\alpha_c = 0.85$ for other concrete cased steel column

For the followings values of δ,

$\delta > 0.9$	Design the column as a non-composite steel column.
$\delta < 0.2$	Design the column as a reinforced concrete column.

13.3.2 The relative slenderness $\bar{\lambda}$ for relevant plane of bending

No reduction factor is applied to the column capacity for $\bar{\lambda}$ is less than 2.0, and its value is calculated from the formula

$$\bar{\lambda} = \sqrt{\frac{N_{pl,Rd}}{N_{cr}}} \tag{13.3}$$

where $N_{pl,Rd}$ = plastic resistance to compression of composite cross section with γ_a, γ_c and γ_s taken to be equal to 1.0. N_{cr} is the elastic critical normal compression force for the *relevant buckling mode* and is defined as

$$N_{cr} = \frac{\pi^2 (EI)_{eff}}{l^2} \tag{13.4}$$

$$(EI)_{eff} = E_a I_a + K_e E_{cm} I_c + E_s I_s \tag{13.5}$$

where

$(EI)_{eff}$ = the effective flexural stiffness for relevant plane of buckling
l = buckling effective length
$E_a I_a$ = the flexural stiffness of the steel section alone
E_a = elastic modulus of steel section
I_a = second moment of area of the steel section
$E_{cm} I_c$ = the flexural stiffness of the concrete
I_c = second moment of area of the concrete section
$E_s I_s$ = the flexural stiffness of reinforcement
E_s = elastic modulus of steel bars
I_s = second moment of area of steel bars
K_e = correction factor taken = 0.6, Clause 6.7.3 (3) BS EN 1992-1-1:2004(E)

For E_{cm}, see Table 3.1 of BS EN 1992-1-1:2004(E), where

$$E_{cm} = 22[(f_{ck} + 8)/10]^{1/3} \tag{13.6}$$

E_{cm} in M_{pa}

Where long-term loading is predominant, then $E_{c,eff}$ is given by

$$E_{c,eff} = E_{cm}/[1 + \phi_t N_{G,Ed}/N_{Ed}] \quad (13.7)$$

where

ϕ_t = the creep coefficient; see BS EN 1992-1-1:2004(E)
N_{Ed} = design compression normal force
$N_{G,Ed}$ = permanent part of N_{Ed}

13.4 AXIAL COMPRESSION RESISTANCE OF COMPOSITE COLUMN CROSS SECTION: DESIGN RESISTANCE ($\chi N_{pl,Rd}$)

The design condition for axial compression is that the design resistance $\chi N_{pl,Rd}$ should exceed the applied load design N_{Ed}, i.e.

$$\chi N_{pl,Rd} > N_{Ed}$$

The buckling coefficient χ is determined using a non-dimensionalised slenderness ratio $\bar{\lambda}$, where

$$\bar{\lambda} = \sqrt{\frac{N_{pl,Rd}}{N_{cr}}}$$

The χ value can be obtained from BS EN 1993-1-1, Clause 6.3.1.2, in combination with the buckling curve and the curves associated defined in Table 6.5: Buckling curves and member imperfections for composite columns, BS EN 1994-1-1:2004, EN 1994-1-1:2004(E).

ρ_s = % of the reinforcement ratio A_s/A_c

Selection of buckling curves a, b, c and d is dependent on the type of composite column and its axis of buckling; see Table 6.5, BS EN 1994-1-1.

13.5 RESISTANCE OF COMPOSITE COLUMN CROSS SECTION IN COMBINED COMPRESSION AND BENDING

In a similar way to reinforced concrete columns, the interaction force–moment diagram between the normal compression resistance and bending moment resistance shown in Figure 13.2 is used to initially calculate the

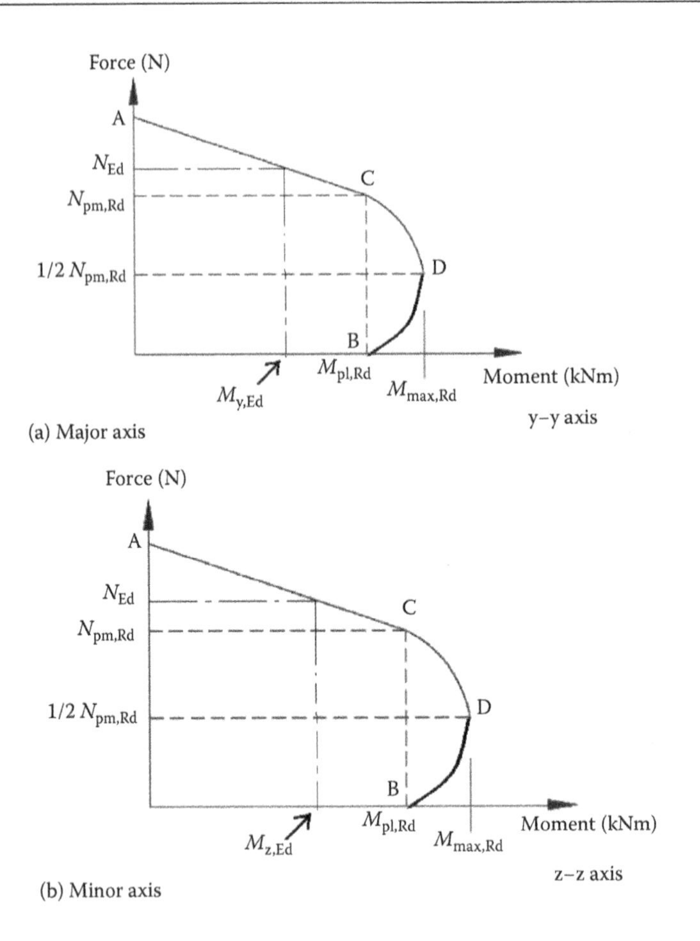

Figure 13.2 Interaction diagram for steel composite columns; (a) y–y axis; (b) z–z axis.

resistance of the composite cross section, where $N_{pl,Rd}$ = the characteristic plastic resistance to compression (point A) and $M_{pl,Rd}$ = the plastic moment resistance (point B).

Comments: Normally the interaction diagram is curved, similar to the interaction diagrams for reinforced concrete columns. However, it may be approximately represented by a series of straight lines.

13.6 CASE STUDY 1: COMPOSITE COLUMN: COMPRESSION RESISTANCE OF CROSS SECTION

A design engineer decided to use a 2000 × 200 × 10 Grade S355 rolled hollow steel section filled with Grade C30/35 normal-weight concrete and an effective height of 3.95 m to carry an axial design compression load of

400 kN. Check if the column axial compression resistance is adequate to carry the design load.

Steps of checking are as follows:

13.6.1 Check h/t (actual) < 52ε (allowable), Clause 9.7.9, Table 6.3 BS EN 1994-1-1 2004(E)

h/t (actual) = 2000/10 = 20

h/t (allowable)

$52\varepsilon = 52 \times (235/355)^{1/2} = 42.3$

Therefore h/t (actual 20) < 42.3 (satisfactory).

13.6.2 Plastic resistance of cross section, $N_{pl,Rd}$

With all materials taken, partial safety factors = 1.0, and $\alpha_c = 1.0$:

$$N_{pl,Rd} = A_a \frac{f_y}{\gamma_a} + A_c \frac{\alpha_c f_{ck}}{\gamma_c} + A_s \frac{f_{sk}}{\gamma_s}$$

$$= 7490 \frac{355}{1.0} + [(200-20)^2 - (4-\pi)10^2] \frac{30}{1.0}$$

$$= 3628.374\,\text{kN}$$

$$\delta = \frac{A_a \frac{f_y}{\gamma_a}}{N_{pl,Rd}} = \frac{7940 \frac{355}{1.0} \times 10^{-3}}{3628.374} = 0.732$$

$0.2 < \delta < 0.9$; the column may be designed as a composite steel column.

Effective stiffness, $(EI)_{eff}$, takes $E_s I_s = 0.0$; no steel bars are used.

$$E_{cm} = 22\left(\frac{f_{ck}+8}{10}\right)^{1/3} = 22\left(\frac{30+8}{10}\right)^{1/3} = 34.3\ \text{kN/mm}^2$$

$$(EI)_{eff} = E_a I_a + 0.6 E_{cm} I_c$$

$$= 210 \times 10^3 \times 4470 \times 10^4 + 0.6 \times 34.3 \times 1000 \times \frac{(200-20)^4}{12}$$

$$= 9388.8\ \text{kNm}^2$$

$$N_{cr} = \frac{\pi^2 (EI)_{eff}}{l^2} = \frac{\pi^2 \times 9388.8}{3.95^2} = 5939.031\,\text{kN}$$

N_{cr} is the elastic critical normal force–Euler critical load.

$$\bar{\lambda} = \sqrt{\frac{N_{pl,Rk}}{N_{cr}}} = \sqrt{\frac{3628.374}{5939.031}} = 0.781$$

This is less than the critical value of 2.0 (satisfactory).

The strength reduction factor is determined using buckling curve (a); see Table 6.5 BS EN 1994-1-1, $\chi = 0.8$; see BS EN 1993-1-2005(E).

$$N_{Rd} = \chi N_{pl,Rd}$$

$$N_{pl,Rd} = A_a \frac{f_y}{\gamma_a} + A_c \frac{\alpha_c f_{ck}}{\gamma_c} + A_s \frac{f_{sk}}{\gamma_s}$$

$$= 7490 \frac{355}{1.0} + [(200-20)^2 - (4-\pi)10^2]\frac{30}{1.5}$$

$$= 3305.25\,\text{kN}$$

$N_{Rd} = \chi N_{pl,Rd} = 0.8 \times 3305.25 = 2644.2$ kN > 400 kN (acting design loads)

therefore satisfactory.

13.7 CASE STUDY 2: RESISTANCE OF COMPOSITE COLUMN IN COMBINED COMPRESSION AND BENDING ABOUT THE MAJOR AXIS

Use EC4 to check the adequacy of the RHS steel column fabricated from 160 × 80 × 8 Grade S355 and filled with Grade C30/35 normal-weight concrete having a system length 3.95 m to carry a design normal compression force of 600 kN and a design moment about the major axis of 14 kNm at the top of the column and (0.0 kNm) moment at the bottom of the column.

13.7.1 Section properties

$A_a = 3520\ \text{mm}^2$ (from table of steel section properties)

$W_{pl,y} = 175 \times 10^3\ \text{mm}^3$ (from table of steel section properties), $I_a = 1090 \times 10^4\ \text{mm}^4$

$$W_{pc,y} = \frac{(b-2t)(h-2t)^2}{4} - \frac{2}{3}r^3 - (4-\pi)\left(\frac{h}{2} - t - r\right)r^2$$

$$= \frac{(80-2\times8)(160-2\times8)^2}{4} - \frac{2}{3}8^3 - (4-\pi)\left(\frac{160}{2} - 8 - 8\right)8\times8$$

$$= 327{,}039.62 \text{ mm}^3$$

Area of the concrete A_c is given by

$A_c = (b-2t)(h-2t) - (4-\pi)r^2 = (80 - 2 \times 8)(160 - 2 \times 8) - (4-\pi)8^2 =$ 9161.0 mm^2

13.7.2 Plastic compression resistance, $N_{pl,Rd}$

$$N_{pl,Rd} = A_a \frac{f_y}{\gamma_a} + A_c \frac{\alpha_c f_{ck}}{\gamma_c} + A_s \frac{f_{sk}}{\gamma_s} = 3520 \frac{355}{1.0} + 9161 \frac{30}{1.5} = 1432.82 \text{ kN}$$

13.7.3 Steel contribution ratio, δ

$$\delta = \frac{A_s f_{yd}}{N_{pl,Rd}} = \frac{3520 \frac{355}{1.0} \times 10^{-3}}{1432.82} = 0.872$$

$0.2 < \delta < 0.9$, within the limits for design as a composite column.

13.7.4 Maximum moment capacity, $M_{max,Rd}$

$$M_{max,Rd} = W_{pa} \frac{f_y}{\gamma_a} + 0.5 W_{pc} \frac{f_{ck}}{\gamma_c}$$

$$= 175{,}000 \frac{355}{1.0} \times 10^{-3} + 0.5 \times 327{,}039.62 \frac{30}{1.5} \times 10^{-3}$$

$$= 65.395 \text{ kNm}$$

13.7.5 Resistance of concrete to compression normal force, $N_{pm,Rd}$

From

$$N_{pm,Rd} = A_c \frac{f_{ck}}{\gamma_c} = 9161 \frac{30}{1.5} \times 10^{-3} = 183.22 \text{ kN}$$

$$\frac{N_{pm,Rd}}{2} = 91.61 \text{ kN}$$

13.7.6 Neutral axis position, h_n

$$h_n = \frac{A_c f_{cd} - A_{sn}(2f_{sn} - f_{cd})}{2b\frac{f_{ck}}{\gamma_c} + 4t_w\left(2\frac{f_y}{\gamma_a} - \frac{f_{ck}}{\gamma_c}\right)} = \frac{183.22 - 0.0}{2\times 80\times\frac{30}{1.5} + 4\times 8\left(2\frac{355}{1.0} - \frac{30}{1.5}\right)} = 7.247 \text{ mm}$$

13.7.7 Calculate W_{pcn}, $M_{n,Rd}$, $M_{pl,Rd}$

$$w_{pcn} = (b - 2t)h_n^2 = (80 - 2\times 8)7.247^2 = 3361.216 \text{ mm}^3$$

$$w_{pan} = bh_n^2 - w_{pcn} = 80\times 7.247^2 - 3361.216 = 840.30 \text{ mm}^3$$

$$M_{n,Rd} = W_{pan}\frac{f_y}{\gamma_a} + 0.5W_{pcn}\frac{f_{ck}}{\gamma_c}$$

$$= 840.3\frac{355}{1.0}\times 10^{-6} + 0.5\times 3361.216\frac{30}{1.5}\times 10^{-6}$$

$$= 0.332 \text{ kNm}$$

$$M_{pl,Rd} = M_{max,Rd} - M_{n,Rd} = 65.395 - 0.332 = 65.063 \text{ kNm}$$

The values required to plot the interaction diagram given in Figure 13.3 are given in Table 13.1.

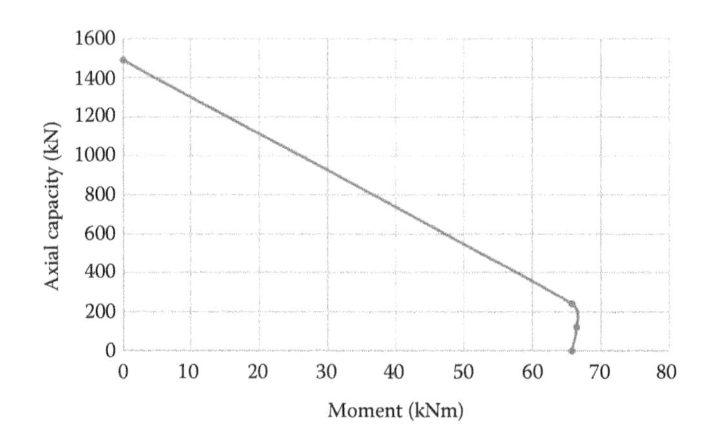

Figure 13.3 Axial compression resistance versus moment capacity curve: case study 2.

Table 13.1 Major axis interaction diagram–force and moment data for case study 2

Point		Moment capacity (kNm)	Axial capacity (kN)
A	$(0, N_{pl,Rd})$	0	1432.82
B	$(M_{pl,Rd}, 0)$	65.063	0
C	$(M_{pl,Rd}, N_{pm,Rd})$	65.063	183.22
D	$(M_{max,Rd}, 0.5\ N_{pm,Rd})$	65.395	91.61

13.7.8 Resistance of composite column for axial buckling about the major axis

$E_aI_a = 210 \times 10^3 \times 1090 \times 10^4 = 2289\ \text{kNm}^2$

$$E_{cm} = 22\left(\frac{(f_{ck} + 8)}{10}\right)^{1/3} = 34.31\ \text{kN/mm}^2$$

$$I_c = \frac{(b-2t)^3(b-2t)}{12} = 15.925 \times 10^{-6}\ \text{m}^4$$

$E_{cm}I_c = 34.31 \times 15.925 = 546.386\ \text{kNm}^2$

Effective flexural strength, $(EI)_{eff,II}$

$(EI)_{eff,II,y} = 0.9\ (E_aI_a + 0.5\ E_{cm}I_c) = 0.9(2289 + 0.5 \times 546.386)$
$= 2305.973\ \text{kNm}^2$.

Elastic critical buckling load, $N_{cr,eff}$*;*

$$N_{cr,eff,y} = \frac{\pi^2 (EI)_{eff,II,y}}{l^2} = \frac{\pi^2 \times 2305.973}{3.95^2} = 1458.679\ \text{kN}$$

$N_{Ed} = 600\ \text{kN} > N_{cr,eff,y}/10$, and therefore second-order effects need to be considered.

$N_{pl,Rk} = (3520 \times 355 + 9161 \times 30) \times 10^{-3} = 1524.43\ \text{kN}.$

$$\bar{\lambda} = \sqrt{\frac{N_{pl,Rk}}{N_{cr,eff}}} = \sqrt{\frac{1524.43}{1458.679}} = 1.02 < 2.0$$

< 2.0 (satisfactory)

$\chi = 0.68$ and $\chi N_{pl,Rd} = 0.68 \times 1432.82 = 974.317\ \text{kN} > 600\ \text{kN}$ (satisfactory).

Check compression resistance based on second-order linear elastic analysis.

Member imperfection, $e_{o,y} = L/300 = 3.95 \times 1000/300 = 13.166$ mm; see Table 6.5, BS EN 19994-1-1 2004(E).

$M_{i(imperfection)} = N_{Ed} \times e_{o,y} = 600 \times 13.166 = 7.896$ kNm

K_i (amplication factor) for $\beta_i = 1.0$ (conservative)

$$K_i = \frac{\beta_i}{1 - \frac{N_{Ed}}{N_{cr,eff,y}}} = \frac{1.0}{1 - \frac{600}{1458.679}} = 1.698$$

$M_{Ed,II,i} = 1.698 \times 7.896 = 13.407$ kNm

For $0.9M_{Ed,II,i} = 12.06$ kNm, use simplified interaction curve (Figure 13.3); therefore N_{Rd} (at $M_i = 12.06$) = 1201.14 kN > 600 kN acting axial compression with member imperfections.

Note: $\chi N_{pl,Rd}$ is also >600 kN; for χ, use Table 6.4, BS EN 1993-1-1 2005(E).

13.7.9 Compression and bending resistance of column based on second-order linear elastic analysis

$r = 0.0$, so $\beta = 0.66 + 0.44r$, $\beta = 0.66$.

$$K_m = \frac{\beta}{1 - \frac{N_{Ed}}{N_{cr,eff}}} = \frac{0.66}{1 - \frac{600}{1458.679}} = 1.121$$

$$M_{Ed} = k_m M_{Ed,max} + k_i N_{Ed} \times e_{o,y} = 1.121 \times 14 + 1.698 \times 7.896$$
$$= 29.10 \text{ kNm} > 14 \text{ kNm}$$

Therefore, $M_{y,max,Ed,II} = 29.1$ kNm.

From the interaction diagram in Figure 13.3, the moment M_{600} corresponding to an axial load of 600 kN is given by

$$M_{600} = M_{pl,Rd} \frac{N_{pl,Rd} - N_{Ed}}{N_{pl,Rd} - N_{pm,Rd}} = 65.063 \frac{1432.82 - 600}{1432.82 - 183.22} = 43.362 \text{ kNm}$$

and $\mu_d = 43.362/65.063 = 0.67 < (\alpha_m = 0.9)$

$$\frac{M_{Ed}}{\mu_d M_{pl,Rd}} = \frac{M_{Ed}}{M_{600}} = \frac{29.1}{43.362} = 0.671 < 0.9$$

The column is satisfactory; see 6.7.3 (1) EN 1994-1-1:2004(E). The coefficient α_m is 0.9 for steel grades between S235 and S355 inclusive, and 0.8 for steel grades S420 and S460.

13.8 CASE STUDY 3: COMBINED COMPRESSION AND BIAXIAL BENDING

A 160 × 80 × 8 Grade S355 RHS steel column filled with normal-weight concrete of Grade C30/35 has an effective height of 3 m and is selected to carry the following design loadings:

Axial design load = 350 kN
Design moment about the major axis y–y = 14 kNm at the top of the column, 0.0 kNm at the bottom of the column
Design moment about the minor axis z–z = 11 kNm at the top of the column, 0.0 kNm at the bottom of the column

Check the suitability of the column.

13.8.1 Interaction diagrams for y–y axis and z–z axis

Hint: For y–y axis (major axis), the interaction diagram is as in case study 2.

$A_a = 3520\ \text{mm}^2$, $W_{pl,y} = 175 \times 10^3\ \text{mm}^3$, $W_{pl,z} = 106{,}000\ \text{mm}^3$,
$I_{yy} = 1090 \times 10^4\ \text{mm}^4$, $I_{zz} = 356 \times 10^4\ \text{mm}^4$

Minor z–z axis

$$W_{pc} = \frac{(h-2t)(b-2t)^2}{4} - \frac{2}{3}r^3 - (4-\pi)(0.5b - t - r)r^2$$

$$= \frac{(160-2\times 8)(80-2\times 8)^2}{4} - \frac{2}{3}8^3 - (4-\pi)\left(\frac{160}{2} - 8 - 8\right)\times 8 \times 8$$

$$= 143{,}598.63\,\text{mm}^3$$

$A_c = 9161\ \text{mm}^2$, $N_{pl,Rd} = 1432.82$ kN and $\delta = 0.872$ (see case study 2)

$$M_{max,Rd} = W_{pa}\frac{f_y}{\gamma_a} + 0.5\,W_{pc}\frac{f_{ck}}{\gamma_c}$$

$$= 106{,}000\frac{355}{1.0}\times 10^{-6} + 0.5\times 143{,}598.63\frac{30}{1.5}\times 10^{-6}$$

$$= 39.066\ \text{kNm}$$

$N_{pm,Rd}$ = 183.22 kN and $0.5N_{pm,Rd}$ = 91.61 kN (see case study 2)

$$h_n = \frac{N_{pm,Rd}}{2b\frac{f_{ck}}{\gamma_c} + 4t_w\left(2\frac{f_y}{\gamma_a} - \frac{f_{ck}}{\gamma_c}\right)} = \frac{183,220}{2\times 80\times\frac{30}{1.5} + 4\times 8\left(2\frac{355}{1.0} - \frac{30}{1.5}\right)} = 7.247 \text{ mm}$$

$$w_{pcn,z} = (b-2t)h_n^2 = (80-2\times 8)7.247^2 = 3361.216 \text{ mm}^3$$

$$w_{pan,z} = bh_n^2 - w_{pcn,z} = 80\times 7.247^2 - 3361.216 = 840.304 \text{ mm}^3$$

$$M_{n,Rd,z} = W_{pan,z}\frac{f_y}{\gamma_a} + 0.5W_{pcn,z}\frac{f_{ck}}{\gamma_c}$$

$$= 840.3\frac{355}{1.0}\times 10^{-6} + 0.5\times 3361.216\frac{30}{1.5}\times 10^{-6} = 0.332 \text{ kNm}$$

$M_{pl,Rd,z} = M_{max,Rd} - M_{n,Rd,z}$ = 39.066 − 0.332 = 38.734 kNm.

The values required to plot the interaction diagram in Figure 13.4 are given in Table 13.2.

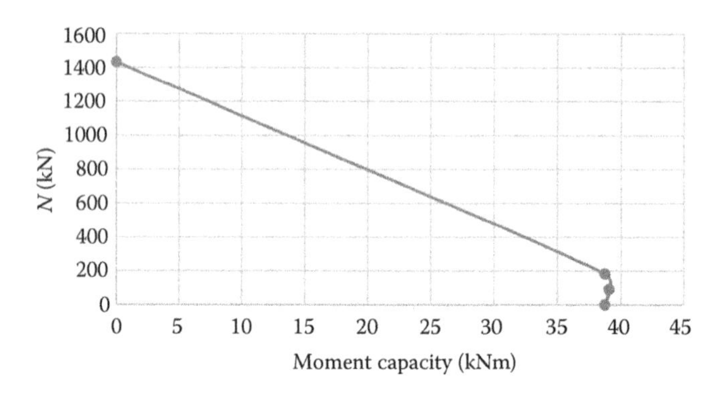

Figure 13.4 Interaction curve: case study 3.

Table 13.2 Normal force and moment data–minor axis interaction diagram: case study 3

Point		*Moment capacity (kNm)*	*Axial capacity (kN)*
A	$(0, N_{pl,Rd})$	0	1432.82
B	$(M_{pl,Rd}, 0)$	38.734	0
C	$(M_{pl,Rd}, N_{pm,Rd})$	38.734	183.22
D	$(M_{max,Rd}, 0.5N_{pm,Rd})$	39.066	91.61

13.8.2 Resistance to axial buckling about the minor axis z–z

$E_a I_{a,z} = 210 \times 10^3 \times 356 \times 10^4 = 747.6 \text{ kNm}^2$

$E_{cm} = 34.31 \text{ GPa}$

$$I_{c,z} = \frac{(80-16)^3(160-16)}{12} = 3.145 \times 10^{-6} \text{m}^4$$

$E_{cm} I_{c,z} = 34.31 \times 10^6 \times 3.145 \times 10^{-6} = 107.9 \text{ kNm}^2$

$(EI)_{eff,II,z} = 0.9(E_a I_a + 0.5E_{cm} I_c) = 0.9(747.6 + 0.5 \times 107.9)$

$= 721.395 \text{ kNm}^2.$

$$N_{cr,eff,z} = \frac{\pi^2 (EI)_{eff,II}}{l^2} = \frac{\pi^2 \times 721.395}{3.0^2} = 791.0 \text{ kN}$$

$$N_{cr,eff,y} = \frac{\pi^2 (EI)_{eff,II}}{l^2} = \frac{\pi^2 \times 2305.973}{3.0^2} = 2528.782 \text{ kN, for } (EI)_{eff,II,y}$$

$= 2305.973$ see case study 2

$N_{Ed} = 350 \text{ kN} > N_{cr,eff,z}/10$, and therefore second-order effects need to be considered.

$N_{pl,Rk} = A_a f_y + A_c f_{ck} = 3520 \times 355 \times 10^{-3} + 9161 \times 30 \times 10^{-3} = 1524.43 \text{ kN}.$

$$\overline{\lambda z} = \sqrt{\frac{N_{pl,Rk}}{N_{cr,eff,z}}} = \sqrt{\frac{1524.43}{791.0}} = 1.388 < 2.0 \qquad \text{(satisfactory)}$$

$$\overline{\lambda y} = \sqrt{\frac{N_{pl,Rk}}{N_{cr,eff,y}}} = \sqrt{\frac{1524.43}{2528.782}} = 0.6 < 2.0 \qquad \text{(satisfactory)}$$

13.8.3 Second-order effects (major y–y axis) – based on second-order linear elastic analysis

(a) $K_{m,y}$

$r = 0.0$, so $\beta = 0.66$.

$$K_{m,y} = \frac{\beta}{1 - \frac{N_{Ed}}{N_{cr,eff}}} = \frac{0.66}{1 - \frac{350}{2528.782}} = 0.766$$

(b) $K_{i,y}$ (member imperfection)

Table 6.5 of EN 1994-1-1, for an infilled hollow section, $e_{0,y} = L/300 = 3.95/300$

$\beta_i = 1.0$ (conservative)

$K_{i,y} = \beta_i/[1 - (N_{Ed}/N_{cr,eff,y})] = 1/[1 - (350/2528.782)] = 1.16$

13.8.4 Second-order effects (minor z–z axis) – based on second-order linear elastic analysis

(a) $K_{m,z}$

$r = 0.0$, so $\beta = 0.66$

$$K_{m,z} = \frac{\beta}{1 - \dfrac{N_{Ed}}{N_{cr,eff,z}}} = \frac{0.66}{1 - \dfrac{350}{791.0}} = 1.18$$

(b) $K_{i,z}$

Table 6.5 of EN 1994-1-1, for an infilled hollow section, $e_{0,z} = L/300 = 3/300$

$\beta_i = 1.0$ (conservative)

$K_{i,z} = \beta_i/[1 - (N_{Ed}/N_{cr,eff,z})] = 1/[1 - (350/791)] = 1.793$

13.8.4.1 Major axis, M_{350}, minor axis, M_{350}

From the interaction diagram for the major axis in Figure 13.3

$$M_{350,y} = M_{pl,Rd}\frac{N_{pl,Rd} - N_{ed}}{N_{pl,Rd} - N_{pm,Rd}} = 65.063\frac{1432.82 - 350}{1432.82 - 183.22} = 56.379 \text{ kNm}$$

$$\mu_d = \frac{M_{350}}{M_{pl,Rd}} = \frac{56.379}{65.063} = 0.867 < 0.9$$

Minor axis, M_{350}

From the interaction diagram for the minor axis in Figure 13.4

$$M_{350,z} = M_{pl,Rd}\frac{N_{pl,Rd} - N_{Ed}}{N_{pl,Rd} - N_{pm,Rd}} = 38.734\frac{1432.82 - 350}{1432.82 - 183.22} = 33.564 \text{ kNm}$$

$$\mu_d = \frac{M_{350}}{M_{pl,Rd}} = \frac{33.564}{38.734} = 0.867 < 0.9$$

13.8.4.2 Compression and bending resistance of column based on second-order linear elastic analysis

Major axis, y–y

$$M_{Ed,y} = k_{m,y}M_{Ed,y} + k_{i,y}N_{Ed,e0,y} = 0.766 \times 14 + 1.16 \times 350\frac{3.0}{300}$$

$$= 14.784 \text{ kNm} > 14 \text{ kNm, use } 14.784 \text{ kNm}$$

$$M_{Ed,z} = k_{i,z}M_{Ed,z} = 1.793 \times 11 = 19.723 \text{ kNm}$$

$$\frac{M_{Ed,y}}{\mu_{dy}M_{pl,y,Rd}} = \frac{14.784}{56.379} = 0.262 < 0.9$$

$$\frac{M_{Ed,z}}{\mu_{dz}M_{pl,z,Rd}} = \frac{19.723}{33.564} = 0.587 < 0.9$$

$$\frac{M_{Ed,y}}{\mu_{dy}M_{pl,y,Rd}} + \frac{M_{Ed,z}}{\mu_{dz}M_{pl,z,Rd}} = 0.849 < 1.0$$

Minor axis, z–z

$$M_{Ed,y} = k_{i,y}M_{Ed,y} = 1.16 \times 14 = 16.24 \text{ kNm}$$

$$M_{Ed,z} = k_{m,z}M_{Ed,z} + k_{i,z}N_{Ed,e0,z} = 1.18 \times 11 + 1.793 \times 350\frac{3.0}{300} = 19.255 \text{ kNm}$$

$$\frac{M_{Ed,y}}{\mu_{dy}M_{pl,y,Rd}} = \frac{16.24}{56.379} = 0.288 < 0.9$$

$$\frac{M_{\mathrm{Ed,z}}}{\mu_{\mathrm{dz}} M_{\mathrm{pl,z,Rd}}} = \frac{19.255}{33.564} = 0.573 < 0.9$$

$$\frac{M_{\mathrm{Ed,y}}}{\mu_{\mathrm{dy}} M_{\mathrm{pl,y,Rd}}} + \frac{M_{\mathrm{Ed,z}}}{\mu_{\mathrm{dz}} M_{\mathrm{pl,z,Rd}}} = 0.861 < 1.0$$

Therefore, the minor axis is the critical case in this example.

Exercise: Repeat this case study using the same design moment about the minor axis and the major axis at the top of the column and with $r = -0.5$ (major axis) and $r = -0.3$ (minor axis).

Chapter 14

Steel plate girders

Design to EC3

14.1 GENERAL THEORY, USES AND PRACTICAL EXAMPLES/APPLICATIONS

The main elements of the plate girder are top flange, web and bottom flange; see Figure 14.1a. They are normally selected from the same steel grade and assembled by fillet weld. The stiffeners (see Figure 14.1b) are normally used to help to

- Resist the web buckling
- Provide support to concentrated loads
- Provide supports to reactions

Plate girders are used mainly in steel bridge and building structures where long span beams are required and especially

1. In cases where a rolled section would need to be spliced and as a result may be inefficient.
2. To support heavy loads such as on a bridge structures.
3. In buildings or structures, fewer numbers of internal columns and long span beams are needed.

Rolled steel beams with depth up to 1016 mm are available in the United Kingdom and may be in other countries in the European Union and the world by TATA/Corus, etc.; therefore, the use of plate girder in building structures is less compared to their use in bridge and heavy industrial structures, where fewer numbers of internal columns and long span beams are needed.

In fabrication, normally either continuous automatic electric arc or submerged gas welding is used to form the fillet welds between the flange and web; see Figure 14.1a. Research showed that the welding process may introduce very high residual stress to build in the web and the two flanges.

In this textbook, only uniform straight plate girder with equal flanges and vertical stiffeners will be considered.

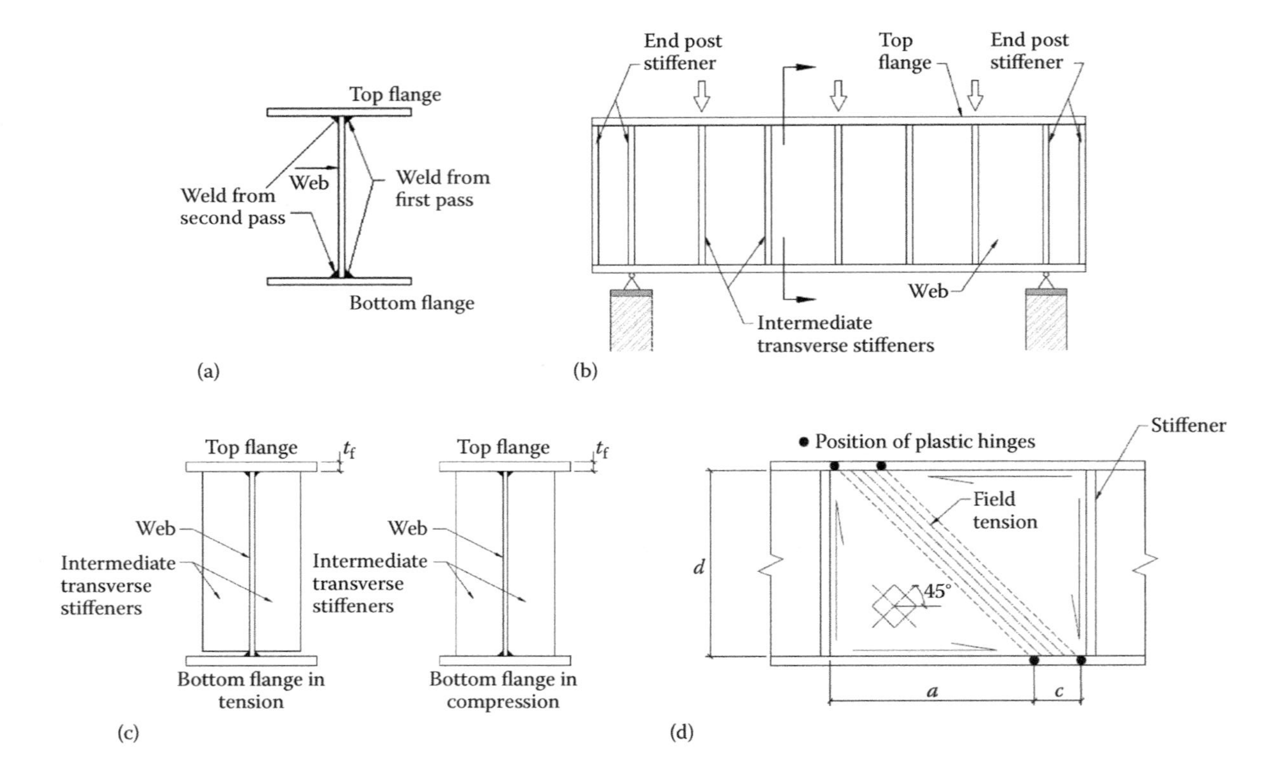

Figure 14.1 (a) Plate girder fabrication; (b) plate girder side elevation; (c) cross section of plate girder; (d) basic geometry of web panel at ultimate limit state just before failure. (*Continued*)

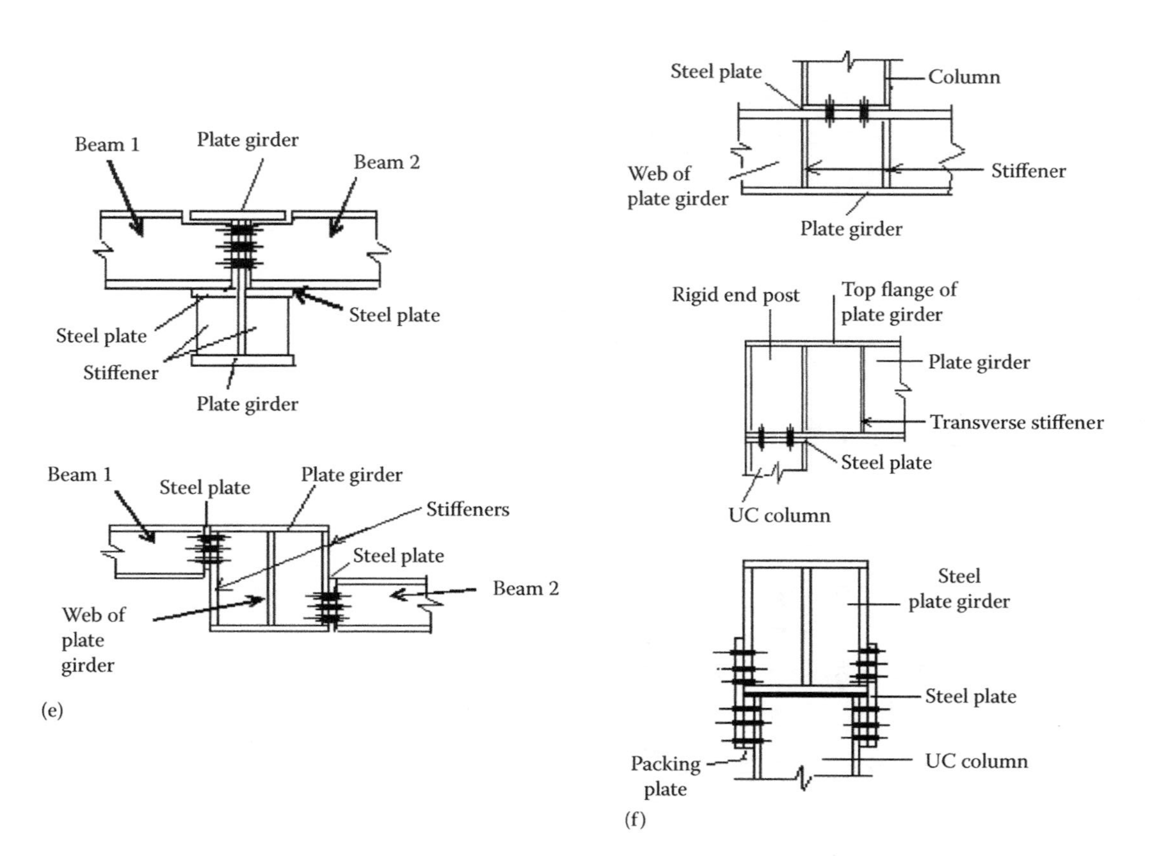

Figure 14.1 (Continued) (e, f) Various schematic examples of connections to plate girder.

Figure 14.1b–d shows plate girder side elevation, section and basic geometry of web panel with position of hinges at the ultimate limit state just before failure, whereas Figure 14.1e and f shows examples of connections to plate girder. This will provide graduate engineers and students with little experience important practical information on fabrication and details of the plate girders.

14.2 DESIGN OF PLATE GIRDER

14.2.1 Critical h_w/t_w ratio for a girder with no web stiffeners

In pure bending test under the action of the applied loads, the flanges of a plate girder are subjected to opposite stresses; see Figure 14.2.

The compression in the top flange tends to cause the top flange to buckle sideways or rotate relative to the web due to the web being unable to support the flange. Researchers call this as flange-induced buckling (BS EN 1993-1-5: 2005, Clause 8).

To prevent the flange undergoing local buckling in the plane of the web, the critical h_w/t_w ratio for a girder is given by (see BS EN 1993-1-5: 2005, Clause 8)

$$\frac{h_w}{t_w} \leq k \frac{E}{f_{yf}} \sqrt{\frac{A_w}{A_c}}$$

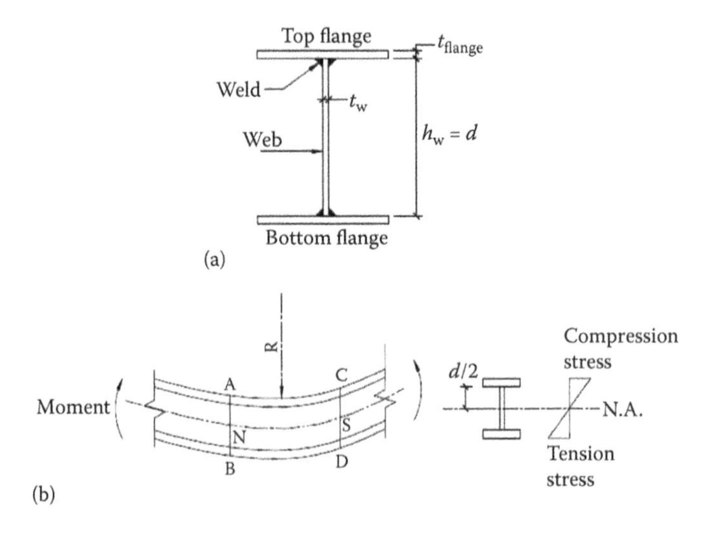

Figure 14.2 (a) Plate girder sectional terminology; (b) plate girder in pure bending where the flanges are subjected to equal and opposite forces.

where

h_w = web depth
t_w = web thickness
f_{yf} = the yield strength of the compression flange
A_c = the effective area of the compression flange
A_w = the area of the web

The parameter k is a constant with the following values:

$k = 0.3$ use when plastic hinge rotation is utilized.
$k = 0.4$ use when plastic resistance is utilized.

For simply supported beams, k may be taken as 0.4 and

$k = 0.55$ use when elastic resistance is utilized.

For researchers, graduates and engineers, the analysis below is of interest to show where the above formula comes from the following.

Force per unit length (in pure bending where the flanges are subjected to equal and opposite forces) can be defined as

$$\sigma_v t_w = \frac{f_{yf} A_{fc}}{R} \tag{14.1}$$

where

σ_v = the vertical stress in the web
A_{fc} = the area of the compression flange
f_{yf} = the yield strength of the compression flange
R = the radius of curvature (see Figure 14.2b), where

$$R = \frac{0.5 h_w}{\varepsilon_f} \tag{14.2}$$

where

h_w = the depth between the centroid of the flanges
$\Delta\varepsilon_f$ = the variation in the strain due to fabrication at mid-depth of the flanges
$\Delta\varepsilon_f = \varepsilon_y + \varepsilon_f$
ε_y = yield strain
ε_f = the residual strain due to fabrication, which can be assumed to equal to $0.5\varepsilon_y$; see relevant research works in this area, and as

$$\Delta\varepsilon_f = \varepsilon_y + \varepsilon_f$$

$$\Delta\varepsilon_f = 1.5\varepsilon_y$$

$$\varepsilon_y = \text{yield strain}$$

since

$$\varepsilon_f = \frac{f_{yf}}{E}$$

From the analysis above, we can find that

$$\sigma_y = 3A_{fc}f_{yf}^2/A_wE \tag{14.3}$$

The stress

$$\sigma_y \leq \sigma_y = \frac{\pi^2 E}{12(1-\upsilon^2)}\left(\frac{t_w}{h_w}\right)^2 \tag{14.4}$$

where

σ_v = the elastic critical buckling stress for a simply supported thin plate.

With $A_c = A_{fc}$ and $\upsilon = 0.3$, Equations 14.3 and 14.4 give

$$\frac{h_w}{t_w} \leq 0.55\frac{E}{f_{yf}}\sqrt{\frac{A_w}{A_{cf}}} \tag{14.5}$$

In BS EN 1993-1-5: 2005, Clause 8, flange-induced buckling, the critical h_w/t_w ratio is given by

$$\frac{h_w}{t_w} \leq k\frac{E}{f_{yf}}\sqrt{\frac{A_w}{A_{cf}}} \tag{14.6}$$

where 0.55 is replaced by k for covering situation where higher strains are required and where

f_{yf} = the yield strength of the compression flange
A_{fc} = the effective area of the compression flange
A_w = the area of the web

Note: Make sure that h_w/t_w is be less than the critical value in Equation 14.6.

For the parameter

$k = 0.3$ use when plastic hinge rotation is utilized.
$k = 0.4$ use when plastic resistance is utilized. For simply supported beams, k may be taken as 0.4.
$k = 0.55$ use when elastic resistance is utilized.

Note: Consult the relevant code of practices for the value of k.

14.2.2 Design methods

Briefly the two methods are as follows.

Method 1 is normally used, where the flanges resist the bending moment. In this case,

- The plate girder is fully laterally restrained and is carrying a UDL.
- The maximum bending moment and maximum shear force are not coincident.
- The shear capacity of the plate girder is based on the shear capacity of the web.

For method 2, the flange and web of the girder carry the moments and forces as an entity. Only method 1 is considered in this textbook.

14.2.3 Plate girder bending moment resistance

The moment resistance of a plate girder with compression flange fully restrained against lateral torsional buckling is given by:

$$M_{f,Rd} = \frac{W_{pl} f_y}{\gamma_{M0}} > M_{Ed}$$

where

$M_{f,Rd}$ = plate girder moment resistance by flanges only
W_{pl} = plate girder cross section plastic modulus
f_y = steel plates design strength
γ_{M0} = partial safety factor
M_{Ed} = design ultimate moment

14.2.4 Plate girder cross section classification

This is determined first in the same manner as rolled sections for beams and columns, and bending resistance is given by $M_{f,Rd}$ as defined above. For section classification, see Table 3.1 EC3. Normally the designer makes sure that the flanges of the girder are classified as plastic section, i.e.

$c_f/t_f \leq 9\varepsilon$, where $\varepsilon = \sqrt{(235/f_y)}$.

The web of the girder is normally slender and needs stiffeners to resist web buckling; see the example in Chapter 15.

14.2.5 Design of the plate girder web sectional dimensions

For estimating the plate girder cross-section dimensions, the moment is assumed to be resisted by the flanges only, whereas the web resists the shear force.

M_{Rd} = flange design strength = $f_{yd}A_f h_w$

Given $A_f = b_f t_f$, then

$$M_{Rd} = f_{yd} b_f t_f h_w \tag{14.7}$$

where

f_{yd} = the design strength of the flanges
t_f and b_f = the thickness and width of the flange plates
h_w = the distance between the internal faces of the flanges

Taken A = cross-sectional area of a plate girder, which is

$$A = 2b_f t_f + h_w t_w \tag{14.8}$$

then from Equations 14.7 and 14.8, A can be defined as

$$A = \frac{2M_{Rd}}{h_w f_{yd}} + h_w t_w \tag{14.9}$$

With $\lambda = h_w/t_w$, then A can be written as

$$A = \frac{2M_{Rd}}{\lambda t_w f_{yd}} + \lambda t_w^2 \tag{14.10}$$

Use mathematics and apply minimum weight solution, i.e. for an optimum solution, where $dA/dt = 0$, Equation 14.11 can be found, which gives the values for t_w and h_w in Equations 14.12 and 14.13, respectively:

$$\frac{dA}{dt_w} = -\frac{2M_{Rd}}{\lambda f_{yd} t_w^2} + 2\lambda t_w \tag{14.11}$$

and

$$t_w = \sqrt[3]{\frac{M_{Rd}}{\lambda^2 f_{yd}}} \tag{14.12}$$

with

$$h_w = \sqrt[3]{\frac{\lambda M_{Rd}}{f_{yd}}} \tag{14.13}$$

and as $A_w = h_w t_w$, then from above, A_w can be defined by

$$A_w = \sqrt[3]{\frac{M_{Rd}^2}{\lambda f_{yd}^2}} \tag{14.14}$$

as $M_{Rd} = f_{yd} b_f t_f h_w$ (see Equation 14.7), and with $A_f = b_f t_f$, then A_f can be calculated from

$$A_f = b_f t_f = \frac{M_{Rd}}{f_{yd}\sqrt[3]{\frac{\lambda M_{Rd}}{f_{yd}}}} = \sqrt[3]{\frac{M_{Rd}^2}{\lambda f_{yd}^2}} \tag{14.15}$$

This shows that the area of a single flange is equal to the area of the web

$$\left(\text{remember } \lambda = \frac{h_w}{t_w}\right) \le k\frac{E}{f_{yf}}\sqrt{\frac{A_w}{A_{cf}}}$$

Note: The reader should be aware that a section of a plate girder may well be classified as Class 4, and in this case, an effective section needs calculating.
Hint:
As

$$\lambda = \frac{h_w}{t_w} \le k\frac{E}{f_{yf}}\sqrt{\frac{A_w}{A_{cf}}}$$

and for $A_w = A_{cf}$, and with $k = 0.4$ (see Section 14.2.1), you can calculate λ; then you can calculate t_w and h_w from Equations 14.12 and 14.13 and A_f from Equation 14.15. Values of f_{yf} and E are normally known or given.

From practical experiences, designers can select an available plate thickness t_f to calculate b_f from above.

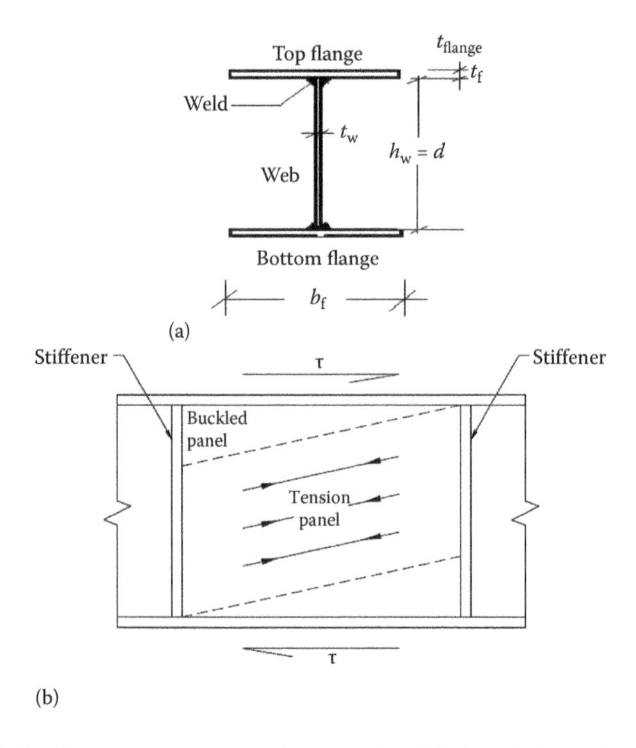

Figure 14.3 (a) Cross-sectional area of plate girder; (b) shear failure of a plate girder panel.

Experimental work showed that after web buckling occurred, a tension field formed in the central diagonal portion of the web (see Figure 14.3), which provides a reserve of strength in the web.

14.2.6 Nondimensional slenderness ration ($\bar{\lambda}_w$)

Shear capacity of web is assessed using the critical shear strength τ_{cr} upon which a shear buckling nondimensional slenderness ration ($\bar{\lambda}_w$) is dependent.

From BS EN 1993-1-5:2005, Clause 5.3 (5), and research literatures in this subject area, $\bar{\lambda}_w$ is defined as follows:

1. For webs with transverse stiffeners at the supports and either intermediate transverse or longitudinal stiffeners, the normalized web slenderness ratio $\bar{\lambda}_w$

$$\bar{\lambda}_w = \frac{h_w}{37.4 \varepsilon t_w \sqrt{k_\tau}} \tag{14.16}$$

The values of k_τ = a shear buckling coefficient are given in Equations 14.17 and 14.18.

t_w = web thickness
h_w = web depth. In this case, $d = h_w$.

The parameter k_τ is dependent on a/h_w and
For

$$a/h_w \geq 1.0 \quad \kappa_\tau = 5.34 + 4.00\left(\frac{h_w}{a}\right)^2 \tag{14.17}$$

For

$$a/h_w < 1.0 \quad \kappa_\tau = 4.00 + 5.34\left(\frac{h_w}{a}\right)^2 \tag{14.18}$$

where a = distance between stiffeners centre to centre.

2. For webs in transverse stiffeners only at the supports, where a/h_w is large, k_τ can be taken as 5.34; thus, from Equations 14.16 and 14.17, $\bar{\lambda}_w$ can be defined by

$$\bar{\lambda}_w = \frac{h_w}{37.4\varepsilon t_w\sqrt{5.34}} = \frac{h_w}{86.4\varepsilon t_w} \tag{14.19}$$

Note: The web plastic shear capacity is its upper limit shear force that may be carried by the web.

14.2.7 Shear capacity of web

14.2.7.1 Nonrigid end post

No post-buckling strength can be generated in this type of end post, and it is strong enough to resist the additional horizontal force from the tension field (Figure 14.4). Design contribution of web to shear buckling resistance is given by

$$V_{bw,Rd} = \chi_w f_{yw} h_w t_w / (\sqrt{3}\gamma_{M1}) \tag{14.20}$$

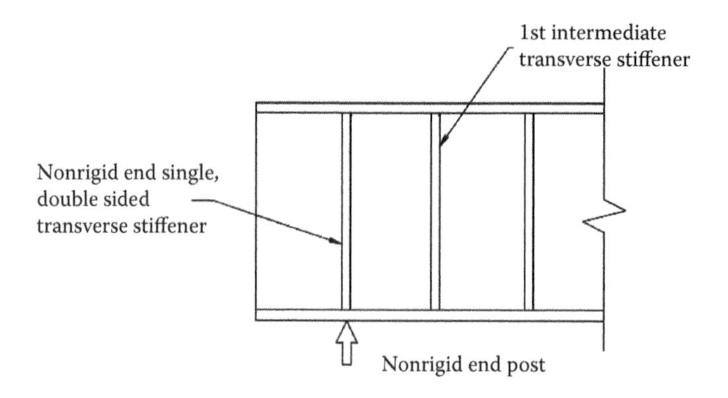

Figure 14.4 Nonrigid end post.

Contribution of web to shear buckling resistance χ_w: In the case of a nonrigid end post, EN 1993-1-5:2006 defines two zones for the contribution of the web to shear buckling resistance χ_w as follows:

(i) $\bar{\lambda}_w \leq 0.83/\eta$

$$\chi_w = \eta$$

(ii) $\bar{\lambda}_w > 0.83/\eta$

$$\chi_w = \frac{0.83}{\bar{\lambda}_w}$$

14.2.7.2 Rigid end post

In this case and in contrast with the nonrigid end post, the rigid end post is strong enough to resist the anchorage force from tension field theory (Figure 14.5). Normalized web slenderness $\bar{\lambda}_w$ and χ_w are given by

(i) $\bar{\lambda}_w \leq 0.83/\eta$

$$\chi_w = \eta$$

(ii) $0.83/\eta \leq \bar{\lambda}_w < 1.08$

$$\chi_w = \frac{0.83}{\bar{\lambda}_w}$$

(iii) $\bar{\lambda}_w \geq 1.08$

$$\chi_w = \frac{1.37}{0.7 + \bar{\lambda}_w}$$

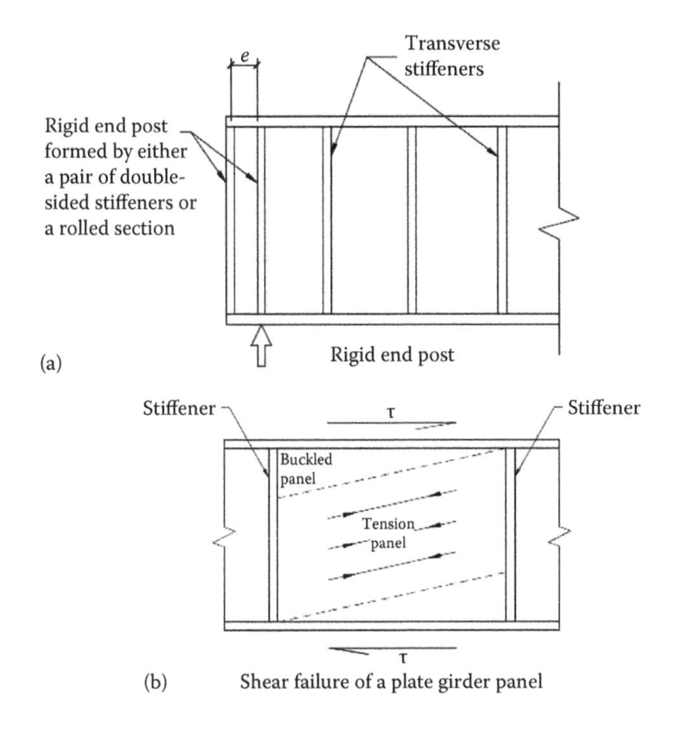

Figure 14.5 (a) Rigid end post; (b) shear failure of panel.

14.2.8 Design shear resistance $V_{b,Rd}$

From Clause 5.2 BS EN 1993-1-5:2006 (E), EN 1993-1-5:2006 and for web with or without stiffeners, $V_{b,Rd}$ is given by

$$V_{b,Rd} = \chi_v h_w t \frac{\frac{f_{yw}}{\sqrt{3}}}{\gamma_{M1}}$$

where shear coefficient χ_v is given by

$$\chi = \chi_w + \chi_f \leq \eta$$

χ_f = the flange buckling contribution factor to shear resistance
χ_w = the web buckling contribution factor to shear resistance

χ_f and χ_w are defined as

$$\chi_f = \frac{b_f t_f^2 f_{yf} \sqrt{3}}{c t_w h_w f_{yw}} \left[1 - \left(\frac{M_{Ed}}{M_{f,Rd}}\right)^2\right]$$

χ_w:
For

(a) Nonrigid end post girder

(i) $\bar{\lambda}_w \leq 0.83/\eta$

$$\chi_w = \eta$$

(ii) $\bar{\lambda}_w > 0.83/\eta$

$$\chi_w = \frac{0.83}{\bar{\lambda}_w}$$

(b) Rigid end post

(i) $\bar{\lambda}_w \leq 0.83/\eta$

$$\chi_w = \eta$$

(ii) $0.83/\eta \leq \bar{\lambda}_w < 1.08$

$$\chi_w = \frac{0.83}{\bar{\lambda}_w}$$

(iii) $\bar{\lambda}_w \geq 1.08$

$$\chi_w = \frac{1.37}{0.7 + \bar{\lambda}_w}$$

[Also see Figure 5.2 of BS EN 1993-1-5:2006, EN 1993-1-5:2006(E).]

where

b_f = the width of the flange
$b_f \leq 15\varepsilon t_f$ on each side of the web
t_f = the flange thickness
V_{Ed} = shear force
$M_{f,Rd}$ = design moment of resistance of the cross section. For $M_{f,Rd}$ use effective flange
c = width of the web between the plastic hinges (see Figure 14.6) and is given by [see BS EN 1993-1-5: 2006(E), Clause 5.4 (1)]

$$c = a\left(0.25 + 1.6\frac{M_{pl,f}}{M_{pl,w}}\right) = a\left(0.25 + \frac{1.6 b_f t_f^2 f_{yf}}{t h_w^2 f_{yw}}\right)$$

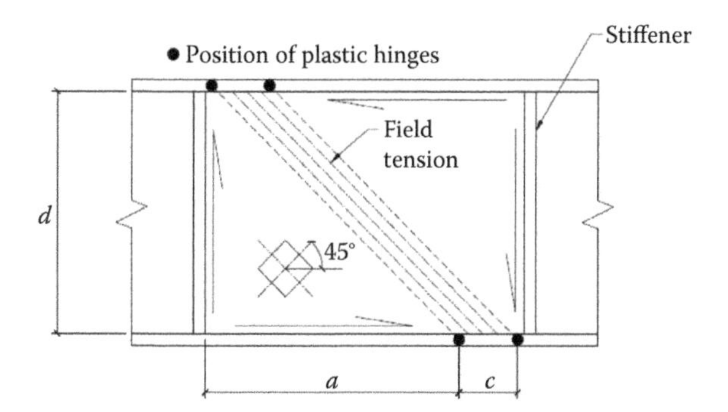

Figure 14.6 Web failure showing position of plastic hinges – post-buckling strength of web.

Note: The shear factors and the applied moment are dependent upon stiffener spacing. Therefore, you may need a few trials. A spreadsheet can be useful for the calculations of shear capacity.

14.2.9 Design of stiffeners

14.2.9.1 Rigid end post (see cl. 9.3.1, EN 1993-1-5)

This may be designed either by

- Using flat steel plates utilising the same steel grade of the web. The plates with a centroid distance, e, welded at the end of the plate girder and above the support.
- Using a rolled section welded at the end of the plate girder. In this case e = the distance between the flange centroids of the rolled section.

14.2.9.2 Minimum cross-sectional area of stiffener of rigid end post, A_{min}

The tension field shown in Figure 14.7a can be approximated by the uniformly distributed q_h loading action as can be seen in Figure 14.7b (Note: the actual tension shield is not uniform). Now if the rigid end post in Figure 14.7b with top flange as the end stiffener of the rigid end post and bottom flange as the second stiffener of the rigid end post is considered as a simply supported steel beam carrying a UDL load of q_h, then the maximum moment acting on the beam is given by

$$M_{max} = \frac{q_h h_w^2}{8}$$

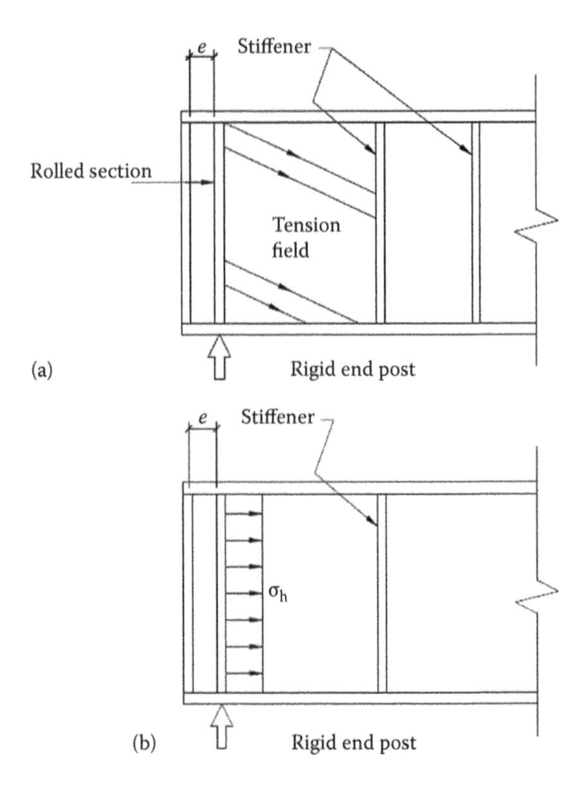

Figure 14.7 Rigid end post horizontal stress σ_h. (a) Tension field; (b) approximate stress.

If this moment is resisted by the beam flanges only, then f_y in the flange can be calculated from

$$f_y = \frac{M_{max}}{W_{pl}(= A_{min}e)}$$

As the horizontal stress σ_h for a panel of high slenderness ratio is given by (Figure 14.8)

$$\sigma_h = \frac{0.43}{\bar{\lambda}_w} f_y$$

and with

$$\bar{\lambda}_w = \frac{h_w}{37.4\varepsilon t\sqrt{k_\tau}}$$

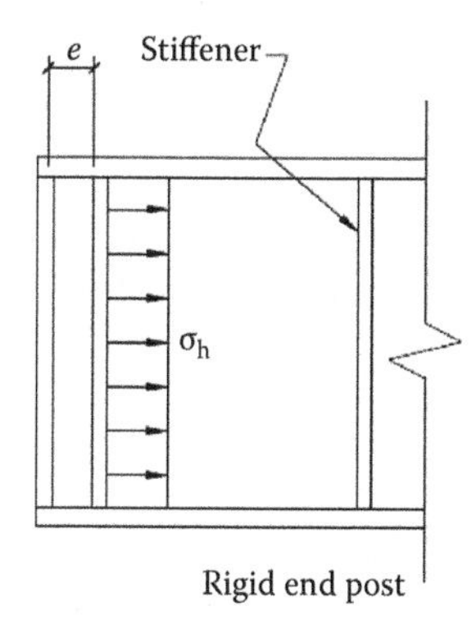

Figure 14.8 Rigid end post considered as a beam carrying σ_h.

and by taking ε = 1.0 and $h_w = a$, then $k_\tau = 9.34$ from

$$\kappa_\tau = 5.34 + 4.00\left(\frac{h_w}{a}\right)^2$$

This gives

$$q_h = \sigma_h t_w = 49.199\frac{t_{fy}^2}{h_w}$$

The value 49.199 is normally replaced by 32 as the stress on the rigid end post from tension field is not uniform.

The value of q_h becomes

$$q_h = \sigma_h t = 32\frac{t_{fy}^2}{h_w}$$

and for rigid end post, therefore,

$$\sigma_{max} = f_y = \frac{M_{max}}{W_{pl}} = \frac{4t^2 h_w f_y}{A_{min} e}$$

from which A_{min} can be defined as

$$A_{min} = \frac{4h_w t^2}{e}$$

To avoid detailing and constructions restrictions, normally e is taken to be greater than $0.1h_w$, i.e.

$e > 0.1h_w$ [see BS EN 1993-1-5:2006(E)]

In summary

(i) *Distance between stiffeners centre-to-centre,* e *(see cl. 9.3.1-4 BS EN 1993-1-5:2005)*

$e > 0.1h_w$ and *normally* less than (a) given by

$$a = h_w\sqrt{\frac{5.35}{k_\tau - 4.00}}$$

where a is the spacing between stiffeners centre to centre. See the example in Chapter 15.

(ii) *Double-sided stiffeners cross-sectional area >* A_{min}, *where*

$$A_{min} = \frac{4h_w t^2}{e}$$

See the example in Chapter 15.

14.2.9.3 Nonrigid end post spacing between stiffeners

In this case, the end panel can be designed as a nonrigid panel carrying the whole applied shear force (Figure 14.9).

$V_{hw,Rd}$ is given by

$V_{hw,Rd} = \chi_w f_{yw} h_w t_w/(\sqrt{3}\gamma_{M1})$, and with $V_{hw,Rd} = V_{Ed}$

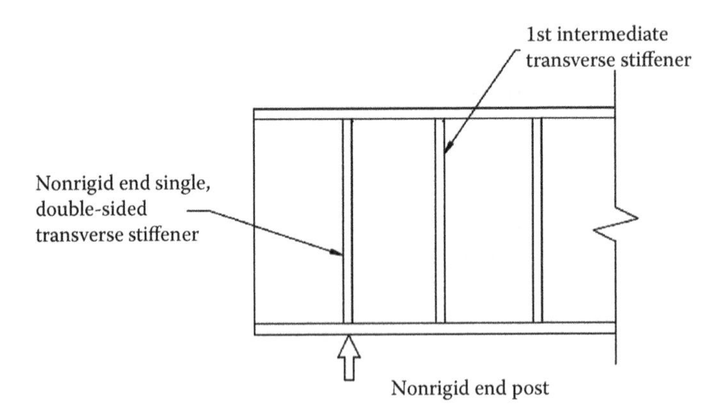

Figure 14.9 Nonrigid end post.

$$\chi_w = \frac{V_{Ed}\sqrt{3}\gamma_{M1}}{f_{yw}h_w t}$$

For web with transverse stiffeners at the supports and either transverse or longitudinal stiffeners, the value of k_τ is given by

$$k_\tau = \left(\frac{h_w}{37.4t\varepsilon\bar{\lambda}_w}\right)^2$$

and as an example for the following range of the non-dimensional normalized slenderness ratio $\bar{\lambda}_w$,

$$0.83/\eta \leq \bar{\lambda}_w < 1.08$$

$$\chi_w = \frac{0.83}{\bar{\lambda}_w}$$

Then for $a/h_w < 1$, the *upper bound* to the required panel width a is equal to

$$a = h_w\sqrt{\frac{5.35}{k_\tau - 4.00}}$$

14.2.9.4 Intermediate transverse stiffeners

The intermediate stiffeners should be able to carry safely a force given by

$$N_{stif,Rd} = V_{Ed} - \chi_w h_w t(f_{yw}/[\sqrt{3}\gamma_{M1}])$$

To calculate the above, the following need to be considered:

1. For web panel between adjacent stiffeners, calculate χ_w assuming the stiffener under consideration is removed.
2. For a plate girder carrying UDL or variable shear, check shear resistance at a distance $0.5h_w$ from the edge of the panel with larger shear force.

See example in Chapter 15 (also Clause 9.3.3, BS EN 1993-1-5:2006, EN 1993-1-5:2006).

14.2.10 Buckling resistance of the stiffener (cl. 9.1, EN 1933-1-3)

14.2.10.1 Buckling resistance of intermediate stiffener

Buckling resistance of intermediate stiffener or end plate stiffener is calculated where a portion of the web of the plate girder in the length equal to $15\varepsilon t$ on either side of the stiffener (making a strut in compression) may be considered into account (Figure 14.10). Also, see Clause 9.1 (1, 2 and 3) BS EN 1993-1-5:2006, EN 1993-1-5:2006.

N_{Rd} (buckling resistance) = $(\chi A f_y/\gamma_{M0})$ > Shear force (N_{Ed}, at the stiffeners location)

χ = *reduction factor*. See Tables 6.1 and 6.2, and Figure 6.4 for the χ *reduction factor* chart provided by BS EN 1993-1-1:2005 and EN 1993-1-1:2005.

A = area of strut section marked black in Figure 14.10; for calculating χ, you need both A_{strut} and I_{strut} (see below).

The minimum second moment of area I_{stif} is given by

for $a/h_w < \sqrt{2}$

$$I_{stif} \geq 1.5\frac{h_w^3 t_w^3}{a^2}$$

for $a/h_w \geq \sqrt{2}$

$$I_{stif} \geq 0.75 h_w t_w^3$$

$$I_{strut} = I_{stif} + 30\varepsilon t_w^4/12$$

$$A_{strut} = A_{stif} + 30\varepsilon t_w^2$$

See example in Chapter 15.

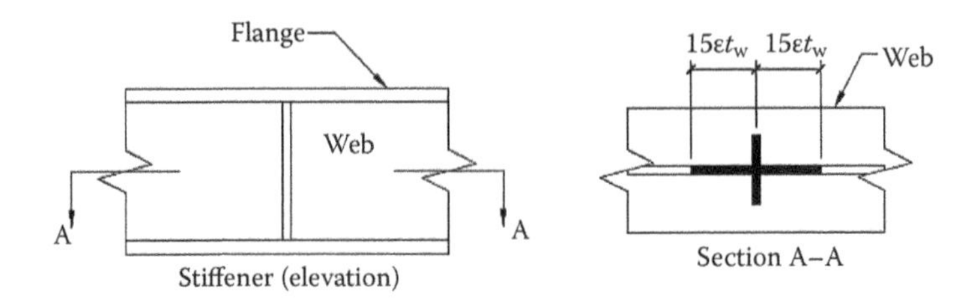

Figure 14.10 Buckling resistance of stiffener.

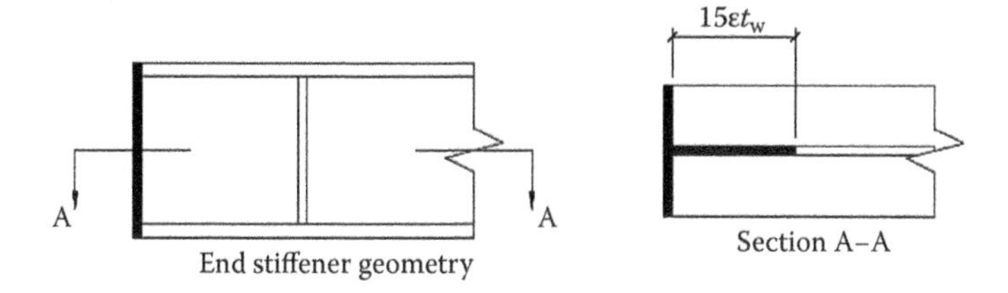

Figure 14.11 End stiffener geometry-buckling resistance.

14.2.10.2 Buckling resistance of the end stiffener

N_{Rd} (buckling resistance) = $(\chi A f_y/\gamma_{M0})$ > Shear force (N_{Ed}, at the stiffeners location)

χ = *reduction factor*. See Tables 6.1 and 6.2, and Figure 6.4 for the χ *reduction factor* chart provided by BS EN 1993-1-1:2005 and EN 1993-1-1:2005.

A = area of strut section marked black in Figure 14.11; for calculating χ, you need both A_{strut} and I_{strut} (see below), where for end stiffeners the coefficient 30 in the equation above is replaced by 15.

$$I_{strut} = I_{stif} + 15\varepsilon t_w^4/12$$

$$A_{strut} = A_{stif} + 15\varepsilon t_w^2$$

The effective length of the stiffener normally taken = $0.75h_w$.

Also, see Clause 9.1 (1, 2 and 3) BS EN 1993-1-5:2006, EN 1993-1-5:2006 and example in Chapter 15.

14.2.11 Web to flange welds

(i) Provided that the following is satisfied

$$V_{Ed} < \chi_w h_w t_w \frac{\frac{f_{yw}}{\sqrt{3}}}{\gamma_{M1}}$$

web to flange welds may be designed for a shear flow of V_{Ed}/h_w.

(ii) For large V_{Ed}, i.e.

$$V_{Ed} > \chi_w h_w t_w \frac{\frac{f_{yw}}{\sqrt{3}}}{\gamma_{M1}}$$

then web to flange welds are designed for a shear flow of $\eta f_{wy} t_w/(\gamma_{M1}\sqrt{3})$.

(iii) For all other cases, please refer to BS EN 1993-1-5:2006, EN 1993-1-5:2006.

Chapter 15

Plate girder: Practical design studies

Use current BS EN codes of practices to design the sectional dimensions together with the dimensional details and spacing of the stiffeners for a simply supported steel plate girder that spans 21.5 m and carrying a UDL characteristics variable load of 120 kN/m. Steel available in grade S355. The plate girder is carrying a cast in situ reinforced concrete floor which provides the compression flange with a full lateral restrained against torsional buckling. Assume a total weight of the plate girder beam = 5.0 kN/m. Architectural consideration is that total depth of the plate girder is ≤ 2.0 m.

15.1 TOTAL DESIGN LOADS ACTING ON THE PLATE GIRDER

Total design loading on the beam = 1.5 × 120 + 1.35 × (5) = 186.75 kN/m

15.2 MAXIMUM BENDING MOMENT ACTING ON THE PLATE GIRDER

$$M_{Ed} = \text{Design loads} \times L/8 = (186.75 \times 21.5) \times 21.5/8 = 10{,}790.648\,\text{kNm}$$
$$= 10.79 \times 10^9\,\text{kNm}$$

15.3 WEB CRITICAL SLENDERNESS RATIO

Assume the flanges resist the bending moment.

Start the design by assuming $A_w = A_f$ (see Chapter 14), with flange-induced buckling, $k = 0.4$ as plastic rotation is not utilized.

The critical h_w/t_w ratio is given by

$$\frac{h_w}{t_w} \le k\frac{E}{f_{yf}}\sqrt{\frac{A_w}{A_c}} = 0.4\frac{210 \times 10^3}{355}\sqrt{1} = 236.619$$

15.4 PLATE GIRDER: DIMENSIONS OF CROSS-SECTION

15.4.1 Web thickness t_w and web depth h_w

Calculate web depth h_w from

$$h_w = \sqrt[3]{\frac{\lambda M_{Rd}}{f_{yd}}} = \sqrt[3]{\frac{236.619 \times 10.79 \times 10^9}{\frac{355}{1.0}}} = 1930\,\text{mm}$$

(The figure does not need rounding up.) The thickness t_w is given as

$$t_w = \frac{h_w}{\lambda} = \frac{1930}{236.619} = 8.156\,\text{mm}, \quad \text{use } t = 9\text{ mm}$$

before starting the design, check if a rolled plate of this thickness is available in the market.

15.4.2 Deciding plate girder flange thickness

Flange outstand/flange thickness = 9ε (maximum value for a class 1 section) where

$\varepsilon = (235/355)^{1/2} = 0.814$

flange outstand/[flange thickness (t_f)] = 9ε

max. flange outstand = $9 \times t_f \times (235/355)^{1/2} = 7.33\, t_f$

A_{flange} = flange width $\times\, t_f = 2\,(7.33\, t_f)\, t_f = 14.66\, t_f^2$.

Assume $M_{pl,Rd} = A$ (flange) × steel design strength × h_w, that is,

$$M_{pl,Rd} = A_{\text{flange}} \frac{f_y}{\gamma_{M0}} h_w$$

$$A_{\text{flange}} = \frac{M_{Ed}}{h_w \frac{f_y}{\gamma_{M0}}} = \frac{10.79 \times 10^9}{1930 \frac{355}{1.0}} = 15,749.323\,\text{mm}^2$$

or

$$14.66 t_f^2 = 15,748.376$$

$t_f = 32.775$ mm

S355, steel plates with thickness; 32 mm, 34 mm, and 36 mm are available.

Use 34 mm thick plate with a width of 500 mm.

Outstand (245.5 mm actual) < 7.33 t_f (7.33 × 34 = 249.22 mm, limit for plastic flange) (satisfactory)

Overall depth, h:

$$h = 2\, t_f + h_w = 2 \times 34 + 1930 = 1998 \text{ mm.}$$

Less than 2.0 m upper limit in this example (satisfactory).

15.5 CHECK THAT ACTUAL h_w/t_w RATIO < ALLOWABLE h_w/t_w

Actual h_w/t_w ratio < allowable h_w/t_w

actual (1930/9 = 214.44 mm)

allowable is given by

$$\frac{h_w}{t_w} \leq k \frac{E}{f_{yf}} \sqrt{\frac{A_w}{A_f}}$$

$$\leq 0.4 \frac{210 \times 10^3}{355} \sqrt{\frac{1930 \times 9}{500 \times 34}} = 239.18$$

214.44 (actual) < 239.18 (allowable) (satisfactory).

The actual value is below the allowable, and therefore is satisfactory.

15.6 PLATE GIRDER CROSS-SECTION CLASSIFICATION

15.6.1 Web classification

Check if $c_w/t_w > 124\varepsilon$, where $c_w = h_w$,

$$c_w/t_w = 239.18 > 124\varepsilon(124 \times 0.814 = 100.94),$$

$$\varepsilon = (235/355)^{1/2} = 0.814$$

Web is class 4, where $h_w/t_w > 124\varepsilon$.

Therefore, the web needs to be checked for shear buckling.

15.6.2 Classification of the flanges

$c_f/t_f = [(500-9)/2]/34 = 7.22 < 9\varepsilon(= 7.326)$,

$\varepsilon = (235/355)^{1/2} = (0.814)$.

Flange is plastic.

Therefore, the flange is plastic and the web is slender.

15.7 CHECK PLASTIC MOMENT OF RESISTANCE OF THE FLANGES $M_{pl,Rd}$

$$M_{pl,Rd} = M_{f,Rd} = A_f \frac{f_y}{\gamma_{M0}}(h - t_f) = 34 \times 500 \frac{355}{1.0}(1998 - 34) = 11.85\,\text{MNm}$$

> acting maximum moment (10.79 MNm) (satisfactory)

15.8 MAXIMUM SHEAR FORCE

$V_{Ed} = 186.75 \times 21.5/2 = 2007.56$ kN (at the support)

15.9 WEB DESIGN

For the benefit of the designer, in this chapter, the web will be designed as a nonrigid end post and rigid end post as follows; calculate panel shear capacity, $V_{b,Rd}$

$$V_{b,Rd} = \chi_v h_w t_w \frac{\frac{f_{yw}}{\sqrt{3}}}{\gamma_{M1}}$$

15.9.1 Determination of χ_v

$\chi_v = x_f + \chi_w$

Calculate x_f.

Check if (on either side of the web) the flange width $\leq 15\varepsilon t_f$

$15\varepsilon t_f = 15 \times (235/355)^{1/2} \times 34 = 414.944$ mm

Actual flange width on one side = 0.5(500 – 9) = 245 mm. Therefore, $15\varepsilon t_f >$ 245 mm. Use actual width of 500 mm.

15.10 PLATE GIRDER WITH NONRIGID END POST

15.10.1 The shear capacity $V_{b,Rd}$ of the first panel from the support is given by

$$V_{b,Rd} = \chi_v h_w t_w \frac{\frac{f_{yw}}{\sqrt{3}}}{\gamma_{M1}}$$

$$\chi_v = \chi_f + \chi_w$$

$$\chi_f = \frac{b_f t_f^2 f_{yf} \sqrt{3}}{c t_w h_w f_{yw}} \left[1 - \left(\frac{M_{Ed}}{M_{f,Rd}}\right)^2\right]$$

Intermediate stiffeners are all equally spaced at 1.25 m between their centers (Figure 15.1). A decision made by the engineer to suit this work. You can try your own spacing.

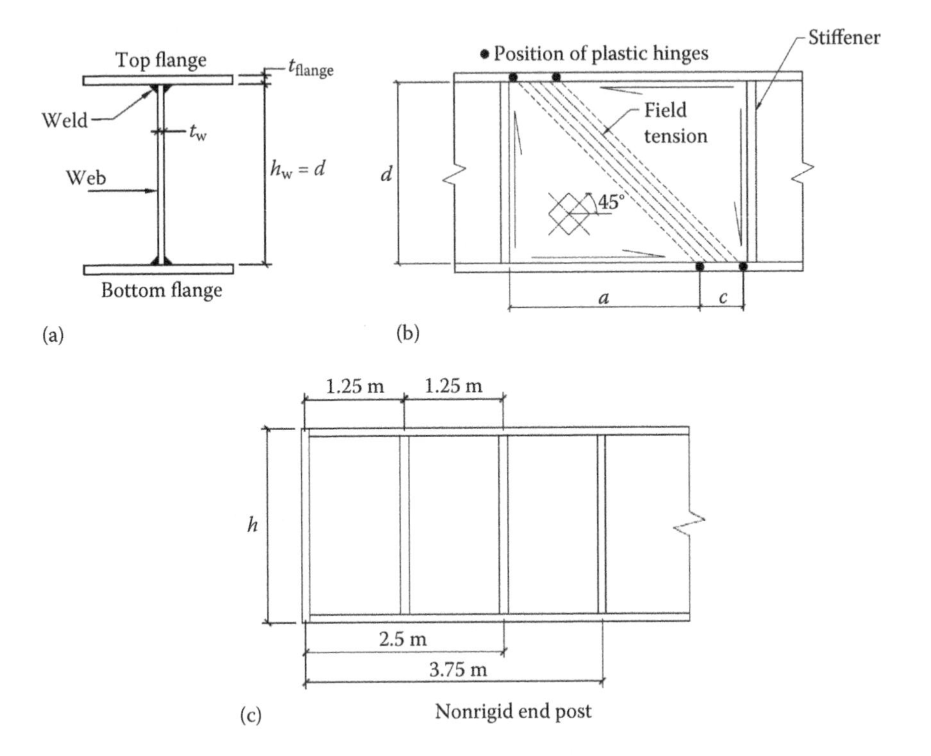

Figure 15.1 (a) Cross-section; (b) nonrigid end post and position of plastic hinge; (c) initial spacing of the stiffeners.

$$c = a\left(0.25 + \frac{1.6b_f t_f^2 f_{yf}}{t_w h_w^2 f_{yw}}\right) = 1250\left(0.25 + \frac{1.6 \times 500 \times 34^2 \times 355}{9 \times 1930^2 \times 355}\right) = 346.982 \text{ mm}$$

χ_f and χ_w, flange and web contribution factors

$$\chi_f = \frac{b_f t_f^2 f_{yf}\sqrt{3}}{ct_w h_w f_{yw}}\left[1 - \left(\frac{M_{Ed}}{M_{f,Rd}}\right)^2\right] = \frac{500 \times 34^2 \times 355 \times \sqrt{3}}{346.982 \times 9 \times 1930 \times 355}\left[1 - \left(\frac{2363.554}{11850}\right)^2\right] = 0.160$$

$M_{Ed} = 2007.562 \times 1.25 - 186.75 \times 1.25^2/2 = 2363.554$ kNm

Calculate χ_w, $\bar{\lambda}_w$ and k_τ see BS EN 1993-1-5 2006(E)
For $a/h_w < 1.0$, k_τ is given by $k_T = 4 + 5.34(h_w/a)^2$

$a/h_w = 1250/1930 = 0.647$

$k_\tau = 4 + 5.34(h_w/a)^2 = 4 + 5.34 \times (1930/1250)^2 = 16.73$

Calculate $\bar{\lambda}_w$ from

$$\bar{\lambda}_w = \frac{h_w}{37.4\varepsilon t_w\sqrt{k_\tau}} = \frac{1930}{37.4\sqrt{\frac{235}{355}} \times 9 \times \sqrt{16.73}} = 1.72$$

For

$$\bar{\lambda}_w > \left(\frac{0.83}{\eta}\right)$$

$$\chi_w = \frac{0.83}{\bar{\lambda}_w} = \frac{0.83}{1.72} = 0.482$$

See Clause 5.3 (1) and Table 5.1, BS EN 1993-1-5 2006(E). Therefore,

$\chi_v = \chi_f + \chi_w = 0.160 + 0.482 = 0.642$

shear capacity $V_{b,Rd}$ of the first panel can be calculated from

$$V_{b,Rd} = \chi_v h_w t_w \frac{\frac{f_{yw}}{\sqrt{3}}}{\gamma_{M1}} = \frac{0.642 \times 1930 \times 9}{1000} \frac{\frac{355}{\sqrt{3}}}{1.0} = 2285.612 \text{ kN}$$

Therefore, $V_{b,Rd}$ > maximum shear force 2007.562 kN.
Use the same procedures above to calculate $V_{b,Rd}$ for the rest of the stiffeners.

15.10.2 Intermediate stiffeners–strength check

15.10.2.1 First intermediate stiffener

Shear force to be carried by the stiffener $N_{Ed,stiff}$ is given by:

$$N_{Ed,stiff} = V_{Ed} - \chi_w h_w t_w \frac{\frac{f_{yw}}{\sqrt{3}}}{\gamma_{M1}}$$

For UDL loading on the girder, calculate V_{Ed} at 0.5 h_w from the stiffener in the panel with the higher shear, see BS EN 1993-1-5 2006(E):

$$V_{Ed,stiff} = 2007.562 - 186.75\ (1.25 - 0.5 \times 1.930) = 1954.338\ \text{kN}$$

Calculate χ_w assuming the stiffener is removed, see BS EN 1993-1-5 2006(E): Therefore,

$$a = 1250 + 1250 = 2.50\ \text{mm}$$

$$\frac{h_w}{a} = \frac{1930}{2500} = 0.772 \quad \frac{a}{h_w} = 1.295$$

For $\frac{a}{h_w} > 1.0$, is given by

$$k_\tau = 5.34 + 4\left(\frac{h_w}{a}\right)^2 = 5.34 + 4 \times 0.772^2 = 7.72$$

Therefore, $\bar{\lambda}$ is from

$$\bar{\lambda}_w = \frac{h_w}{37.4\,\varepsilon t \sqrt{k_\tau}} = \frac{1930}{37.4 \times \sqrt{\frac{235}{355}} \times 9 \times \sqrt{7.72}} = 2.53$$

As $\bar{\lambda}_w > 1.08$, χ_w is given by

$$\chi_w = \frac{0.83}{\bar{\lambda}_w} = \frac{0.83}{2.53} = 0.328$$

$$N_{\text{Ed,stiff}} = V_{\text{Ed}} - \chi_{\text{w}} h_{\text{w}} t_{\text{w}} \frac{\frac{f_{\text{yw}}}{\sqrt{3}}}{\gamma_{\text{M1}}}$$

$$= 1954.338 - 0.328 \times 1930 \times 9 \times \frac{\frac{355}{\sqrt{3}}}{1.0} \times 10^{-3}$$

$$= 786.61\,\text{kN}$$

786.61 kN should be < buckling capacity.
See calculations in the following section, that is, buckling capacity check.

15.10.2.2 Second intermediate stiffener

Calculate axial force $N_{\text{Ed,stiff}}$ on the stiffener (see Figure 15.2), which is given by

$$N_{\text{Ed,stiff}} = V_{\text{Ed}} - \chi_{\text{w}} h_{\text{w}} t_{\text{w}} \frac{\frac{f_{\text{yw}}}{\sqrt{3}}}{\gamma_{\text{M1}}}$$

As in the case for the first intermediate stiffener, and since the load on the girder is a UDL, V_{Ed} is determined at 0.5 h_{w} from the stiffener in the panel with the higher shear: see BS EN 1993-1-1 2005(E): Therefore,

$$V_{\text{Ed}} = (186.75 \times 21.50/2 - 186.75(2.5 - 0.5 \times 1.930)) = 1720.9\ \text{kN}$$

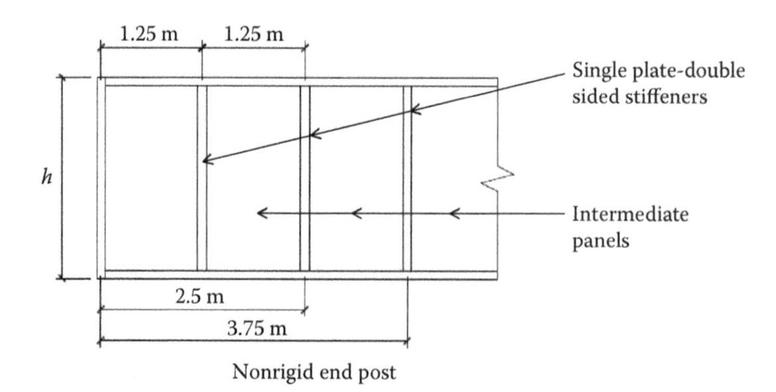

Figure 15.2 Nonrigid end post and position of intermediate stiffeners, first trial.

Calculate χ_w assuming the stiffener is removed, see BS EN 1993-1-5 2006(E): Therefore,

$$a = 1.25 + 1.25 = 2.5 \text{ mm}$$

$$\frac{h_w}{a} = \frac{1930}{2.5} = 0.772$$

$$\frac{a}{h_w} = 1.295$$

For $\frac{a}{h_w} > 1.0$, k_τ is given by

$$k_\tau = 5.34 + 4\left(\frac{h_w}{a}\right)^2 = 5.34 + 4 \times 0.772^2 = 7.72$$

And since $\bar{\lambda}_w$ ratio is given by

$$\bar{\lambda}_w = \frac{h_w}{37.4\varepsilon t\sqrt{k_\tau}} = \frac{1930}{37.4 \times 9 \times \sqrt{\frac{235}{355}}\sqrt{7.72}} = 2.53$$

for $\bar{\lambda}_w > 1.08$, χ_w

$$\chi_w = \frac{0.83}{\bar{\lambda}_w} = \frac{0.83}{2.53} = 0.328$$

$$N_{Ed,stiff} = V_{Ed} - \chi_w h_w t_w \frac{\frac{f_{yw}}{\sqrt{3}}}{\gamma_{M1}}$$

$$= 1720.9 - 0.328 \times 1930 \times 9 \times \frac{\frac{355}{\sqrt{3}}}{1.0} \times 10^{-3}$$

$$= 553.172 \text{ kN}$$

N_{Ed} = 553.172 kN should be < buckling capacity of the stiffener. See the calculations in the following section, that is, buckling capacity check of the stiffener.

For third, fourth, and so on stiffeners, use the same procedures as above.

15.10.3 Dimensions of stiffeners: The minimum stiffness requirement

According to Clause 9.3.3 (3), BS EN 1993-1-5 2006 for $a/h_w < \sqrt{2}$, the minimum stiffness requirement for all stiffeners is given by:

$$I_{stiff} = 1.5\frac{h_w^3 t^3}{a^2} = 1.5\frac{1930^3 \times 9^3}{1250^2} = 5.0 \times 10^6 \text{ mm}^4$$

Taking the same thickness of the web for the stiffeners, that is, 9 mm, then the total breadth of stiffener b is calculates as such:

Since $I_{stiffener} = t_{stiff} \times b^3/12$

$$b = \sqrt[3]{\frac{12 \times 5.0 \times 10^6}{9}} = 188.2 \text{ mm}$$

Since the flange of the plate girder is 500 mm wide, b = 200 mm from each side can be selected, where 200 × 2 + (t_w = 9 mm) = 409 mm < 500 mm

The stiffener can carry shear force, $N_{Rd,stiff}$ of

$$N_{Rd,stiff} = 2 \times 200 \times 9 \times \frac{355}{1.0} / 10^3 = 1278 \text{ kN}$$

Note: This should be greater than the force acting on the stiffeners, also check the buckling strength or sometimes called buckling capacity; see the section below.

15.11 INTERMEDIATE STIFFENER: BUCKLING CHECK

Figure 15.3 shows a nonrigid end post, section A–A for intermediate stiffeners buckling check.

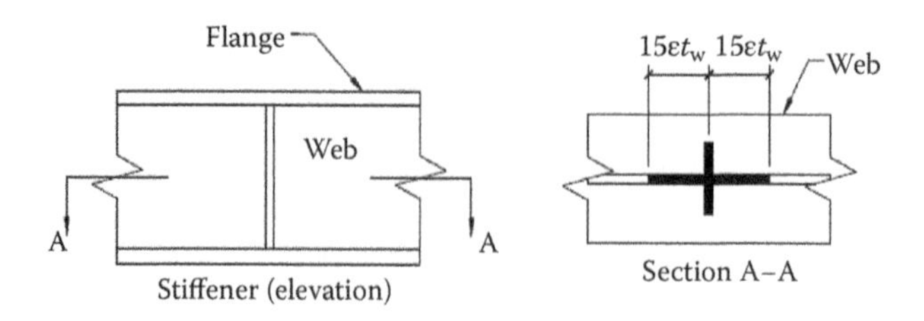

Figure 15.3 Nonrigid end post, section A–A intermediate stiffeners–buckling check.

$L_{eff} = 0.75 \times h_w = 0.75 \times 1930 = 1448$ mm

$$A_{strut} = A_{stiffener} + 40\,\varepsilon t_w^2 = 200 \times 9 \times 2 + 30 \times \sqrt{\frac{235}{355}} \times 9^2 = 5577\ \text{mm}^2$$

A_{strut} = the equivalent area of the strut resisting compression
$A_{stiffener}$ = area of stiffener
I_{strut} = I (the effective area) which is equal to

$$I_{strut} = I_{stiffener} + 30\,\varepsilon t^4 = \frac{1}{12}400^3 \times 9 + \frac{1}{12}30 \times \sqrt{\frac{235}{355}} \times 9^4 = 48.013 \times 10^6\ \text{mm}^4$$

$$N_{cr} = \frac{\pi^2 EI_{strut}}{L^2} = \frac{\pi^2 \times 210 \times 10^3 \times 48.013 \times 10^6}{1448^2} = 47{,}461.4\ \text{kN}$$

$$\bar{\lambda} = \sqrt{\frac{A_{strut}}{N_{cr}} f_y} = \sqrt{\frac{5577 \times 355}{47{,}461.4 \times 10^3}} = 0.204$$

Since $\bar{\lambda} > 0.2$, check if the stiffener buckling capacity is satisfactory (Clause 6.3.1.2 (4): BS EN 1993-1-5:2006, EN 1993-1-5:2006(E)).

N_{Rd} (stiffener buckling capacity) = $(\chi A f_y/\gamma_{M0})$ > shear force ($N_{Ed,stiff}$ at the stiffeners location).

χ = reduction factor. See Figure 6.4 BS EN 1993-1-1:2005 and EN 1993-1-1:2005.

Use curve *c* as $t_f < 40$ mm and buckling about *z-z*.

$N_{Rd} = 0.96 \times 5577 \times 355/1.0 = 1900.64$ kN > ($N_{Ed,stiff} = 786.61$ kN) = the shear force on the first intermediate stiffeners, which is the highest shear force compared with the rest of the intermediate stiffeners of the plate girder in this example.

Thus, the stiffener size is adequate. Use 2 × 200 mm × 9 mm as intermediate stiffeners (Figures 15.2 and 15.3).

15.12 PLATE GIRDER: END STIFFENER – NONRIGID END POST

'Normally' the designer uses for the end stiffener a thicker steel plate and the same steel grade as for the rest of stiffeners. At the time of design, 12 mm, 15 mm, and 20 mm, 24 mm, 25 mm S355 steel plates are available. Let us try a 15 mm thick × 500 mm wide end plate stiffener (Figure 15.4):

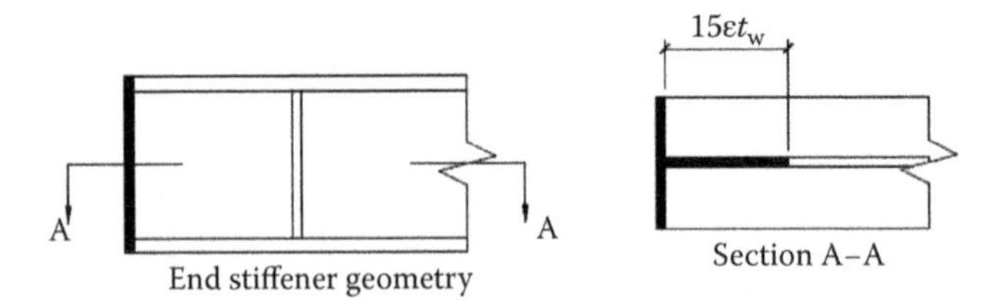

Figure 15.4 Nonrigid end post, section A–A end plate stiffeners – buckling check.

(i) Check buckling resistance of the end stiffener,

Check that

$$N_{Rd} = \chi A_{strut} \frac{f_y}{\gamma_{M0}} > \text{shear force acting on the end stiffener}$$

Check if $\bar{\lambda} < 0.2$

$$\bar{\lambda} = \sqrt{\frac{A_{strut} f_y}{N_{cr}}}$$

a proportion of the web length of ($15\varepsilon t_w$) is taken into account, see BS EN 1993-1-5 2006(E).

Length of the web = $15\varepsilon t_w = 15 \times 9 \sqrt{(235/355)} = 109.84$ mm

$$A_{strut} = A_{stiffener} + A_{web} = 500 \times 15 + 109.84 \times 9 = 8488.6 \text{ mm}^2$$

$$N_{cr} = \frac{\pi^2 E I_{strut}}{L^2}$$

$$I_{strut} = I_{stiffener} + I_{web} = \frac{500^3 \times 15}{12} + \frac{109.84 \times 9^3}{12} = 156.2 \times 10^6 \text{mm}^4$$

Use of an effective length of $L = 0.75\, h_w = 0.75 \times 1930 = 1448$ mm.

$$N_{cr} = \frac{\pi^2 E I_{strut}}{L^2} = \frac{\pi^2 \times 210 \times 10^3 \times 156.2 \times 10^6}{1448^2 \times 1000} = 154.4 \times 10^3 \text{ kN}$$

$$\bar{\lambda} = \sqrt{\frac{A_{strut} f_y}{N_{cr}}} = \sqrt{\frac{8488.6 \times 355}{154.4 \times 10^3 \times 10^3}} = 0.14$$

BS EN 1993-1-1 2005(E), Clause 6.3.1.2 (4), there is no reduction for strut buckling as

$\bar{\lambda} < 0.2.$

So, from buckling curve, see BS EN 1993-1-1 2005(E), $\chi = 1.0$

$$N_{Rd} = \chi A_{strut} \frac{f_y}{\gamma_{M0}} = 1.0 \times 8488.6 \frac{355}{1.0} \times 10^{-3} = 3013.45\,\text{kN} > 2007.562\ \text{kN},$$

the reaction at the end of the plate girder. Therefore, buckling capacity of end stiffener is satisfactory.

(ii) Lateral torsional buckling of end stiffener

Check if $I_T/I_p > 5.3\ f_y/E$ (limiting value to a void lateral torsional buckling), BS EN 1993-1-5 2006(E), Clause 9.2.1 (8)

I_p = Polar second moment of area for the stiffener around the age fixed to the web

$I_p = I_x + I_y$

$$I_y = \frac{bh^3}{12} = \frac{15 \times 500^3}{12} = 156.25 \times 10^6\ \text{mm}^4$$

$$I_x = \frac{bh^3}{12} = \frac{500 \times 15^3}{12} = 140.625 \times 10^3\,\text{mm}^4$$

$I_p = I_x + I_y = 156.25 \times 10^6 + 140.625 \times 10^3 = 156.39 \times 10^6\ \text{mm}^4$

$$I_T = \frac{bh^3}{3} = 500 \times 15^3 / 3 = 0.562 \times 10^6\,\text{mm}^4$$

$$\frac{I_T}{I_p} = \frac{0.562 \times 10^6}{156.39 \times 10^6} = 3.594 \times 10^{-3}\ \text{(Actual value)}$$

This is < limiting value $\left(5.3 \frac{f_y}{E} = 5.3 \frac{355}{210 \times 10^3} = 8.95 \times 10^{-3}\right)$

A 26 mm thick end plate is available and will satisfy the limiting stiffness criterion (I_T/I_P) and will of course satisfy the buckling criteria calculated above.

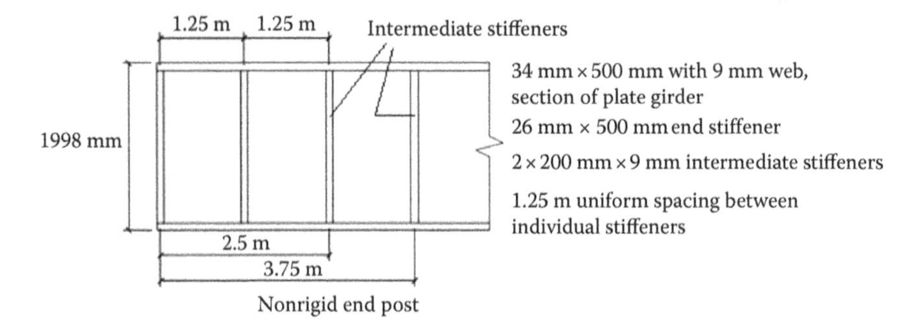

Figure 15.5 Nonrigid end post, the final layout of the girder.

15.13 WELDING BETWEEN FLANGE AND WEBS, CLAUSE 9.3.5-BS EN 1993-1-5 2006(E)

When $V_{Ed} > h_w t_w (f_{yw}/\sqrt{3})/\gamma_{M1}$, provide welds for a shear flow of

$$\eta t_w \frac{\frac{f_{yw}}{\sqrt{3}}}{\gamma_{M1}} = 1.2 \times 9 \frac{\frac{355}{\sqrt{3}}}{1.0} = 2213.56 \frac{\text{N}}{\text{mm}}$$

η = 1.2 for steel grade ≤ S460, see BS EN 1993-1-5 2006 Clause 5.1 (2-note 2)
η = 1.0 for steel grade > S460

Provide welding with strength > 2.22 kN/mm.
Figure 15.5 shows the plate girder layout for a nonrigid end post.

15.14 PLATE GIRDER WITH RIGID END POST

15.14.1 Intermediate stiffener

The intermediate stiffener is 1.5 m from the support. The engineer decided on equal spacing (Figure 15.6). You can decide on your own spacing between the stiffeners.

Check first panel shear capacity, $V_{b,Rd}$ from

$$V_{b,Rd} = \chi_v h_w t_w \frac{\frac{f_{yw}}{\sqrt{3}}}{\gamma_{M1}}$$

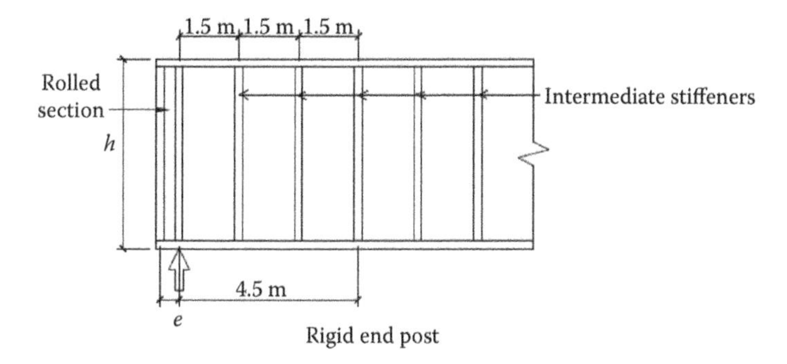

Figure 15.6 Rigid end post, proposed layout of stiffeners.

where

$\chi_v = \chi_f + \chi_w$, and

$$X_f = \frac{b_f t_f^2 f_{yf} \sqrt{3}}{c t_w h_w f_{yw}} \left[1 - \left(\frac{M_{Ed}}{M_{f,Rd}} \right)^2 \right]$$

c and M_{Ed} are given by:

$$c = a\left(0.25 + \frac{1.6 b_f t_f^2 f_{yf}}{t_w h_w^2 f_{yw}} \right) = 1500\left(0.25 + \frac{1.6 \times 500 \times 34^2 \times 355}{9 \times 1930^2 \times 355} \right) = 416.379\,\text{mm}$$

$$M_{Ed} = 2007.562 \times 1.5 - 186.75 \frac{1.5^2}{2} = 2801.249\,\text{kNm}$$

From the above information X_f can be calculated from

$$\chi_f = \frac{b_f t_f^2 f_{yf} \sqrt{3}}{c t_w h_w f_{yw}} \left[1 - \left(\frac{M_{Ed}}{M_{f,Rd}} \right)^2 \right]$$

$$= \frac{500 \times 34^2 \times 355 \times \sqrt{3}}{416.379 \times 9 \times 1930 \times 355} \left[1 - \left(\frac{2801.249}{11{,}850} \right)^2 \right] = 0.139$$

and χ_w is given by

$$\chi_w = \frac{1.37}{0.7 \bar{\lambda}_w}$$

To calculate $\bar{\lambda}_w$, you need to calculate k_τ which is a function of h_w/a where

$$\frac{h_w}{a} = \frac{1930}{1500} = 1.28 \quad \frac{a}{h_w} = 0.77$$

and as $\frac{a}{h_w} < 1.0$, k_τ is given by

$$k_\tau = 4 + 5.34\left(\frac{h_w}{a}\right)^2 = 4 + 5.34 \times 1.28^2 = 12.74$$

therefore $\bar{\lambda}_w$ can be calculated from

$$\bar{\lambda}_w = \frac{h_w}{37.4\,\varepsilon t_w \sqrt{k_\tau}} = \frac{1930}{37.4 \times \sqrt{\frac{235}{355}} \times 9 \times \sqrt{12.74}} = 1.97$$

$\bar{\lambda}_w > 1.08$, therefore χ_w is given by

$$\chi_w = \frac{1.37}{0.7 + \bar{\lambda}_w} = \frac{1.37}{0.7 + 1.97} = 0.513$$

and

$$\chi_v = 0.139 + 0.513 = 0.652$$

Panel capacity $V_{b,Rd}$ can be calculated from

$$V_{b,Rd} = \chi_v h_w t_w \frac{\frac{f_{yw}}{\sqrt{3}}}{\gamma_{M1}} = \frac{0.652 \times 1930 \times 9}{1000} \frac{\frac{355}{\sqrt{3}}}{1.0} = 2321.213\,\text{kN}$$

$(V_{b,Rd} = 2321.21$ kN$) > 2007.562$ kN max. shear force acting on the panel. Check the dimensions of stiffeners

$a = 1.5$ m $< \sqrt{2}h_w$ then,

$$I_{stiffener} = 1.5\frac{h_w^3 t_w^3}{a^2} = 1.5\frac{1930^3 \times 9^3}{1500^2} = 3.493 \times 10^6\ \text{mm}^4$$

Use a plate of 9 mm thick. Calculate the width of the stiffener:

$$b = \sqrt[3]{\frac{12 \times 3.493 \times 10^6}{9}} = 166.9\,\text{mm}$$

As in the case with nonrigid end post, let us try $b = 200$ mm.
Check if the stiffener can carry the shear force acting on it

$$N_{Rd,stiff} = A_{stiff} \times f_y/\gamma_{M0},$$

which should be > shear force acting on the stiffeners at the location considered

$$\therefore N_{Rd,stiff} = 2 \times 200 \times 9\frac{355}{1.0} \times 10^{-3} = 1278\,\text{kN}$$

1278 kN should be > shear force on the stiffeners given by:

$$V_{Ed,stiff} = V_{Ed} - \chi_w h_w t_w \frac{\frac{f_{yw}}{\sqrt{3}}}{\gamma_{M1}}$$

To complete, see the examples on plate girder with nonrigid end post.
Hints: For the following remaining strength (capacity) checks,

- Shear strength $V_{b,Rd}$ of the second, third, ..., and so on panels. Use the same procedures for a plate girder with nonrigid end post – see relevant pages in this example.
- Strength of intermediate stiffeners, that is, first intermediate stiffener, second intermediate stiffener, third intermediate stiffener, ..., and so on. Use the same procedures for a plate girder with nonrigid end post – see relevant pages in this example.
- Buckling capacity of end and intermediate stiffeners. Use the same procedures for a plate girder with nonrigid end post – see relevant pages in this example.

15.15 RIGID END POST

(i) *Distance between stiffeners center-to-center, e (Figure 15.7), see Clause 9.3.1 9 (3) BS EN 1993-1-5:2006(E)*

Distance e > 0.1 h_w and 'normally' less than a, where a, given by

$$a = h_w / \left(\sqrt{\frac{k_\tau - 4}{5.34}}\right)$$

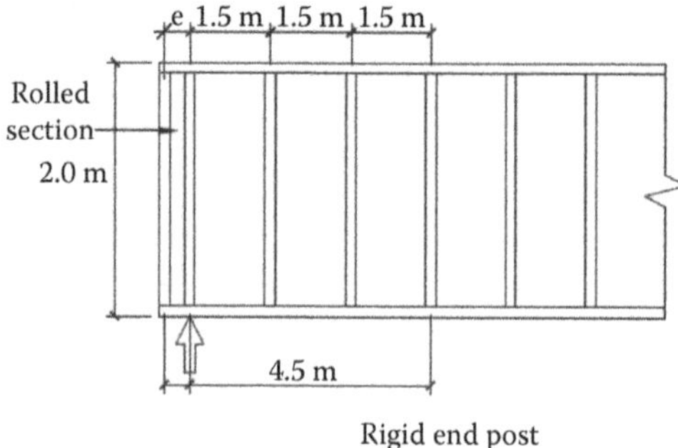

Figure 15.7 Rigid end post, distance (e).

see BS EN 1993-1-5 2006(E) Annex A.3
where k_τ is given by

$$k_\tau = \left(\frac{h_w}{37.4\,t_w \varepsilon \bar{\lambda}_w}\right)^2$$

and

$$\bar{\lambda}_w = \frac{0.83}{\chi_w}$$

where

$$\chi_w = \frac{V_{Ed}\sqrt{3}\gamma_{M1}}{f_{yw}h_w t_w} = \frac{2007.562 \times 10^3 \sqrt{3} \times 1.0}{355 \times 1930 \times 9} = 0.563$$

For rigid end post, the required normalized web slenderness is given by

$$\bar{\lambda}_w = \frac{0.83}{\chi_w} = \frac{0.83}{0.563} = 1.474$$

as for rigid end post $a/h_w < 1.0$, k_τ is given by:

$$k_\tau = \left(\frac{h_w}{37.4 t_w \varepsilon \bar{\lambda}_w}\right)^2 = \left(\frac{1930}{37.4 \times 9 \times \sqrt{\frac{235}{355}} \times 1.474}\right)^2 = 22.858$$

Therefore, a can be determined from:

$$a = h_w / \left(\sqrt{\frac{k_\tau - 4}{5.34}}\right) = 1930 / \left(\sqrt{\frac{22.858 - 4}{5.34}}\right) = 1027.022\,\text{mm}$$

$a = h_w/((\sqrt{k_\tau} - 4)/5.35)$, see BS EN 1993-1-6 Annex A.3

Choose a = 300 mm, center-to-center distance of the pair of double-sided stiffeners.

Therefore, $0.1\ h_w < a < 1027.022$.

(ii) *Double sided stiffeners cross-sectional area* > A_{min}

$$A_{min} = \frac{4h_w t^2}{e} = \frac{4 \times 1930 \times 9^2}{300} = 2084\,\text{mm}^2$$

A_{min} = 2084 = 500 (width of end stiffener plate is taken equal to the width of the flange) × $t_{stiffener}$

$t_{stiffener}$ = 4.168 mm (minimum thickness required)

A (provided for individual double sided stiffener = 26 mm × 500 mm) > 2084 mm², see Figure 15.8, therefore satisfactory.

The work in this book is only a guide and the readers are referred to the current codes of practices to make sure their work is satisfactory.

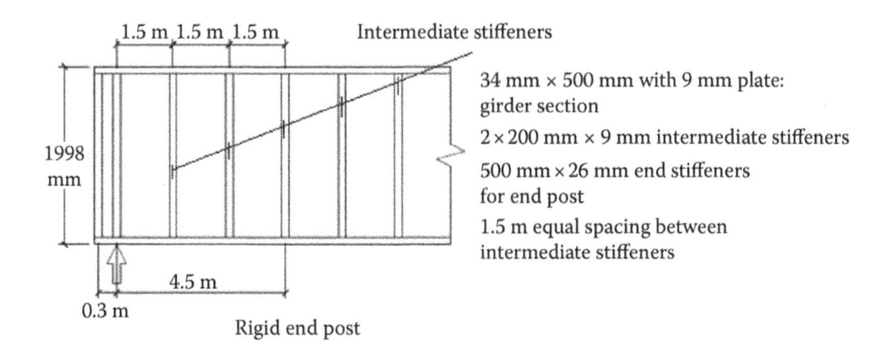

Figure 15.8 Rigid end post, final layout of plate girder.

Chapter 16

Sustainable steel buildings and energy saving

16.1 SUSTAINABLE STEEL BUILDINGS

Sustainable steel buildings are designed to continue to function effectively and be flexible and easier in change of uses. In the long term, they are providing efficiently and economically built assets. In this context, sustainable steel buildings are concerned primarily with

- Minimizing the use of energy
- Using of locally produced materials
- Reducing waste and air pollutions in the construction process
- Using waste recycled materials
- Adapting whole-life-cycle costing schemes in the design process of new build and using schemes to help minimise economic maintenance cost during the life of the building
- Continue to provide sustainable clients attractions
- Continue providing friendly and healthy working environment
- Providing pleasing appearance and no damage to the surrounding environment
- Adapting the policy of minimizing services and providing a pleasing infrastructure

The combination of science, skills and technology currently available in the field of steel fabrication, design and construction provide the designers and construction industry with the means to moderate old buildings and design and built new building/structures with

- Improving physical performance,
- Pleasant working environment,
- Increasing use of natural lightning, and
- Incorporating more energy efficient services.

16.2 ENERGY SAVING AND THERMAL INSULATION

Building services engineering is concerned with providing steel buildings that produce healthy environments for the people to work and or live in, that are economic in use and consumption of energy and meet the challenge of reducing both the impact of the atmosphere on the building and the impact of the building on the atmosphere.

The environmental engineering services will require less energy input if heat losses from the building are minimized. Heat loss from the building is partly through warming up the fresh air that is replacing exhausted and polluted indoor air, and partly by heat flowing out through the structure.

What do you have to do to bring the thermal performance of an external wall or a floor up to date?

To do this it is necessary to know how much heat is flowing through the unit area (each square metre) of the wall. This flow rate is stated as Watts per square metre (W/m^2).

The basic unit of heat in the SI System (*Système Internationale d'Unités*) is the Joule. A Joule is a quantity of heat and a Watt is that quantity of heat used per second, which is the unit of time in the SI System.

As 1 Watt = 1 Joule per second, it can be seen that if there were a flow rate of 10 Watts through each square metre of the wall there would be 10 Joules of heat flowing every second through each square metre.

But heat will only flow if there is a temperature difference between indoor and outdoor. Temperature is a measure of molecular activity and there is more activity as the temperature rises. Excited molecules will pass on their excitement to lesser excited molecules until they are equally excited and are in equilibrium. Therefore, for heat energy to flow there must be a temperature difference.

This brings us to the amount of heat flowing through each square metre of the wall for each unit of temperature difference. In this case, the unit of temperature is the Kelvin. This can be expressed as $W/m^2\,K$ (earlier on in the use of SI units the degree Celsius was used and some older textbooks used the expression $W/m^2 °C$, but this is incorrect).

Kelvin is the thermal transmittance coefficient of the complete wall, otherwise known as the *U*-value. Currently domestic external walls must have a *U*-value less than 0.35 $W/m^2\,K$. This will be reduced in the future to 0.25 $W/m^2\,K$ or even 0.18 $W/m^2\,K$.

Check for current *U*-value requirements in:

- Approved Document L of the Building Regulations 2000 and TI One stop. Also see Approved Document L1A: conservation of fuel and power in new dwellings, 2013 edition with 2016 amendments, Ref: ISBN 978 1 85946 324 6.

Designers in the United Kingdom may use a 20°C temperature difference with a –1°C as an external design temperature. (This switch to °C may seem confusing, but just 'go with it' at this point. A temperature of –1°C is easier to comprehend than 272 K.)

To find out how much heat will flow through a particular wall, it is necessary to know how effectively each material used in the wall prevents the passage of heat.

The thermal conductivity of each material is determined, and this data is used to work out the total thermal resistance of the wall. A material's thermal conductivity value is known as its *k*-value (sometimes the Greek letter lambda (λ) is used instead), see Sections 16.8 and 16.9.

But there is more to heat flow through the wall than just how the materials conduct heat. The surfaces have a resistance to flow, both receiving and transmitting heat. This also affects heat flow across cavities, as transmission and receipt of thermal energy is involved.

It is necessary to look at the methods of heat flow, which are

- Conduction
- Radiation
- Convection

Conduction is involved by conducting heat through materials. As a general rule, heavy materials conduct heat more quickly than light materials.

Radiation is involved as heat is transmitted across cavities. This also involves the phenomena of absorptivity, resistivity and emissivity.

Absorptivity is the ability of a material to absorb radiation at its surface. As radiated heat arrives at a material's surface, it is partly absorbed (and passed through the material by conductivity), and partly resisted (turned back). Resistivity is the reciprocal of absorptivity.

A material's emissivity determines its ability to emit heat from its surface by radiation. An example of a material that would resist the passage of heat is aluminium foil. If placed in a cavity, it would be very effective at resisting the passage of heat for two reasons:

- Its absorptivity is slow (its resistivity is high);
- Its emissivity is low.

If the foil were in contact with another material on either or both sides, these effects would be lost as heat would be conducted from one material to the other.

To be able to calculate a *U*-value, we need

- The internal surface resistance;
- The *k*-value of any materials and their thickness;
- The external surface resistance;
- The resistance values of any cavities.

Convection is involved with heat being lost from the external surface.

16.3 THE *U*-VALUE

The *U*-value, or thermal transmittance coefficient, is a measure of the structure's ability to transfer heat. It is the air-to-air heat transfer coefficient, whereas thermal conductance is from surface-to-surface.

$$U\text{-value} = \frac{1}{\text{total thermal resistance of the structure}}$$

$$U\text{-value} = \frac{1}{R_{SI} + R_1 + R_2 + R_3 + R_4 + \cdots + R_N + R_{SO}}$$

where

R_{SI} = inside surface resistance
$R_1, R_2, \ldots, R_N$ = thermal resistances of each part of the structure
R_{SO} = outside surface resistance.

Note that R_{CAV}, R_A or R_{AIR} may be used to denote thermal resistance of air spaces.

The thermal resistance (m^2 K/W) of a particular material of given thickness is obtained by dividing the thickness of the material (in metres) by its thermal conductivity (*k*-value; W/m K).

Resistance (R) = thickness (L) × resistivity (r),

$$\text{but, resistivity} = \frac{1}{\text{conductivity } (k)}$$

$$\therefore \text{Resistance} = \frac{1}{k} \times L$$

$$\therefore \text{Resistance} = \frac{L}{k}$$

As mentioned above, four things are required to calculate the *U*-value of any structure:

- The thickness of each material;
- The thermal conductivity (*k*-value) of each material;
- The surface resistances;
- The resistance values of any enclosed air-spaces.

16.4 RESISTANCES OF SURFACES

The values of resistances of most surfaces have been computed and those normally used in the United Kingdom are set out in the CIBSE guide.

In well-insulated buildings, the effect of different external exposures is minimal and the following values are generally used:

Outside surface resistance for walls = 0.06 m^2 K/W
Inside surface resistance for walls = 0.12 m^2 K/W

16.5 RESISTANCES OF AIR SPACES

With traditional building materials (brick, stone, concrete – high emissivity) it is normal practice to take the resistance value of an unventilated cavity over 19 mm as 0.18 m^2 K/W.

16.6 EXAMPLE CALCULATION

Part	*Thickness (m)*	*Thermal conductivity (W/m K)*	*Thermal resistance (m^2 K/W)*
R_{SO}	–	–	0.06
Brickwork	0.103	0.840	0.12
R_A	–	–	0.18
Insulation	0.020	0.035	0.57
Blockwork	0.100	0.230	0.43
Plaster	0.013	0.460	0.03
R_{SI}	–	–	0.12
Total thermal resistance			1.51

$$U\text{-value} = \frac{1}{1.51} = 0.66\ \text{W/m}^2\text{k}$$

$$\textbf{Note:}\ \text{The resistance} = \frac{\text{thickness (m)}}{\text{thermal conductivity (K)}}$$

16.7 SOME MAXIMUM *U*-VALUES

Residential buildings:

Roof with loft 0.25 W/m^2 K
Exposed walls 0.35 or 0.45 W/m^2 K
Exposed floor 0.45 W/m^2 K
Ground floor 0.45 W/m^2 K

See Approved Document L of Building Regulations (2000). Also, latest versions in the Document L1A: conservation of fuel and power in new dwellings, 2013 edition with 2016 amendments.

16.7.1 Example calculation 1

Solid brick wall: 215 mm thick brickwork (standard clay brick units and mortar, density 1700 kg/m^3) and 13 mm plaster (gypsum) inside.

Part	*Thickness (m)*	*Thermal conductivity (W/m K)*	*Thermal resistance (m^2 K/W)*
R_{SO}	–	–	0.06
Brickwork	0.215	0.840	0.26
Plaster	0.013	0.460	0.03
R_{SI}	–	–	0.12
Total thermal resistance			0.47

$$U\text{-value} = \frac{1}{0.47} = 2.13 \text{ W/m}^2\text{k}$$

16.7.2 Example calculation 2

A cavity brick wall: 102.5 mm brick outer leaf (1700 kg/m^3), 50 mm cavity, 102.5 mm brick inner leaf (1700 kg/m^3 – protected, 1% moisture) and 13 mm plaster (gypsum) inside.

Part	*Thickness (m)*	*Thermal conductivity (W/m K)*	*Thermal resistance (m^2 K/W)*
R_{SO}	–	–	0.06
Brickwork	0.103	0.840	0.12
R_A	–	–	0.18
Brickwork	0.103	0.620	0.17
Blaster	0.013	0.460	0.03
R_{SI}	–	–	0.12
Total thermal resistance			0.68

$$U\text{-value} = \frac{1}{0.68} = 1.47\ \text{W/m}^2\text{k}$$

16.7.3 Example calculation 3

A cavity brick wall: 102.5 mm brick outer leaf (1700 kg/m^3), 50 mm cavity, 100 mm lightweight concrete block inner leaf (600 kg/m^3) and 13 mm plaster (gypsum) inside.

Part	*Thickness (m)*	*Thermal conductivity (W/m K)*	*Thermal resistance (m^2 K/W)*
R_{SO}	–	–	0.06
Brickwork	0.103	0.840	0.12
R_A	–	–	0.18
Blockwork	0.100	0.190	0.53
Plaster	0.013	0.460	0.03
R_{SI}	–	–	0.12
Total thermal resistance			1.04

$$U\text{-value} = \frac{1}{1.04} = 0.96\ \text{W/m}^2\text{k}$$

16.7.4 Example calculation 4

A cavity brick wall: 102.5 mm brick outer leaf (1700 kg/m^3), 50 mm cavity, 100 mm lightweight concrete block inner leaf (600 kg/m^3) and 13 mm lightweight plaster (vermiculite) inside.

Part	*Thickness (m)*	*Thermal conductivity (W/m K)*	*Thermal resistance (m^2 K/W)*
R_{SO}	–	–	0.06
Brickwork	0.103	0.840	0.12
R_A	–	–	0.18
Blockwork	0.100	0.190	0.53
Plaster	0.013	0.200	0.07
R_{SI}	–	–	0.12
Total thermal resistance			1.08

$$U\text{-value} = \frac{1}{1.08} = 0.93\ \text{W/m}^2\text{k}$$

16.7.5 Example calculation 5

A cavity brick wall: 102.5 mm brick outer leaf (1700 kg/m³), 25 mm cavity, 25 mm heavy duty expanded polystyrene board, 100 mm lightweight concrete block inner leaf (600 kg/m³) and 13 mm plaster (perlite) inside.

Part	*Thickness (m)*	*Thermal conductivity (W/m K)*	*Thermal resistance (m^2 K/W)*
R_{SO}	–	–	0.06
Brickwork	0.103	0.840	0.12
R_A	–	–	0.18
Insulation	0.025	0.035	0.71
Blockwork	0.100	0.190	0.53
Plaster	0.013	0.190	0.07
R_{SI}	–	–	0.12
Total thermal resistance			1.79

$$U\text{-value} = \frac{1}{1.79} = 0.56\ \text{W/m}^2\text{k}$$

16.7.6 Example calculation 6

A cavity wall: 102.5 mm brick outer leaf (1700 kg/m³), 50 mm urea formaldehyde foam (k = 0.032 W/m K), 100 mm lightweight block inner leaf (600 kg/m³) and 15 mm lightweight plaster (vermiculite).

Part	*Thickness (m)*	*Thermal conductivity (W/m K)*	*Thermal resistance (m^2 K/W)*
R_{SO}	–	–	0.06
Brickwork	0.103	0.840	0.12
Foam	0.050	0.032	1.56
Blockwork	0.100	0.190	0.53
Plaster	0.013	0.200	0.07
R_{SI}	–	–	0.12
Total thermal resistance			2.46

$$U\text{-value} = \frac{1}{2.46} = 0.41\ \text{W/m}^2\text{k}$$

To return to the question of how to improve a wall's U-value to a specific value, consider a wall with a U-value of 0.86 W/m^2 K.

How thick would extra blockwork need to be for the wall to have a U-value of 0.3 W/m^2 K, if the k-value of the blockwork is 0.23 W/m K and the blockwork is already 150 mm thick?

For $U = 0.3$, total thermal resistance (R) required = $1/U$ = 3.33 m^2 K/W.

Existing $R = 1.16$, therefore extra $R = 3.33 - 1.16 = 2.17$ m^2 K/W.

Extra thickness of blockwork required = $2.17 \times 0.23 = 0.499$ m.

The blockwork would have to be 150 + 499 = 649 mm thick, which is impractical.

If the blockwork remained at 150 mm, and an insulation material having a k-value of 0.02 W/m K was added to the wall construction instead, what would be the required thickness to achieve a U-value of 0.45 W/m^2 K?

For $U = 0.45$, total thermal resistance (R) required = $1/U$ = 2.22 m^2 K/W.

Existing $R = 1.16$, therefore extra $R = 2.2 - 1.16 = 1.06$ m^2 K/W.

Thickness of insulation required = $1.06 \times 0.02 = 0.021$ m.

The insulation would have to be 21 mm thick.

It may be necessary to specify a material of 25 mm thickness.

16.8 THERMAL CONDUCTIVITIES OF COMMONLY USED INSULATING MATERIALS

Material	*Thermal conductivity (k) (W/m K)*
Expanded polystyrene board	
Heavy duty	0.034
Glassfibre wall bat	0.034
Glassfibre quilt	0.040
Polyisocyanurate board	0.022
Polyurethane board	0.022
Vermiculite loose fill	0.065
Wood-wool slab	0.085

16.9 SOME TYPICAL *k*-VALUES (W/m K)

Carbon steel	50
1:2:4 concrete	1.5
Water	0.6
Aerated concrete	0.2
Cellular polystyrene	0.035
Air	0.025

Thermal conductivities of building materials

Material	*Thermal conductivity (k) (W/m K)*
Aluminium	160.0
Asphalt	0.58
Brickwork	
Commons	
Light	0.806
Average	1.21
Dense	1.47
Lightweight bricks	0.374
Engineering bricks	1.15
Concrete	
Dense	1.44
Cellular concrete	0.26
Vermiculite concrete	0.2
Corkboard	0.042
Cork slab (damp) 160	–
Glass	1.05
Glass-fibre	
Quilt (40% RH)	0.032
Quilt	0.04
Wall bat	0.034
Glass wool	
Quilt	0.035
Blanket	0.042
Mild steel	50.0
Mineral wool	
Felt	0.04
Rigid slab	0.05
Plaster	
Gypsum	0.46
Perlite	0.08
Cement: lime: sand	0.48
Gypsum: sand (1:3)	0.65
Plasterboard	
Gypsum	0.16
Plywood	0.138
Polyisocyanurate board	0.022

(Continued)

Material	*Thermal conductivity (k) (W/m K)*
Polystyrene	
Cellular	0.034
Expanded board: HD	0.034
Polyurethane	
Board	0.022
Cellular	0.026
Purlboard	0.02
Rendering: cement: sand	0.532
Roofing felt	0.19
Stone	1.3
Artificial	
Granite	2.5
Marble	2.0
Sandstone	1.3
Slate	1.9
Tiles	
Clay	0.85
Concrete	1.1
Cork	0.085
Timber	
Softwood (across grain)	0.13
Hardwood (across grain)	0.15
Wall board	0.16
Water	
Fresh at 20°C	0.60
Sea at 20°C	0.58
Ice at −1°C	2.24
Wood chipboard	500
Woodwool slab	0.08

Note: The information above is extracted from the CIBSE guide, Table A3.22 and other sources (see Table A3.23 for correction factors for moisture content in masonry). Also, see the latest versions in the Document L1A: conservation of fuel and power in new dwellings, 2013 edition with 2016 amendments for up date information.

16.10 THERMAL INSULATION

To meet part L of the current *UK Building Regulations* in terms of thermal efficiency, roofs and external walls are required to achieve a certain *U*-value.

Floor finishes

Screed floor

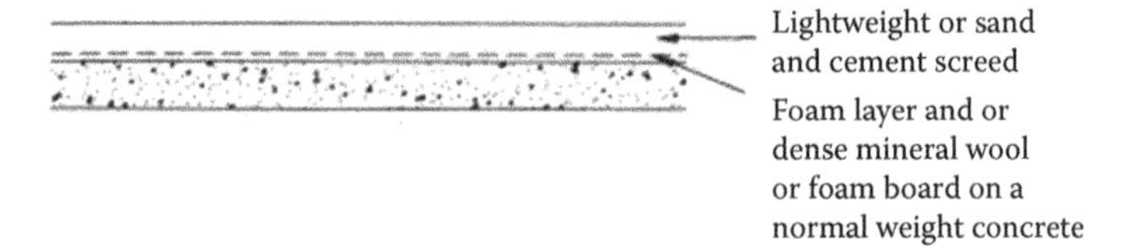

Platform floor

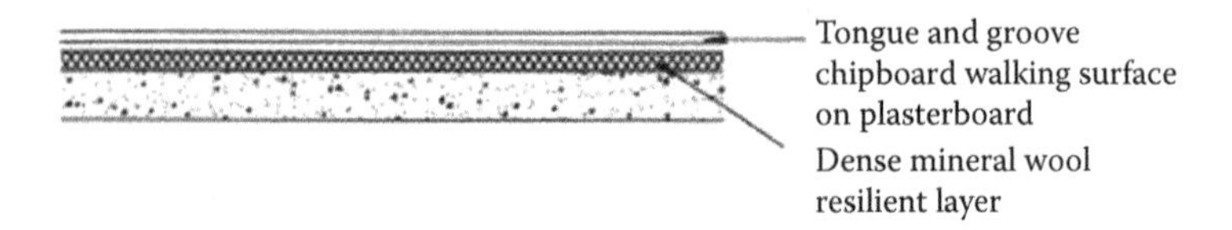

Raft floor

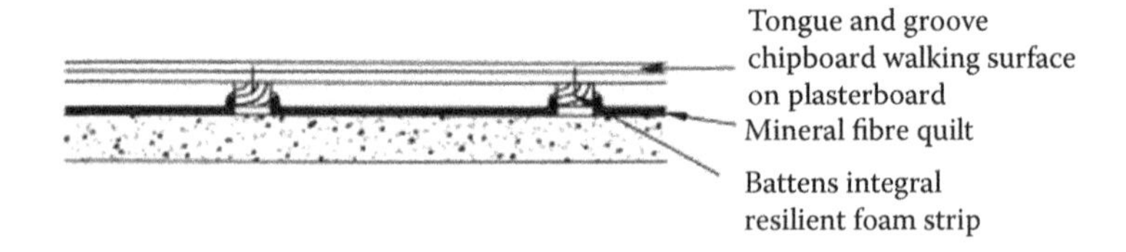

Cradle floor

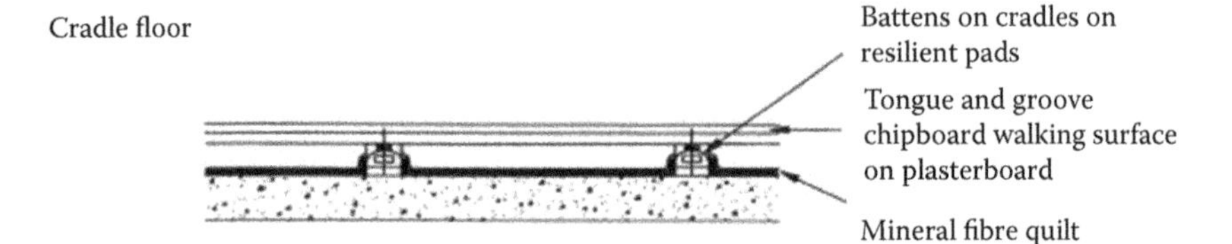

Floor finishes on slimdeck slab. Slab normally made from 280 ASB beams with SD225 composite decking supporting a normal weight concrete. Total depth approx. 300 mm

Figure 16.1 Alternative slimdek floor finishes (see SCI P321 Acoustic performance of slimdek, SCI, 2003).

A value of 0.2 $W/m^2{}^\circ C$ can be achieved by using steel 'warm frame' construction (SCI Publication 174, 1999), where the insulation is placed in a closed cell insulation with mineral wool insulation between the light steel wall studs. Using the warm frame principle, the majority or all the insulation is placed externally to the frame. This creates a continuous layer of insulation and highly reduces the risk of condensation (see Figure 16.1).

In recent years, using light steel framing, in commercial, low rise and residential buildings, insulated walls, roofs and floors has shown good insulated building materials with *U*-value satisfying the requirements of the current *UK Building Regulations*.

16.11 ACOUSTIC INSULATION

To meet the new requirements of part E of the current *UK Building Regulations* (2003), the performance of separating floors and walls between dwellings is required to achieve the minimum performance standard values of $D_{nT,w} + C_{tr}$ and $L_{nT,w}$ (for the separating floors) and $D_{nT,w} + C_{tr}$ (for separating walls). For more details, see the following websites and relevant publications:

www.steel-sci.org
www.Slimdek.com
www.coruspanelsandprofiles.co.uk
Case studies on slimdek, P309, SCI (2002)
Multi-storey residential buildings using Slimdek, P310 SCI (2002)
Slimdek-Engineering floor solutions, Corus Construction Centre (2002)
Acoustic performance of light steel framing, P320, SCI (2003)

Approved Document L of the Building Regulations 2000 and TI One stop. Also see Approved Document L1A: conservation of fuel and power in new dwellings, 2013 edition with 2016 amendments.

Bibliography

BRITISH STANDARDS AND EUROCODES

BRE-Ecohomes (2006) The environmental rating for homes, The Guidance, 2006, Issue 1.2.
British Constructional Steelwork Association/The Steel Construction Institute: Joints in Steel Construction. Simple Connections (2010) BCSA, SCI, London.
British Standard Institution (1994) Eurocode 4 Design for Composite Steel and Concrete Structures, General Rules for Buildings. DD ENV 1994-1-1, BSI, London.
British Standards Institution, BS EN 1994-1-2:2005 EC 4 – Design of composite steel and concrete structure Part 1-2: General rules; Structural fire design. London.
BS 449: Part 2 (1969) The use of structural steel in building.
BS 476: Part 3, Part 8 (1968–75) Fire tests on building materials and structures.
BS 4848 Specification for Hot-Rolled Structural Steel Sections.
BS 5493 (1977) Protective Coatings of Iron and Steel Structures Against Corrosion.
BS 5950 Structural Use of Steelwork in Building.
Part 1 (1990) Code of practice for design in simple and continuous construction: Hot rolled sections.
Part 3 (1990) Code of practice for design of simple and continuous composite beams.
Part 4 (1994) Code of practice for design of composite slabs with profiled steel sheeting.
Part 6 Code of practice for design of light gauge sheeting, decking and cladding.
Part 8 (1990) Code of practice for fire resistant design.
BS 5950-1-2000 Structural Use of Steelwork in Building.
Part 1: Code of practise for design in simple and continuous construction. Hot rolled sections.
BS 5950-2003: Structural use of steelwork in building: Part 8: Code of practice for fire resistant design. BSI, London.
BS 5950-4 (1992) Structural use of steelwork in buildings, Part 4: Code of practice for design of composite slabs using profiled steel sheeting. BSI, London.

BS 5950-8-2003: Structural use of steelwork in buildings. Part 8: Code of practice for fire resistant design. BSI, London.

BS 6399 Loading for Buildings.
Part 1 (1984) Code of practice for dead and imposed loads.
Part 2 (1995) Code of practice for wind loads.

BS 6399-1 Loading for Buildings.
Part 1 (1996) Code of practice for dead and imposed loads.

BS 6399-2 Loading for Buildings.
Part 1 (1997) Code of practice for wind loads.

BS 6399-3 Loading for Buildings.
Part 1 (1988) Code of practice for imposed roof loads.

BS 648 Schedual of Weight of Building Materials.

BS 8110 (1997) Structural Use of Concrete.
Part 1: Code of practice for design and construction.

BS 8110-1997: Structural use of concrete Part 1: Code of practice for design and construction. BSI, London.

BS EN 10 080-2005: Steel for reinforcement of concrete-weldable reinforcing steels-General. BSI, London.

BS EN 1990:2002 Eurocode – Basis of structural design. London.

BS EN 1990-2002: Eurocode – Basis of structural design. CEN. London, BSI.

BS EN 1991-1-1:2002 EC 1: Actions on structures – Part 1-1: General actions – Densities, self-weight, imposed loads for buildings. London.

BS EN 1991-1-1-2002: EC 1 – Actions on structures – Part 1-1: General actions – Densities, self-weight, imposed loads for buildings. CEN. BSI, London.

BS EN 1991-1-3-2003: EC 1 – Actions on structures – Part 1-3: General actions – Snow loads. CEN. BSI, London.

BS EN 1991-1-4-2005: EC 1 – Actions on structures – Part 1-4: General actions – Wind actions. CEN. BSI, London.

BS EN 1991-1-6-2005: Actions on structures. General Actions, Actions during execution. London, BSI.

BS EN 1992-1-1-(2005) EC 2 Design of concrete structures. Part 1-1 General rules and rules for buildings. London.

BS EN 1993-1-1:2005 EC 3: Design of steel structures – Part 1-1: General rules and rules for buildings, BSI, London.

BS EN 1993-1-1-2005 EC 3 Design of steel structures. Part 1-1 General rules and rules for buildings. London.

BS EN 1993-1-3-2006-EC 3: Design of steel structures Part 1.3. General rules. Supplementary rules for cold formed steel members and sheeting. London, BSI.

BS EN 1993-1-5-2006, EC 3 – Design of steel structures – Part 1.5: Plated structural elements. CBS EN. London.

BS EN 1993-1-8-2005: EC 3: Design of steel structures – Part 1-8: Design of joints. CEN. London, BSI.

BS EN 1994-1-1-2004 EC 4 Design of composite steel and concrete structures. Part 1-1: General rules and rules for buildings. London.

BS EN 1994-1-1-2004: Eurocode 4: Design of composite steel and concrete structures – Part 1-1: General rules and rules for buildings: The European Union.

BS EN 1994-1-2 EC 4: Design of composite steel and concrete structures. Part 1-1: General rules – Structural fire design, BSI, London.

BS EN 1994-1-2-2005: Design of composite steel and concrete structures, Part 1-2 Structural fire design. London, BSI.
BS EN ISO 13918-2008: Welding studs and ceramic ferrules for arc stud welding, BSI, London.
CP3: Chapter V, Part 2 (1972) *Loading. Wind Loads.*
EN 1944-2008: Design of composite steel and concrete structures. Part 1-1: General rules and rules for buildings, BSI, London.
NA to BS EN 1993-1-8:2005, UK National Annex to EC 3: Design of steel structures Part 1.8 Design of joints, 2008, BSI, London.
SCI (2004) Comparative Structure Cost of Modern Commercial Buildings, 2nd ed., SCI Publication P137.
SCI and BCSA (2009) Steel Building Design: Concise Eurocodes. SCI Publication No. P362. Ascot. London.
SCI and BCSA (2012) Steel Building Design: Design Data – In accordance with Eurocodes and the UK National Annexes. SCI Publication No. P363.
Steel Construction Institute (1997) Design for Construction, SCI Publication P178. Ascot.
Steelwork Design Guide to BS 5950-1: 2000 (2001) *Section Properties and Member Capacities*, 6th ed., SCI & BCSA, Steel Construction Institute.
The Institution of Structural Engineers (2010) UK: Manual for the design of steelwork building structures to EC 3. London, ICE.

BUILDING REGULATIONS

Building Regulations (2010) Part B, Fire Safety, HMSO, London.
Building Regulations (2013) Statutory Guidance Structure: Approved Document A, Department for Communities and Local Government, HMSO, London.

REFERENCES AND FURTHER READING

Al Nageim, H. and MacGinley, T.J. (2005) *Steel Structures Practical Design Studies*, Spon Text, Taylor & Francis, London.
Bowles, J.E. (1988) *Foundation Analysis and Design*, McGraw Hill Book Company, New York.
Brettle, M. (2009) Steel Building Design: Worked Examples – Open Sections – In accordance with Eurocodes and the UK National Annexes. SCI Publication No. P364. Ascot, The SCI, London.
Cauvin, A. and Stagnitto, G. (1998) La progettazione delle strutture in CA e l'elaboratore, in Malerba (ed.), *Analisi limite e non lineare di strutture in CA*, CISM, Udine.
Cauvin, A., Stagnitto, G. and Passera, R. (1998) Integration of expert system in a structural design office, in Ian Smith (ed.), *Artificial Intelligence in Structural Engineering*, Springer Verlag, Berlin.
Chung, K.F. and Narayanan, R. (1994) *Composite column design to Eurocode 4*, SCI, London.

Coates, R.C., Coutie, M.G. and Kong, F.K. (1988) *Structural Analysis*, Van Nostrand Reinhold, Wokingham, UK.
Couchman, G.H., Mullet, D.L. and Rackham, J.W. (2000) *Composite Slabs & Beams using Steel Decking: Best Practice for Design & Construction*. Metal Cladding & Roofing Manufacturers, Association & SCI, Ascot, Berks.
Council on Tall Buildings (1985) *Planning and Design of Tall Buildings*, 5 vols., American Society of Civil Engineers.
Curtin, W.G., Shaw, G., Beck, J.K. and Bray, W.A. (1991) *Structural Masonry Designers' Manual*, 2nd ed., BSP Professional Books, Oxford.
Dowling, P.J., Knowles, P.R. and Owens, G.W. (eds.) (1988) *Structural Steel Design*, Butterworths, London.
Galambos, T.V. (ed.) (1988) *Guide to Stability Design Criteria for Metal Structures*, 4th ed., John Wiley, New York.
Gardner, L. (2011) *Stability of Steel Beams and Columns*. SCI Publication No. P360, Escot.
Gardner, L. and Grubb, P.J. (2010) Eurocode load combinations for steel structures. London, BCSA Publication No. 53/10, BCSA.
Ghali, A. and Neville, A.M. (1989) *Structural Analysis*, 3rd ed., Chapman & Hall, London.
Hart, F., Henn, W. and Sontag, H. (1978) *Multistorey Buildings in Steel*, Granada Publishing, London.
Hicks, S.J. and Smith, A. (2009) *Design of floors for vibrations – A new approach*. The Steel Construction Institute P354. Ascot, SCI.
Horne, M.R. (1971) *Plastic Theory of Structures*, Nelson, London.
Horne, M.R. and Morris, L.J. (1981) *Plastic Design of Low Rise Frames*, Collins, London.
Hughes, A.S. and Malik, A. (2010) *Steel Building Design: Combined bending and torsion*. SCI Publication 385.
Johnson, B.G. el (1976) *Guide to Stability, Design Criteria for Metal Structures*, 3rd ed., John Wiley, New York.
Johnson, R.P. and Molenstra, N. (1991) *Partial shear connection in composite beams for buildings*, Proc Inst Civil Engineers Vol. 91 No. 4 Dec, (1991), 679–704.
King, C.M. (1995) Plastic Design of Single-Story Pitched-Roof Portal Frames to Eurocode 3, SCI P147, Steel Construction Institute.
King, C.M. (2001) Design of Steel Portal Frames for Europe, SCI P164.
Kirby, P.A. and Nethercot, D.A. (1979) *Design for Structural Stability*, Collins, London.
Lee, G.C., Ketter, R.L. and Hsu, T.L. (1981) *The Design of Single Storey Rigid Frames*, Metal Manufacturers Association, Cleveland, OH.
Leonard, J.W. (1988) *Tension Structures–Behaviour and Analyses*, McGraw Book Company, New York.
Lothers, J.E. (1960) *Advanced Design in Structural Steel*, Prentice Hall, Englewood Cliffs, NJ.
MacGinley, T.J. (2002) *Steel Structures Practical Design Studies*, 2nd ed., E&FN Spon. London and New York.
MacGinley, T.J. and Ang, P.T.C. (1992) *Structural Steelwork–Design to Limit State Theory*, Butterworths/Heinemann, Oxford.
Makowski, Z.S. (ed.) (1984) *Analysis, Design and Construction of Braced Domes*, Granada Technical Books, London.

Martin, L. and Purkiss, J. (2008) *Structural design of steelwork to EN 1993 and EN 1994*, 3rd ed., Butterworth-Heinemann, Oxford.
Morris, L.J. and Plum, D.L. (1988) *Structural Steelwork Design*, Nichols Publishing, New York.
Narayanan, R. (ed.) (1985) *Steel Framed Structures–Strength and Stability*, Elsevier Applied Science Publishers, London and New York.
Newberry, C.W. and Eaton, K.J. (1974) *Wind Loading Handbook*, Building Research Establishment, HMSO, London.
Orton, A. (1988) *The Way We Build Now – Form, Scale and Technique*, Van Nostrand Reinhold UK, Wokingham.
Pask, J.W. (1982) *Manual on Connections for Beam and Column Construction*, British Constructional Steelwork Association, London.
Rackham, J.W., Couchman, G.H., and Hicks, S.J. (2009) *Composite slabs and beams using steel decking: Best practice for design and construction*, SCI Publication 300, Ascot, SCI, London.
Salter, P.R. (2002) *Design of Single-Span Steel Poprtal Frames*, SCI P252.
Schueller, W. (1977) *High-rise Building Structures*, John Wiley, New York.
Schueller, W. (1983) *Horizontal-span Building Structures*, John Wiley, New York.
Smith, A.L., Hicks, S.J. and Devine, P.J. (2009): Design of floors for vibration: A new approach, SCI Publication 354-2009. Ascot, SCI.
Steel Designers Manual, 4th ed., (1986) BSP Professional Books, Oxford.
Steel Designers Manual, 5th ed., (1994) Blackwell Scientific Publications, Oxford.
Structural Safety 1997–1999: Review and Recommendations. 12th report of SCOSS. The Standing Committee on Structural Safety, February 1999.
Subramanian, N. (2008) *Design of Steel Structures*, OUP India, New Delhi.
Taranath, B.S. (1988) *Structural Analysis and Design of Tall Buildings*, McGraw Hill, New York.
Timoshenko, S.P. and Gere, J.M. (1961) *Theory of Elastic Stability*, 2nd ed., McGraw Hill, New York.
Trahair, N.S. and Bradford, M.A. (1988) *The Behaviour and Design of Steel Structures*, Chapman & Hall, London.
Trahair, N.S., Bradford, M.A., Nethercot, D.A. and Gardner, L. (2008) *The Behaviour and Design of Steel Structures to EC3*, 4th ed., London and New York, Taylor and Francis.
Woolley, T., Kimmins, S., Harrison, P. and Harrison, R. (1997) *Green Building Handbook*, E&FN Spon. London.
Yam, L.C.P. (1981) *Design of Composite Steel – Concrete Structures*, Surrey University Press, London.

TECHNICAL PAPERS – JOURNALS AND CONFERENCES

Bucholdt, H.A. (1984) Cable roofs, in *Symposium–Long Span Roofs*, Institution of Structural Engineers.

Cauvin, A. (1995) Use of neural networks for preliminary structural design. *Proceedings of the Sixth International Conference on Computing in Civil and Building Engineering*, Berlin, Balkema, Rotterdam.

Cauvin, A. and Stagnitto, G. (1995) A 'top down' procedure for design of tall buildings structures using expert systems. *Proceedings of the Fifth World Congress of the 'Council on Tall Buildings and Urban Habitat'*, Amsterdam.

Cauvin, A. and Stagnitto, G. (2001) Structural modelling of complex structures, *Proceedings of the Cacquot Conference*, Ecole Nationale des Ponts et Chaussées, Paris.

Davies, J.M. (1990) Inplane stability of portal frames. *The Structural Engineer*, 68(8).

Dickie, J.F. (1984) Domes and Vaults, in *Symposium–Long Span Roofs*, Institution of Structural Engineers.

Dowling, P.J., Harding, J.E. and Bjorhovde, R. (ed.) (1987) *Journal of Constructional Steel Research – Joint Flexibility in Steel Frames*, Elsevier Applied Science, London and New York.

Fraser, D.J. (1980) Effective lengths in gable frames, sway not prevented. *Civil Engineering Transactions, Institution of Engineers, Australia*, CE22(3).

Horridge, J.F. Design of Industrial Buildings, Civil Engineering Steel Supplement, November 1985.

Jenkins, W.M., Tong, C.S. and Prescott, A.T. (1986) Moment transmitting endplate connections in steel construction and a proposed basis for flush endplate design. *The Structural Engineer*, 64(5).

Khan, F.R. and Amin, N.R. (1973) Analysis and design of framed tube structures for tall concrete buildings. *The Structural Engineer*, 51(3).

Masterton, G.T. (2008) Power and industry centenary special edition. *Journal of the Institution of Structural Engineers*, 86(14).

Skilling, J.B. (1988) Advances in high rise building. Annual lecture. *Singapore Structural Steel Society*, 3(3).

Way, A.G.J. and Salter, P.R. (2003) Introduction to steelwork design to BS 5950-1:2000 (P325), The Steel Construction Institute, Ascot.

Weller, A.D. (1993) An introduction to EC3. *The Structural Engineer*, 71(18).

HANDBOOKS AND TECHNICAL LITERATURE

British Steel Corporation Tubes Division, Corby (1984) Construction with hollow sections.

British Steel Corporation Tubes Division, Corby. Nodus Space Frame Grids Part 1 – Design and construction.

Conder International Ltd, Winchester. Steel framed buildings.

Constradd. Profiled steel cladding and decking for commercial and industrial buildings.

Constradd: Elliott, D.A. (1983) Fire protection for structural steel in buildings.

Eastern Partek, Singapore. Hollow-core prestressed slabs. Section properties, load/span chart.

European Convention for Constructional Steelwork, Brussels (1991) Essentials of Eurocode 3, Design manual for steel structures in building.

Gardner, L. and Nethercot, D.A. (2005) *Designers' Guide to EN 1993-1-1-2005: EC 3: Design of Steel Structures*. Thomas Telford Publishing, London.

Hendy, C.R. and Murphy, C.J. (2007) *Designers' Guide to EN 1993-2.*
John Lysaght, Australia. Galvanised 3W and 2W steel decking.
Mero space frames. Mero-Raumstruktur. GmbH & Co., Wurzburg, Germany.
Nippon Steel Corporation, Tokyo (1986) NS space truss system. Design manual. Otis lifts, Newcastle-upon-Tyne. Passenger lift planning guide.
Precision Metal Forming Limited, Cheltenham (1993) P.M.F. composite floor decking systems.
Space Deck Limited, Somerset. Space frame grids. Part 1. Design and Construction. Part 2. Analysis.
Steel Construction Institute, Ascot (1987) Steelwork design guide to BS 5950: Part 1. Vol. 1 Section properties, member capacities; Vol. 2 Worked examples.
Trade ARBED, Luxembourg. Structural shapes. Histar rolled section.
Ward Building Components Ltd, Matton (1986) Multibeam purlin and cladding rail systems.

STRUCTURAL DESIGN AND ANALYSIS BY COMPUTER

Computer educational package supplied to by EdSoft Ltd. You are also encouraged to see their web site www.Reel.co.uk
QSE Steel Designer BS 5950+EC3
QSE-Section Wizard (all properties, stress)
QSE-Space 2/3D analysis
Staad-Pro (FE, 2nd order, seismic + other advanced feature with QSE interface)
www.access-steel.com. access-steel design of fixed column base joints
www.access-steel.com. access-steel design of simple column bases with shear nibs
www.access-steel.com. NCCI SN002 Access, Steel Determination of non-dimensional slenderness of I- and H-sections
www.access-steel.com. NCCI; Vibrations, SN036a_EN-EU Vertical and horizontal deflection limits for multi-story buildings SN034a-EN-EU
www.coruspanelsandprofiles.co.uk
www.planningportal.gov.uk: Building Regulations
www.planningportal.gov.uk: Code for Sustainable Homes
www.planningportal.gov.uk: The Building Regulations (2000), Approved Document E – Resistance to the passage of sound, 2003 ed., incorporating 2004 amendments. RIBA Bookshops, London
www.Slimdek.com
www.steel-ncci.co.uk: NCCI SN031a (2009) Effective lengths of columns and truss elements in truss portal frame construction
www.steel-ncci.co.uk: NCCI; Vertical and horizontal deflection limits for multi-story buildings SN034a-EN-EU
www.steel-sci.org

Index

Page numbers followed by f and t indicate figures and tables, respectively.

N

O

Q

R